# Data, Models and Statistical Analysis

# Data, Models and Statistical Analysis

R A COOPER & A J WEEKES

BARNES & NOBLE BOOKS
TOTOWA, NEW JERSEY

*First published in the USA 1983 by*
BARNES & NOBLE BOOKS
81 ADAMS DRIVE
TOTOWA, NEW JERSEY, 07512



ISBN: 0-389-20382-3 Cloth
IBSN: 0-389-20383-1 Paper

**Library of Congress Cataloging in Publication Data**
Cooper, R. A. (Ron A.)
Data, models, and statistical analysis.
Bibliography: p.
Includes index.
1. Social sciences—Statistical methods.
2. Statistics. I. Weekes, A. J. (Tony J.) II. Title
HA29.C7714 1983 519.5 82-25110
ISBN 0-389-20382-3
ISBN 0-389-20383-1 (pbk.)

Printed in Great Britain

# Contents

# Preface

## BACKGROUND

Our purpose has been to write a non-specialist text which will offer the reader a feel for statistical analysis of real data at a relatively elementary level and at the same time provide an introduction to useful advanced topics, particularly in multivariate statistical analysis.

We have taken advantage of three relatively recent developments in the subject. The first is the increasing availability of computer program packages for analysis of statistical data. This sometimes leads to misuse of statistical techniques because it is so easy to carry out quite inappropriate analyses. We hope that this text, by relating underlying theory to computer output, will make a small contribution to correcting this situation.

The second is an increasing concern amongst many statisticians that formal and uncritical application of hypothesis testing procedures is over-emphasised in introductory courses. We share this unease and consider it important that even (or especially) in an introductory text, readers should be introduced to problems in use and interpretation.

Finally, there is the growing interest among statisticians in informal and visual methods of data analysis. Many of these methods are far from new and they are part of the working tools of any applied statistician. However, in many applied statistics texts, they are ignored or treated in a dismissive fashion.

The body of the text assumes that the reader is familiar with the content of a typical non-specialist introductory course. However, Chapter 0 (as we have called it) summarises this introductory material and so in this sense the book is self-contained. The mathematical level goes little further than that required for introductory courses.

## THE APPROACH TAKEN

The discussion of statistical methods is based upon the analysis of real data sets. There are, in our view, a number of advantages which flow from this. For many readers, the purpose of a particular technique is more clearly understood when presented in numerical as well as algebraic terms. It is also possible to establish a clear link between statistical method and computer output. Finally, the reader is made aware of the fact that real data rarely yield the neat diagrams and tidy conclusions which discussions based upon contrived data sets might suggest.

This is essentially an applied text. However, informed application depends upon an understanding of theoretical principles. So, although we have tried to cut theoretical development and proof to a minimum, we have been concerned to set out the essence of the theoretical background (even though sometimes in an intuitive fashion) and to stress the assumptions of underlying statistical models.

The text is directed to the analysis of non-experimental data of the kind found in economics and business studies, in sociology and the behavioural sciences generally, and in many miscellaneous areas—archaeology, literature analysis, history. We exclude any consideration of the analysis of data from designed experiments and sample survey methodology.

## HOW TO USE THIS BOOK

To help readers who wish to study selectively, we indicate at the head of each chapter the more relevant earlier sections on which the chapter depends. For example, Chapters 3, 4, 7, 8 and 9 form a fairly self-contained body of material leading up to a discussion of multiple regression. Chapters 10 to 13, dealing with various multivariate methods, are also self-contained. The final chapter on categorial variables concentrates upon contingency table analysis and depends little upon earlier discussion. As already remarked, the preliminary

Chapter 0 is intended simply to offer a summary account of introductory statistics drawn upon in the text.

At the end of each chapter, we offer two kinds of work for the reader. The first, headed **Examples**, consists of conventional examples based upon routine application of material in the chapter; these are prefixed with the letter **E**. Answers to selected examples are at the end of the book. The second, headed **Further Work** (and prefixed **F**), is more open-ended and suggests areas for further empirical analysis. In the later chapters these will usually require access to appropriate computer programs.

The References and Bibliography section is mainly intended as a guide to further reading. We hope this will be particularly useful for those areas where, because of the limited technical level assumed, we have inevitably done little more than introduce some basic ideas. The section also contains references to sources cited within the body of the text.

Finally, a word about computation. An understanding of the statistical analyses which we present does not depend upon the reader having access to the same programs which we have used. There are two reasons why this is so. First, the differences between programs as regards standard procedures are often little more than presentational. Second, since we invariably set out the full data set, it should be relatively easy to compare our output with that available to the reader, thus facilitating interpretation of the latter.

## ACKNOWLEDGEMENTS

We are grateful to several colleagues at the University of York who have provided advice and encouragement at various stages. In particular, we must record our debt to John Byrne, Les Godfrey, Peter Lee, Don Poskitt and Andy Tremayne. Needless to say, they are not responsible for any errors, random or otherwise, which remain. We should like to pay our compliments to Ann Hirst for the excellent work she has put into proof reading and correction in its widest sense.

July 1982

Ron Cooper
Tony Weekes

# 0

# Recapitulation of Introductory Statistics

## 0.1 INTRODUCTION

In this chapter we summarise those areas of basic statistical theory and method which will be freely drawn upon from time to time in the rest of the book. The recapitulated material is fairly typical of that included in a one-year introductory course for non-specialist students. Readers with some prior statistical knowledge may find it convenient to use this chapter for reference as the need arises. Readers with minimal experience may find it useful as a link and guide to a more thorough exposition to be found elsewhere (see Bibliography).

The methods are presented in a neutral fashion with little comment or evaluation. It will however be clear from the body of the text that we take, and would encourage others to take, a sceptical view of some methods — and the emphasis accorded to them — embodied in the more traditional elementary textbooks.

Much of this text draws only partially upon the material in this chapter. To take an extreme example, some of the chapters on multivariate analysis require as a prerequisite little more than a knowledge of descriptive statistics (Sections 0.2 — 0.9 below) and a familiarity with the mathematics of simple graphical analysis.

## 0.2 UNIVARIATE ANALYSIS: DESCRIPTIVE STATISTICS

Throughout statistical analysis, we are concerned with the notion of a *variable*, such as the height of individuals or the population of countries. A *numerical variable* is a measurable property and different numerical measurements are associated with different individuals or objects. These measurements are referred to as the *values of the variable*. A *continuous variable* can assume any numerical value (decimal or otherwise) in its relevant range (e.g. the variable 'height'). A *discrete variable* can assume only a limited set of numerical values (e.g. the variable 'size of households').

Variables will be labelled by upper case letters such as $X$, $Y$, ... Individual values will be denoted by the corresponding lower case letters $x_1$, $x_2$, ... $x_n$, where: (i) the numerical subscripts relate to the individuals or objects for which the measurements are given; and (ii) the letter $n$ will generally stand for the total number of individuals. This set of $n$ values will be spoken of as the *data set* or the *data values* (or more simply as the *data* when the analysis is concerned with only one variable).

For some purposes it is useful to rearrange the data in ascending order of size. (The original order $x_1$, ... $x_n$ will usually be arbitrary, or related to some incidental ordering such as an alphabetical arrangement of individuals.) This process is known as *ranking* and the values are said to be in *rank order*. The ranked data will be denoted as

$$x_{(1)}, x_{(2)}, \dots x_{(n)},$$

where the subscript in brackets is called the *rank value*.

### EXAMPLE 0.1

Variable $X$ is 'population of capital cities' (measured to nearest million). The data comprises the populations of 5 cities, i.e. $n = 5$, where $x_1$, ... $x_5$ respectively equal 8, 11, 9, 6, 12. Then $x_{(1)}$, ... $x_{(5)}$ respectively equal 6, 8, 9, 11, 12, and, for example, the city with population value 9 has the rank value 3.

## 0.3 SUMMARY STATISTICS

Introductory texts lay emphasis on summarising and describing data by means of summary statistics. These are arithmetical calculations based upon the data. The more important can be categorised as measures either of central tendency or of dispersion.

## *MEASURES OF CENTRAL TENDENCY*

A measure of central tendency, or *centrality* as we shall call it, is a number which is somewhere near the 'middle' of the data values. There are a number of different centrality measures including:

(i) *Arithmetic mean* (or *mean* for short) This is the ordinary average and is defined as $\Sigma x/n = (x_1 + \ldots + x_n)/n$. This is often denoted by $\bar{x}$, although an alternative symbol used is $m$. Similarly, if the data were to be denoted by $y_1, \ldots, y_n$, the arithmetic mean would be denoted by $\bar{y}$.

(ii) *Median* The median is defined as that number such that the same number of data values are less than or equal to it as are greater than or equal to it. This definition has the following implication. If the number of data values, $n$, is odd, then the median equals that value whose rank order is $(n + 1)/2$. If $n$ is even, the median is some value between the two data values having rank orders $(n/2)$ and $[(n/2) + 1]$. In this case the median may be conventionally defined as the average of these two values.

(iii) *Mode* The mode is that value which occurs most frequently in the data. It need not be unique.

The *quartile points* are related to the median although they are useful as indicators of dispersion rather than of centrality. If the data are in rank order, then the median divides the data set into two equal parts (equal in the sense of having the same number of values in each part). Roughly speaking, the lower quartile point and upper quartile point divide the lower and upper half of the ranked data into equal parts; an exact rule will be found in Section 2.7. In effect, the rank ordered data are divided into four approximately equal parts (in terms of number of data values), and the lower quartile point ($Q_1$), the median ($M$) and the upper quartile point ($Q_3$), mark their boundaries. The terms *first quartile, second quartile,* etc. refer collectively to the

**Table 0.1** Examples of Calculations of Central Tendency

| *Data values* | *Rank order* | *Mean* | *Median* | *Mode* |
|---|---|---|---|---|
| 8,11,8,6,12 | 6,8,8,11,12 | 9 | 8 | 8 |
| 7,2,7,4,6,7,3,6 | 2,3,4,6,6,7,7,7 | 5.25 | 6 | 7 |
| 4,7,3,2,7,2, | 2,2,3,4,7,7 | 4.2 | 3 to 4 (or conventionally 3.5) | 2 and 7 |

data values in the four parts. Other partitions of the ranked data can be made: ten parts yield 'deciles' and one-hundred parts 'percentiles'.

## *MEASURES OF DISPERSION*

By 'dispersion' is meant the extent to which the data values differ from each other overall. Thus the data values 2.1, 2.7, 1.8, 2.4 have less dispersion than the data values 1, 2, 3, 5, 7. There are a number of different measures of dispersion including:

(i) *Range* The range is based upon the lowest and highest data values, that is upon $x_{(1)}$ and $x_{(n)}$, either by stating the difference between them, $x_{(n)} - x_{(1)}$, or simply the two values themselves. The latter is preferable.

(ii) *Inter-quartile range* This is the difference between the upper quartile point and the lower quartile point and equals $Q_3 - Q_1$. Again, it is preferable to state the two figures separately. The quartile deviation (also known as the semi-interquartile range) is defined as one-half of this difference.

(iii) *Variance* This is based upon an average of the squared differences between each data value and the mean. There are two alternative difinitions, one using a divisor of $n$ and the other a divisor of $(n - 1)$, indicated respectively by $S^2$ and $s^2$:

$$S^2 = \frac{\Sigma(x_r - \bar{x})^2}{n} = \frac{\Sigma x_r^2}{n} - \bar{x}^2$$

and

$$s^2 = \frac{\Sigma(x_r - \bar{x})^2}{n - 1}$$

A notation for the variance of a variable $X$ is var $(X)$.

(iv) *Standard deviation* This is defined as the square root of the variance, that is,

$$S = \sqrt{\left[\frac{\Sigma(x_r - \bar{x})^2}{n}\right]}$$

and

$$s = \sqrt{\left[\frac{\Sigma(x_r - \bar{x})^2}{n - 1}\right]}$$

(v) *Mean absolute deviation* This is the average of the absolute differences (that is, ignoring any negative signs) between each data value and the mean.

**EXAMPLE 0.2**

*Data values (arranged in rank order):*
1 3 4 4 5 7 10 10 12 14 18
*Measures of centrality:*
Mean = 8, Median = 7, Modes = 4, 10
*Measures of dispersion:*
Quartile points: $Q_1 = 4$, $Q_3 = 12$.

$$\text{Semi-interquartile range} = \frac{12 - 4}{2} = 4$$

Range = 18 − 1 = 17, or can be written (1, 18)
Variance: $S^2 = 276/11 = 25.1$; $s^2 = 276/10 = 27.6$
Standard deviation: $S = \sqrt{25.1} = 5.0$; $s = \sqrt{27.6} = 5.3$
Mean average deviation = 48/11 = 4.4

## 0.4 FREQUENCY DISTRIBUTION AND HISTOGRAM

A method of partial summarisation of data, particularly when there is a relatively large number of data values and the variable is continuous, is provided by the frequency distribution and its visual counterpart, the *histogram*. A frequency distribution of data values may be set up in two stages. First, a set of non-overlapping consecutive intervals is set up which taken together include all the data values within its range. Secondly, a count is made of the number of data values which fall within each of the intervals. The intervals are called 'class intervals' and the counts 'frequencies'. A table of the results of this procedure is called a *frequency distribution.*

If the frequencies are replaced by the *proportion* of data values falling within each class interval, then a relative frequency distribution emerges.

**EXAMPLE 0.3**

*Data values (arranged in rank order for convenient calculation)*
2.09 4.02 4.78 4.93 5.61 5.98
7.16 7.63 8.24 8.69 9.15 9.86

The frequency distribution and relative frequency distribution of the data are set out in Table 0.2, where the class intervals are:

2.00–3.99, 4.00–5.99, 6.00–7.99, 8.00–9.99.

**Table 0.2** Frequency and Relative Frequency Distribution

| *Class interval* | *Frequency* | *Relative frequency* |
|---|---|---|
| 2 – | 1 | 1/12 = 0.08 |
| 4 – | 5 | 5/12 = 0.42 |
| 6 – | 2 | 2/12 = 0.17 |
| 8 – | 4 | 4/12 = 0.33 |
| Total | 12 | 1.00 |

The histogram (Figure 0.1) is a visual method of presenting the information in a frequency distribution or a relative frequency distribution. In a histogram, the horizontal axis shows the range of values of the variable and the class intervals. A histogram is always drawn such that the *area* of each rectangular block above a class interval is proportional to the *frequency* (or relative frequency) for that class interval. (In the example, the class intervals are of equal magnitude and so the heights of the columns are similarly proportional to the frequencies or relative frequencies.) In the case of a histogram of a relative frequency distribution, it is useful to think of the unit of area measurement being defined so that the total area of all blocks is unity. The area of any one block in terms of this unit of measurement then equals the corresponding relative frequency — for example, the area of the left-hand block equals 0.08. Also, there is the important property that areas of blocks are additive. Thus the relative frequency of occurrence of data values between 2.00 and 8.00 = the sum of the areas of blocks in the range 2.00 to 8.00 = 0.08 + 0.42 + 0.17 = 0.67.

If the data are values of a discrete variable, then a frequency

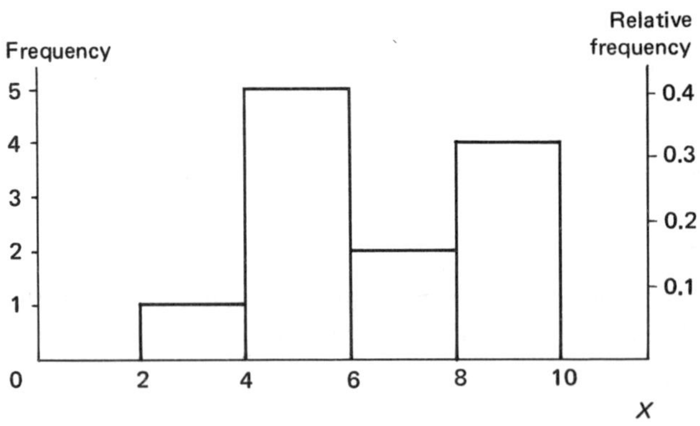


**Figure 0.1** Histogram of Example 0.3 Data

**Table 0.3** Summary Statistics for Grouped Data. Notation for Frequency Distribution

| *Mid-point of class interval* | *Frequency* | *Relative frequency* |
|---|---|---|
| $x_1$ | $f_1$ | $p_1$ |
| ... | ... | ... |
| $x_r$ | $f_r$ | $p_r$ |
| ... | ... | ... |
| $x_n$ | $f_n$ | $p_n$ |
| Total | $n = \Sigma f_r$ | 1 |

distribution can be simply set up by counting the number of occurrences of each of the possible variable values.

It may be necessary to calculate summary statistics of a data set already expressed as a frequency distribution, that is, 'grouped data', where data values have been grouped into particular class intervals. Table 0.3 sets out a general notation for this situation. The formulae for the mean and variance are:

$$\bar{x} = \Sigma f_r x_r / n \quad \text{and} \quad S^2 = \Sigma f_r (x_r - \bar{x})^2 / n$$

## 0.5 CUMULATIVE FREQUENCY DISTRIBUTION

The cumulative frequency distribution is a simple development of the frequency distribution. The cumulative frequency at any particular value is the total number of data values less than or equal to that value; there is a corresponding cumulative distribution based upon relative frequencies. Table 0.4 shows the cumulative relative frequency

**Table 0.4** Cumulative Relative Frequency Distribution of Example 0.3 Data

| *Class interval boundary* | *Cumulative frequency* | *Cumulative relative frequency* |
|---|---|---|
| 2.00 | 0 | 0 |
| 4.00 | 1 | 0.08 |
| 6.00 | 6 | 0.50 |
| 8.00 | 8 | 0.67 |
| 10.00 | 12 | 1.00 |

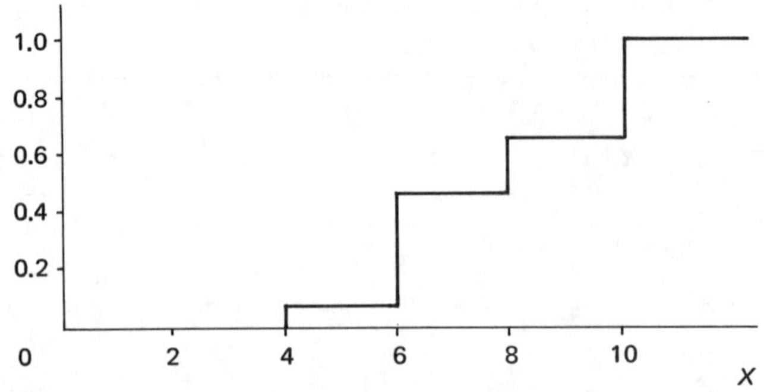


**Figure 0.2** Cumulative Histogram based upon Figure 0.1

distribution of the data in Example 0.3 and Figure 0.2 is the visual version, which we shall call the cumulative histogram. Table 0.4 and Figure 0.2 are based upon the data as grouped into class intervals; a 'finer' cumulative frequency distribution could also be compiled from the individual data values.

## 0.6 CATEGORICAL VARIABLES

The discussion so far has concerned values of a *numerical variable* (sometimes called a quantitative variable). A *categorical variable* is a property whose 'values' are specified in descriptive terms. (Another term is qualitative variable.) Examples of categorical variables are: sex (values—male, female) and marital status (values—single, married, widowed, etc.)

## 0.7 BIVARIATE ANALYSIS

Bivariate analysis is concerned with the relationship between values of two variables, a pair of values being known for each of a number of individuals or entities. The two variables may be conventionally labelled $X$ and $Y$ and the values for each individual $x_1, y_1; x_2, y_2; \ldots x_n, y_n$. The term *data point* will be used for the pair of values of the variables of any one individual. Three cases may be distinguished: (i) both variables are numerical; (ii) both variables are categorical: or (iii) one variable is numerical and the other is categorical. Each of these cases will be reviewed.

## 0.8 BIVARIATE ANALYSIS OF TWO NUMERICAL VARIABLES

Examples are: (i) $X$ is height of a person, $Y$ is weight of a person, and (ii) $X$ is national income per head of a country, $Y$ is number of people per physician in a country. For (i), it would be expected that there would be a *positive* relation between the two variables, that is to say, that taller people tend to weigh more. For (ii), a negative relation would be likely, that is, the higher the national income per head, the more doctors per head and the fewer the number of people per doctor. Table 0.5 sets out two illustrative data sets of these two examples.

The *scatter diagram* or *scatterplot* gives a quick visual appreciation of the nature of the relationship. In a scatterplot, each data point, that is each pair of values of $X$ and $Y$, is plotted as a point on a conventional graph. (Note that for this purpose the $X$ and $Y$ scales need not be the same.) Figure 0.3 shows the scatterplots for the two data sets of

**Table 0.5**

| *Data set (i)* | | *Data set (ii)* | |
|---|---|---|---|
| *x* | *y* | *x* | *y* |
| 18 | 16 | 15 | 15 |
| 10 | 11 | 20 | 19 |
| 6 | 8 | 13 | 22 |
| 14 | 12 | 11 | 25 |
| 22 | 20 | 12 | 17 |

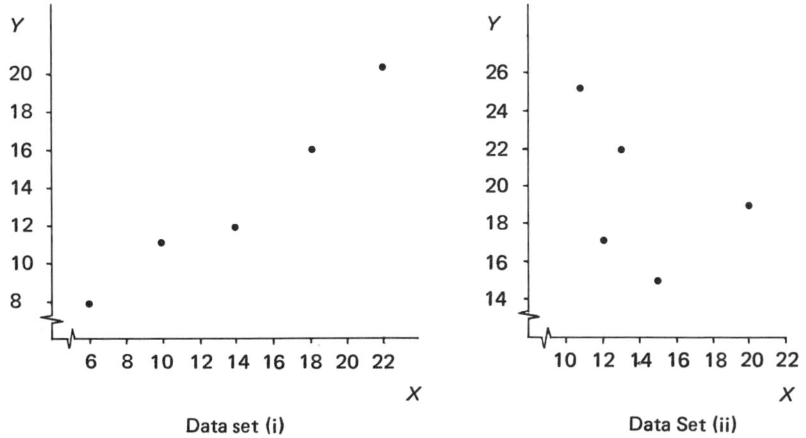


**Figure 0.3** Scatterplots for Table 0.5 Data

Table 0.5. The first shows a very strong positive linear relationship. In other words, the points would very closely fit a straight line. The second relationship is much more tenuous.

The *correlation coefficient* measures how closely the data points lie to a straight line. It is conventionally labelled $r$ and lies between $-1$ and $+1$. A positive sign to $r$ indicates a positive relation (as for data set (i)) and a negative sign a negative relation (as for data set (ii)). Values of $r$ near $+1$ or $-1$ indicate a good fit and values around zero a poor fit. For data set (i), $r = 0.98$ and for data set (ii), $r = -0.38$.

*Regression analysis* is concerned with the calculation of the line or mathematical curve which best fits the data points. In the present summary only linear regression analysis is discussed; the line of best fit is called the regression line. Figure 0.4 shows the regression line fitted to data set (ii); its equation is $y = 25.6 - 0.4x$. The regression line is defined as that line which minimises the sum of squares of the 'residuals' in the direction of the $Y$-axis. That is to say, referring to Figure 0.4, if (a) the distance in the $Y$-axis direction is measured from each data point to the regression line (these distances being known as the residuals), and (b) these residuals are each squared and added up, then (c) this sum of squared residuals will be smaller for the regression line than if the same procedure had been repeated for any other line. This method of fitting is known as applying the principle of *least squares* because its regression line is determined by minimising the sum of the squared residuals. The regression procedure is not symmetrical as between the two variables. One variable needs to be identified as the '$y$-variable' (called the dependent variable) and the other as the $x$-variable (called the independent or the explanatory variable). The choice is based upon the presumed direction of causality.

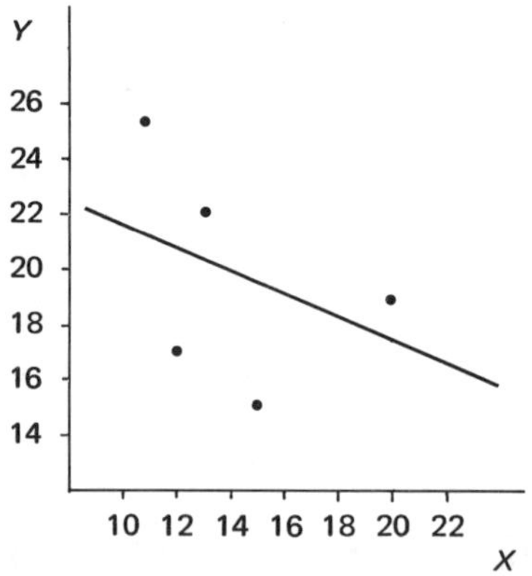


**Figure 0.4** Regression Line and Scatterplot

## 0.9 CORRELATION AND REGRESSION ANALYSIS: FORMULAE

The *correlation coefficient* equation is

$$r = \frac{[\Sigma(x - \bar{x})(y - \bar{y})]/n}{\sqrt{[\Sigma(x - \bar{x})^2/n)(\Sigma(y - \bar{y})^2/n]}}$$

The numerator in this expression is known as the *covariance* of the two variables, or $\text{cov}(X, Y) = [\Sigma(x - \bar{x})(y - \bar{y})]/n$. The covariance has the same sign as the correlation coefficient.

The *regression equation* is

$$y = a + bx$$

where $b = \text{cov}(X, Y)/\text{var}\, X$, ($b$ is known as the regression coefficient), and $a$ is such that $\bar{y} = a + b\bar{x}$. These formulae for $a$ and $b$ are derived by solving what are called the *normal equations,* which are

$$na + b\Sigma x = \Sigma y$$
$$a\Sigma x + b\Sigma x^2 = \Sigma xy$$

Having calculated the regression line, it is possible to calculate the residuals which often play an important part in analysis. These are calculated in two stages as in Table 0.6. First, for each data point, the $x$-value is substituted in the regression equation and a corresponding $y$-value calculated; this is the $y$-value of the point on the regression line below or above the data point. These calculations are shown in the column headed $\hat{y}$ and they are called the predicted or fitted $y$-values. The residuals are then given by subtracting from the original $y$-values the calculated $y$-values. The residuals are conventionally denoted by the letter $e$; with $n$ data points there are associated the $n$ residuals $e_1, e_2, \ldots e_n$. The residuals have the property that their average value is always zero and also that their correlation coefficient with the $x$-values is zero.

**Table 0.6** Regression Calculations: Data Set (ii)
Regression equation $y = 25.6 - 0.4x$

| $x$ | $y$ | $\hat{y}$ | $e$ |
|---|---|---|---|
| 11 | 25 | 21.0 | 4.0 |
| 12 | 17 | 20.5 | −3.5 |
| 13 | 22 | 20.1 | 1.9 |
| 15 | 15 | 19.3 | −4.3 |
| 20 | 19 | 17.1 | 1.9 |

## 0.10 BIVARIATE ANALYSIS OF TWO CATEGORICAL VARIABLES

We consider the simple case where each categorical variable is dichotomous, that is, described by only two categorical values. They might, for example, be sex (values: male, female) and marital status (values: single, not-single). Then, if data are available on these two variables for a number of individuals each individual will fall into one of four categories: male and single, female and single, male and non-single, female and non-single. By counting the number of individuals in each of these four categories, what is known as a *contingency table* can be formulated. An example is shown in Table 0.7. The final row and final column show what are known as the marginal totals—the total number of individuals who are male or female etc. The overall total number is 12.

**Table 0.7** Contingency Table

| | *Male* | *Female* | *Total* |
|---|---|---|---|
| Single | 2 | 7 | 9 |
| Non-single | 2 | 1 | 3 |
| Total | 4 | 8 | 12 |

For data of this kind, one may ask how far males differ from females as regards marital status or, similarly, how far single people differ from non-single ones as regards their sex. One way of answering this question is to calculate relevant proportions. Thus, of all the individuals, 1/3 are male whereas 2/3 of the non-single people are male and only 2/9 of the single people are male. When these proportions differ from each other as in the example, the two variables are said to be *associated*.

Another related approach is to calculate a table of 'equiproportionate values' as in Table 0.8. The table has two properties. The marginal row and column totals are those of the original data as in Table 0.7. But the four cell counts are replaced by numbers such that the proportion of males who are single (3/4) equals the proportion of females who are single (6/8) equals the total proportion who are single (9/12). This equiproportionate property applies to all rows and columns. The table is compiled by (i) copying down the marginal totals from Table 0.7, (ii) computing from them one cell frequency—e.g. the top left-hand $3 = (4 \times 9)/12$, and (iii) writing in the remaining cell frequencies by subtraction from the marginal totals.

**Table 0.8** Equiproportionate Values

| | *Male* | *Female* | *Total* |
|---|---|---|---|
| Single | 3 | 6 | 9 |
| Non-single | 1 | 2 | 3 |
| Total | 4 | 8 | 12 |

The table of equiproportionate values may be looked on as a numerical description of a situation where there is no association between the two variables sex and marital status. It therefore offers a standard of comparison against which to view the data frequencies of Table 0.7. For this rudimentary data, such a comparison shows that it would have required only one male and one female to have had a different marital status to make the two tables identical. At a superficial level, one might say the degree of association is not significant.

The analysis is readily extended to non-dichotomous variables. If one variable has, say, $p$ categories and the other $q$ categories, then a contingency table can be drawn up with $p$ rows and $q$ columns, apart from the marginal row and column totals. A corresponding table of equiproportionate values can similarly be calculated. The cell frequencies in $(p - 1)(q - 1)$ cells would be calculated proportionately and the remaining row and column filled in by subtraction.

## 0.11 BIVARIATE ANALYSIS OF ONE CATEGORICAL AND ONE NUMERICAL VARIABLE

An example of this situation would be data where, for each individual, the sex is stated (a dichotomous categorical variable) and the height (a numerical variable). Analysis would then be concerned with a comparison of heights of males with heights of females. Essentially, univariate analysis is applied to each group defined by the categorical variable.

## 0.12 DESCRIPTIVE STATISTICS AND STATISTICAL THEORY

The synopsis thus far has been about the methods of descriptive statistics. They take the data as it stands and describe and summarise it. The methods are, however, limited in scope. This may be illustrated in terms of three examples drawn from bivariate analysis:

(i) Suppose a correlation coefficient of $r = 0.2$ has been calculated from some data. This is fairly near to zero and one may wonder whether it is large enough to be meaningful or significant.
(ii) The contingency table data of Table 0.7 were little different from the related table of equiproportionate values in Table 0.8. Could this small difference have arisen by chance?
(iii) The heights of two groups of people are observed. The mean height of group $A$ is greater than that of group $B$. But if the individual heights vary very much, it may be asked how reliable a guide is the difference of the two means.

Questions of this nature cannot be tackled without making use of statistical theory which itself is founded upon probability concepts. The elementary theory is summarised below.

## 0.13 PROBABILITY

Probability is a concept with various meanings both in everyday life and in the subject of statistics. Two principal interpretations may be illustrated by reference to the following examples.

Consider the two statements: 'it is likely he will be home early tomorrow', and 'the missing aircraft has almost certainly crashed in the mountains'. They are examples of the use of *subjective probability;* different people would arrive at different numerical measures of the probability. Contrast these with the following: 'the probability of drawing two aces at random from a pack of cards is 0.005', and 'in a random sample size 6, from a population having equal numbers of men and women, the probability that there will be more men than women is 0.34'. These probabilities can be calculated 'objectively' by mathematics. So-called 'classical statistics' makes exclusive use of the second interpretation and this approach of classical statistics is adopted in most introductory texts and will also be followed in the present text.

A *random event* is one of a number of possible 'outcomes' or results of a *random experiment*. Examples are:

(i) Random experiment — choosing 6 people at random from a population with equal sex distribution. Possible random events — 0, 1, 2, 3, 4, 5 or 6 men are chosen.
(ii) Random experiment — two cards are drawn (without replacement) from a pack of cards. Possible random events — any two differing cards in the pack.

A list of all possible outcomes of a random experiment is called the *sample space*, that is, the possible random events collectively make up the sample space. A random event has the characteristic that it cannot be predicted though, in principle, the probability of its occurrence may be calculated.

The probability of occurrence of a random event is usually interpreted as being equal to the *long-run relative frequency* of its occurrence. Thus, to say that the probability of throwing a 1 or a 2 with a dice is 1/3 is taken to be equivalent to the statement that in the long run, one-third of throws will show a 1 or a 2. Therefore, a probability value lies between 0 and 1.

## 0.14 SAMPLING FROM A POPULATION

In a social survey, a *random sample* of say, 1000 persons is selected from a group of people under study, called the *population.* A random sample is such that each person in the population has a particular probability (often equal probabilities) of being selected for the sample. The random event here is the personal property being observed: the income, the sex, etc. Similarly, in industrial quality control, a random sample of products from the week's output may be selected for testing; the population here is the week's output and the random event is whether or not the product selected is satisfactory.

This terminology of random sampling from a population is often used in situations where no actual sampling takes place. Thus, the single tossing of a coin may be thought of as taking a random sample from a 'population' containing equal numbers of heads and tails.

## 0.15 RANDOM VARIABLES AND DISCRETE PROBABILITY DISTRIBUTIONS

Consider again the example of a random sample of size 6 from a population with equal sex distribution, where the random event to be observed is the number of men in the sample. It is possible to calculate the probabilities of each of the possible random events 0 to 6 (see binomial distribution, Section 0.19) and they are shown in Table 0.9. Such a table constitutes a *probability distribution.* A probability distribution specifies the possible outcomes of the random experiment together with their respective probabilities. The numbers 0 to 6 are known as values of a *random variable,* that is, they arise as the outcome of a random experiment. In this case it is a discrete numerical

**Table 0.9** Probability Distribution of Number of Men in Sample

| *Value of random variable* | *Probability* |
|---|---|
| 0 | 1/64 |
| 1 | 6/64 |
| 2 | 15/64 |
| 3 | 20/64 |
| 4 | 15/64 |
| 5 | 6/64 |
| 6 | 1/64 |
| Total | 1 |

random variable. This is consistent with the terminology for ordinary statistical variables. Similarly, it is possible to have a probability distribution of a categorical random variable. Table 0.10 shows such a distribution relating to the outcome of tossing a biased coin having a probability of 0.6 for a head and 0.4 for a tail. The values of this random variable are the categories Head, Tail.

There is a close correspondence between a probability distribution and a corresponding relative frequency distribution due to probability being interpreted as long-run relative frequency. Thus, Table 0.9 can also be looked upon as the relative frequency distribution of the number of men turning up in a long-run sequence of samples. The mean of this frequency distribution is 3. When the 'mean' of a probability distribution is likewise calculated, it is known as the *expected value* of the random variable. That is, the expected value, 3, is the long-run average value of the random variable. Similarly, the variance of a probability distribution can be calculated.

**Table 0.10** Probability Distribution for the Tossing of a Biased Coin

| *Value of random variable* | *Probability* |
|---|---|
| Head ($H$) | 0.6 |
| Tail ($T$) | 0.4 |
| Total | 1.0 |

## 0.16 MATHEMATICS OF PROBABILITY

First, we need some notation. $P(A)$ will mean the probability that an event, denoted by $A$, occurs. The information in Table 0.10 would be given as $P(H) = 0.6$ and $P(T) = 0.4$. If $A$ and $B$ are two events, $P(A$ or $B)$ indicates the probability that either $A$ or $B$ occurs. (In set theory notation this is $P(A \cup B)$.) $P(A$ and $B)$ indicates the probability that both $A$ and $B$ occur. (The set theory notation is $P(A \cap B)$.)

Two events $A$ and $B$ are *mutually exclusive* if the occurrence of one precludes the occurrence of the other. Thus, on one toss of a coin, the events head and tail are mutually exclusive. Two events $A$ and $B$ are said to be *independent* if the occurrence of one event does not affect the probability of occurrence of the other event. An example of this is the tossing of a coin twice where event $A$ is the occurrence of a head on the first toss and event $B$ is the occurrence of a tail on the second toss. If two events are not independent, they are said to be *dependent*.

*Addition theorem* If $A$ and $B$ are mutually exclusive, then $P(A \text{ or } B) = P(A) + P(B)$.

### EXAMPLE 0.4

(i) For one toss of a coin $P(H \text{ or } T) = P(H) + P(T) = 1$
(ii) Referring to Table 0.9, the selection of a sample with 0, 1 or 2 men are mutually exclusive events. By an obvious extension to the addition theorem, $P(0 \text{ or } 1 \text{ or } 2) = 22/64$.

The probability distribution for the values of any one random variable lists mutually exclusive values of the random variable. And so the probability that any one of a subset of these values occurs is given by adding the relevant probabilities. The addition theorem shows why the total of all probabilities of a probability distribution must sum to unity.

*Multiplication theorem* If $A$ and $B$ are *independent* events, then $P(A \text{ and } B) = P(A) \times P(B)$

### EXAMPLE 0.5

This theorem is best interpreted in the context of two random experiments where $A$ is a particular event observed at the first experiment and $B$ at the second experiment. A random sample of two persons is chosen from a large population with 60% men and 40% women. Event $A$ is 'the first choice is a woman' and event B 'the second choice is a woman'. Then the probability of choosing two women in the sample $= P(A \text{ and } B) = P(A) \times P(B) = 0.4 \times 0.4 = 0.16$.

The theorem extends simply to the occurrence of more than two events.

*Conditional probability* is concerned with describing the probability of occurrence of two or more random events when they are not independent. Suppose that in the last example, the population had consisted of only 5 people — three men and two women. Then, with the same notation, $P(A) = 0.4$ as before. But $P(B)$, the probability of choosing a women at the second sample choice, is now different. $P(B) = 1/4$ (not $2/5$) if a woman has been chosen initially. So the probability of choosing two women becomes $2/5 \times 1/4 = 1/10$. Because the probability of event $B$ is affected by whether or not event $A$ occurs, $A$ and $B$ are said to be *dependent events.* The probability of $B$ given that $A$ occurs, is written as $P(B \mid A)$. The multiplication theorem extends to dependent events thus:

*Multiplication theorem* If $A$ and $B$ are *dependent* events, then $P(A \text{ and } B) = P(A) \times P(B \mid A)$

This second version is consistent with the first in the sense that if two events are independent, $P(B \mid A) = P(B)$.

## 0.17 CONTINUOUS FREQUENCY DISTRIBUTIONS

The frequency distribution and its pictorial representation, the histogram, were reviewed in Section 0.4. The frequency distribution was *discrete* in the sense that the frequencies were counts based upon discrete and finite class intervals. Figure 0.5 illustrates how the concept of a *continuous* frequency distribution can be developed from the discrete frequency distribution. Diagram (a) shows the histogram of a frequency distribution[1] which, for illustrative purposes, is assumed to be a distribution of income of a large number of individuals. The total area of the histogram is unity and so the shaded area (see Section 0.4) equals the proportion of incomes between 500 and 1000. The second histogram shown in diagram (b) is presumed to be based upon the same data, but, in setting up the class intervals, a finer gradation has been used. Again, the shaded area equals the proportion of incomes between 500 and 1000. In diagram (c), a large number of class intervals has been imposed upon the data and it has been presumed that the histogram is such that the tops of its columns present a smooth outline

1. Henceforward, in this chapter, frequency distribution will invariably mean *relative* frequency distribution.

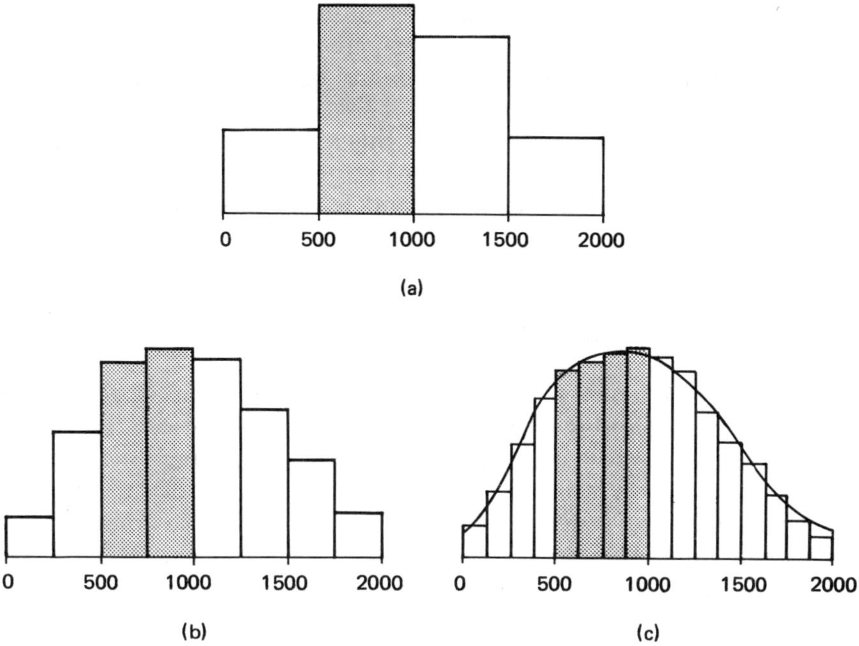


**Figure 0.5** Discrete and Continuous Frequency Distributions

through which a curve has been drawn. Such a curve is called a *frequency curve* Again, the shaded area has the same property as before.

In Figure 0.6, the frequency curve has been copied without the histogram. The shaded area is now the area under the frequency curve (almost identical to the corresponding histogram area). In fact, the frequency curve embodies all the information in the histogram about the distribution of incomes. In general, a frequency curve has the following properties:

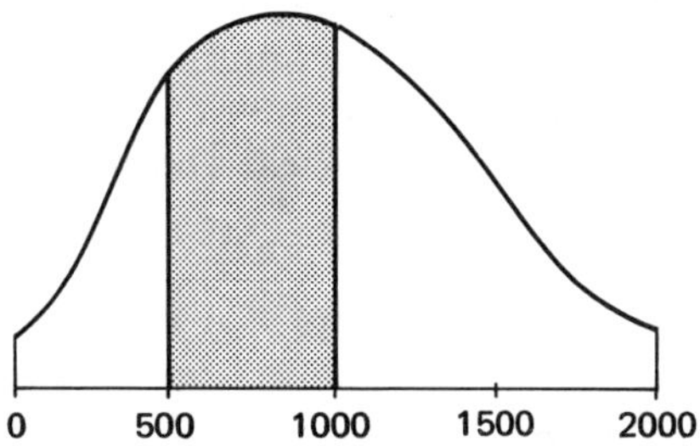


**Figure 0.6** Frequency Curve

(i) the total area under the curve is unity; and
(ii) the area under the curve between any two values of the variable equals the proportion of data values between those two values.

An empirical frequency curve of this kind can exist only if (a) the variable is continuous, (b) there is a large number of data values to allow many fine class intervals, and (c) the histogram presents a smooth outline.

## 0.18 THEORETICAL DISTRIBUTIONS

So far in this chapter we have summarised the various aspects of frequency distributions derived from data: 'empirical frequency distributions' as we shall call them. Depending upon the kind of variable, three kinds have been identified: (i) for categorical variables, a count of the frequency of occurrence of each category; (ii) for discrete numerical variables, a count of the frequencies of each discrete value; or (iii) for continuous numerical variables, either a frequency distribution based upon class intervals or the frequency curve derived from such a distribution. These kinds of empirical frequency distributions are mirrored by corresponding probability distributions again depending upon whether the random variable is categorical, discrete numerical or continuous numerical. Discrete probability distributions were considered in Section 0.15 and continuous probability distributions will be reviewed in Section 0.21.

The theoretical distributions are idealised distributions to which empirical distributions might approximate. When this is so, a theoretical distribution is said to be a 'model' for the empirical data. Theoretical distributions are exactly specified in mathematical terms and their statistical properties, such as the mean and the variance, can be calculated mathematically. Depending upon the context, a theoretical distribution may be regarded either as a frequency or as a probability distribution. Two theoretical distributions normally included in introductory courses are summarised below. The binomial distribution relates to a discrete numerical variable and the normal distribution to a continuous numerical variable.

## 0.19 BINOMIAL DISTRIBUTION

The binomial distribution is most easily interpreted as a theoretical probability distribution. Consider a random experiment (e.g. rolling a dice) where the probability of a particular event $A$ (throwing a five or a

**Table 0.11** Binomial Probability Distribution Example

| *Number of fives or sixes* | *Proportion of fives or sixes* | *Probability* |
|---|---|---|
| 0 | 0 | 0.20 |
| 1 | $\frac{1}{4}$ | 0.40 |
| 2 | $\frac{1}{2}$ | 0.30 |
| 3 | $\frac{3}{4}$ | 0.10 |
| 4 | 1 | 0.01 |
| Total | | 1.00 |

six) has a constant value (in this case 1/3). Suppose the experiment is repeated $n$ number of times (say 4). The binomial distribution describes the probability distribution of the random variable 'number of times $A$ occurs in $n$ experiments' (i.e. number of fives or sixes in 4 rolls). The random variable can alternatively be defined as the proportion rather than the number of occurrences. Table 0.11 shows the probability distribution for this example and Figure 0.7 is a visual representation. Table 0.11 can also be regarded as a frequency distribution: it sets out the relative frequencies of occurrence in the long run if the experiment of 4 rolls of the dice is repeated very many times.

The binomial distribution has two constants or *parameters* ($\pi$ and $n$) which are given numerical values in any particular application. These parameters appear in the formula for the probabilities of the binomial distribution. This is:

$$\binom{n}{x} \pi^x (1 - \pi)^{n-x}$$

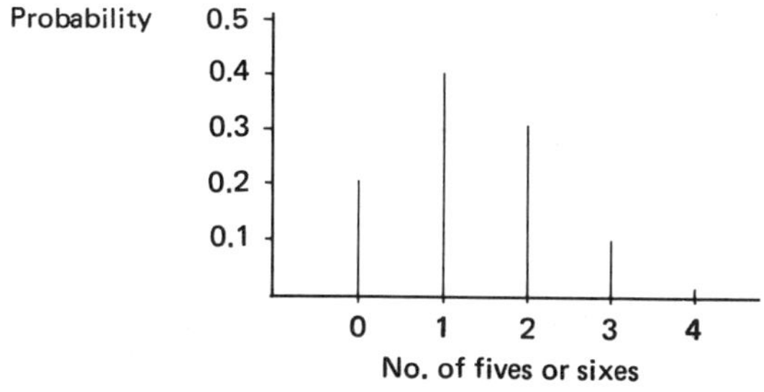


**Figure 0.7** Line Histogram of Binomial Distribution at Table 0.11

**Table 0.12** Mean and Variance of Binomial Distribution

| | *Mean* (*Expected Value*) | *Variance* |
|---|---|---|
| Variable: | | |
| Number of occurrences | $n\pi$ | $n\pi(1-\pi)$ |
| Proportion of occurrences | $\pi$ | $\dfrac{\pi(1-\pi)}{n}$ |

where $x$ is the possible number of occurrences from 0 to $n$. This formula is known as the *density function* of the binomial distribution. As in Table 0.11, these probabilities can be set in correspondence either with the number of occurrences of the event or the proportion of occurrences ($p$). In the latter case, $p$ can take the values 0, $1/n$, $2/n$, . . . 1. Table 0.12 shows the mean and variance of the binomial distribution for each of these two cases.

The expected value (Section 0.15) of the number of fives or sixes in the dice experiment equals 4/3 — this is the long-run average number of fives or sixes in 4 rolls.

## 0.20 NORMAL DISTRIBUTION

The normal distribution relates to a continuous numerical variable. It is specified by the mathematical formula for its frequency curve. This is:

$$\frac{1}{\sqrt{(2\pi)}\sigma} \exp\left[-\frac{1}{2}\left(\frac{x-\mu}{\sigma}\right)^2\right]$$

The exact form of this expression is not important in practice. Its essential features are:

(i) $\pi$ is the mathematical number 3.14..; exp is the exponential function and indicates that the mathematical number $e$ (= 2.718..) is to be raised to the power $-\frac{1}{2}[(x-\mu)/\sigma]^2$. (See Section 0.28.)

(ii) $\mu$ and $\sigma$ are the *parameters* of the distribution. That is to say, any particular pair of values of $\mu$ and $\sigma$ determines a particular normal distribution. Such a pair of values can be substituted in the density function and its value (that is, the height of the frequency curve) can be calculated for any value of the variable, $X$. Figure

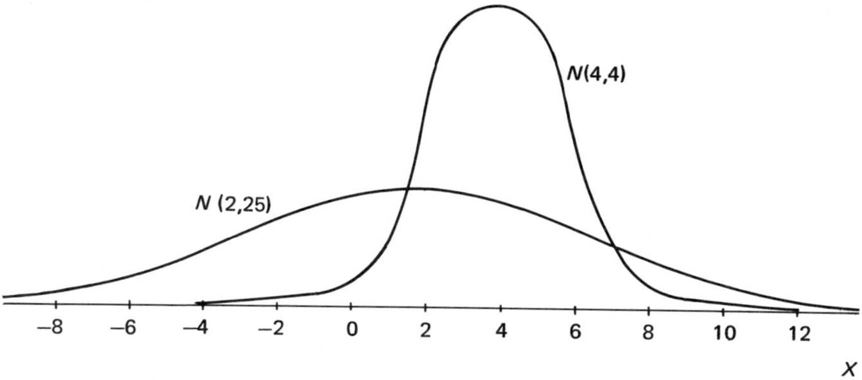


**Figure 0.8** Normal Distribution Frequency Curves

0.8 shows the frequency curves of two normal distributions, one with $\mu = 4$ and $\sigma = 2$, and the other with $\mu = 2$ and $\sigma = 5$.

All normal frequency curves are symmetrical. The value of $\mu$ determines the *mean* of the distribution and the value of $\sigma$, the *standard deviation*; the *variance* is consequently $\sigma^2$. A conventional notation for stating that a distribution is normal with mean $\mu$ and variance $\sigma^2$ is to write that the distribution is $N(\mu, \sigma^2)$. Thus, the two frequency curves of Figure 0.8 relate to the distributions $N(4,4)$ and N(2,25) (or $N(4,2^2)$ and $N(2,5^2)$).

The *standard normal distribution* has a mean of zero and a standard deviation (and also variance) of unity — it is the distribution N(0,1). *Normal distribution tables* allow the calculation of the area under the standard normal frequency curve between any two values of the variable $x$. A convenient property of all normal distributions allows these tables to be used to calculate areas and therefore relative frequencies for any normal distribution curve. This property states that the area under any normal curve $N(\mu, \sigma^2)$ between the mean and the value $x$ expressed in the form $\mu + z\sigma$ depends only upon $z$. As a consequence, by putting $\mu = 0$ and $\sigma = 1$, this area equals the area under the standard normal curve between the values 0 and $z$. Hence we have the usual procedure for area calculations: (i) for the area under the curve $N(\mu, \sigma^2)$ between the particular pair of values, $x_1$ and $x_2$, calculate $z_1$ and $z_2$ where $z_1 = (x_1 - \mu)/\sigma$ and $z_2 = (x_2 - \mu)/\sigma$. (These are usually known as the *z-scores* of $x_1$ and $z_2$.) (ii) Use the standard normal tables to calculate the area under N(0,1) between the values $z_1$ and $z_2$. An extract from the normal distribution table is in Appendix A.

**EXAMPLE 0.6**

(i) *Calculation of area under* N(2,9) *between* $x_1 = -1$ *and* $x_2 = 8$.
$z_1 = [(-1) - 2]/3 = -1$ and $z_2 = (8 - 2)/3 = 2$. From tables of N(0,1), area over range (−1,0) is 0.34 and over range (0,2) is 0.48. Hence area required = 0.34 + 0.48 = 0.82.

(ii) *Calculation of that value, for N(3,4), which determines a right-hand tail area of 0.05.*
From tables of N(0,1), the value 1.65 determines a right-hand tail area of 0.05. This is z-score of required value, $x$ say. Hence $1.65 = (x - 3)/2$ and $x = 3.3 + 3.0 = 6.3$.

Among the theoretical distributions, the normal distribution has a number of unique properties, two of which are particularly important. The first is that, if a set of values $x_1, x_2, \ldots$ collectively has the normal distribution $N(\mu, \sigma^2)$, then the derived set of values $x_1 + k, x_2 + k, \ldots$, obtained by adding the constant $k$ to each value, also has a normal distribution, $N(\mu + k, \sigma^2)$. The second result states that if a set of values $x_1, x_2, \ldots$ has the normal distribution $N(\mu, \sigma^2)$, then the derived set of values $x_1/\sigma, x_2/\sigma, \ldots$ obtained by dividing each value by the standard deviation $\sigma$, has the normal distribution $N(\mu/\sigma, 1)$. It is a consequence of these two results that if a set of values has the distribution $N(\mu, \sigma^2)$, and if the set is standardised by subtracting the mean $\mu$ and dividing by the standard deviation $\sigma$, then the standardised values have the distribution N(0,1).

Depending upon the context, the normal distribution can be regarded as either a continuous frequency distribution or a continuous probability distribution. The latter interpretation is reviewed within the following section.

## 0.21 CONTINUOUS PROBABILITY DISTRIBUTIONS

Section 0.15 drew attention to the close connection between a discrete relative frequency distribution and a discrete probability distribution. There is a similar kind of connection between a continuous frequency distribution and a continuous probability distribution.

Consider, for example, the general normal distribution $N(\mu, \sigma^2)$. If it is being used to describe a frequency distribution, then an area under the normal curve within a range of any two values equals the relative frequency of values in that range. Thus, the area under the normal curve between $\mu - 1.96\sigma$ and $\mu + 1.96\sigma$ being 0.95 implies that 95% of the values having a normal distribution are in the range $\mu \pm 1.96\sigma$.

Now suppose the normal distribution is being interpreted as a probability distribution. This could arise in the following way. Suppose a set of values of a continuous variable, say, the heights of a large number of children of the same age, has a normal frequency distribution, $N(\mu, \sigma^2)$, where the parameters $\mu$ and $\sigma^2$ would have known numerical values. Then 95% of the children will have heights in the range $\mu \pm 1.96\sigma$. If a child is chosen at random, the height of that child constitutes the value of a random variable and statements can be made about the probability of this random variable having a value in any specified range. For example, the probability is 0.95 that the height will be in the range $\mu \pm 1.96\sigma$, because 95% of all the children have heights in that range. In other words, an area under the normal curve $N(\mu, \sigma^2)$ specifies the probability that the random variable has a value within the range corresponding to the area. The random variable is then said to have the probability distribution $N(\mu, \sigma^2)$. In general, the probability distribution of a value chosen at random is mathematically identical to the frequency distribution of the set of values from which it comes.

## 0.22 BINOMIAL DISTRIBUTION APPROXIMATED BY NORMAL DISTRIBUTION

The shape of the binomial distribution approaches that of the normal distribution when the binomial parameter, $n$, is large (say, greater than 20) so long as $\pi$ is not close to 0 or 1. In these circumstances, a normal curve based upon the same mean and variance as that of the binomial will closely fit the histogram of the binomial distribution. The number of occurrences (of a particular event) will have the distribution $N\{n\pi, n\pi(1 - \pi)\}$ and the proportion of occurrences the distribution $N\{\pi, [\pi(1 - \pi)]/n\}$.

## 0.23 SAMPLING THEORY

Sampling theory is concerned with the relation between, on the one hand, the properties of a frequency distribution of a large set of values of a variable (known as the 'population' values), and, on the other hand, the properties of a random sample of values taken from the population. As an example, suppose that we know the income of everyone in a country; the income would then constitute the population values which would have a particular frequency distribution, mean and standard deviation. And suppose a random sample of say

100 individuals is taken giving 100 sample values. In the nature of things, no *definite* statement can be made about the sample values. Sampling theory is concerned with making *probability* statements such as: (i) there is a probability of 0.95 that the mean of the sample values will be within £100 of the population mean, or (ii) if the sampling procedure is repeated many times, the sample mean values will constitute a particular distribution.

Sometimes, as in this example, the population may be real and the sampling procedure a practical one, as is the case in economic and social sample surveys. But often, the population-sample relation may be introduced in the form of a statistical model of data which itself has not been obtained by a sampling procedure. This kind of model building is introduced substantively in Chapter 3.

## 0.24 SAMPLING DISTRIBUTIONS

Table 0.13 sets out some basic results. It is concerned with two quite different situations. The first, under the heading I, deals with the case where some property is either present or absent for each individual. In the second case, under heading II, each individual has the value of a numerical variable — height, income, distance and so on. The table follows the usual convention of using the letter $n$ for the size of the sample.

Part I of Table 0.13 is concerned with the relation between $p$, the proportion of individuals in the sample possesing a particular property, with $\pi$, the corresponding proportion in the population. $p$ has a binomial distribution (see Section 0.19) but the binomial distribution approximates to a normal distribution with the same mean and variance if $n$ is large (see Section 0.22). This implies probability statements such as:

the probability is 0.95 that $p$ is in the range $\pi \pm 1.96 \sqrt{\left(\frac{\pi(1-\pi)}{n}\right)}$

or

the probability is 0.99 that $p$ is in the range $\pi \pm 2.58 \sqrt{\left(\frac{\pi(1-\pi)}{n}\right)}$

Part II of Table 0.13 sets out the distribution results for the case where the observed variable is a continuous numerical variable. The distribution of the sample mean, $\bar{x}$, is described in terms of the population mean $\mu$ and the population variance $\sigma^2$. Again, the distributional

**Table 0.13** Sampling Distributions

| *Population parameters* | *Sample summary statistic* | *Distribution of sample summary statistic* | *Comment* |
|---|---|---|---|
| I. *Categorical variable* | | | |
| $\pi$, the proportion of individuals in population possessing specified property | $p$, the corresponding proportion in sample | $p$ is $N\left(\pi, \frac{\pi(1-\pi)}{n}\right)$ | Valid for 'large' samples, i.e. $n > 20$ if $\pi$ not too near 0 or 1. |
| II. *Numerical variable* | | | |
| $\mu$, the mean value<br>$\sigma^2$, the variance | $\bar{x}$, the sample mean | $\bar{x}$ is $N(\mu, \frac{\sigma^2}{n})$ | Valid if (i) population has normal distribution or (ii) if $n$ is large, whatever the population distribution |

result leads to probability statements such as:

the probability is 0.95 that $\bar{x}$ is in the range $\mu \pm 1.96\frac{\sigma}{\sqrt{n}}$ or

the probability is 0.99 that $\bar{x}$ is in the range $\mu \pm 2.58\frac{\sigma}{\sqrt{n}}$

## 0.25 SAMPLING INFERENCE

The distributional results of Table 0.13 are not directly useful in practice, for they describe the behaviour of, for example, the sample mean in terms of the usually unknown characteristics of the population. The typical practical question is: what can be said about the unknown population mean given information only from a single random sample? It is answered by the results of inference or estimation theory which itself is derived from distribution theory. *Estimation theory,* which is concerned with estimating population parameter values, falls under two heads: *point estimation* and *interval estimation*. Point estimation is concerned with the calculation of a single number as a 'best' estimate, whereas interval estimation involves the calculation of a range of values within which the unknown population parameter is likely to lie.

## 0.26 POINT ESTIMATION

Table 0.14 sets out the results for the sampling situations covered in Table 0.13. It will be noticed that two formulae are given for the population variance: the sample variance based upon a divisor of $(n-1)$, $s^2$, and that based upon a divisor of $n$, $S^2$. This complication is resolved in Chapter 4. When the sample size is large, there is very little difference between $S^2$ and $s^2$.

## 0.27 CONFIDENCE INTERVAL ESTIMATION: LARGE SAMPLE THEORY

Confidence interval estimation is a method of calculating a range of values within which a population parameter can be expected to lie. The essential structure of basic confidence interval estimation is set out in Table 0.15. In this section, we are concerned with so-called large

**Table 0.14** Point Estimates

| *Population parameter* | *Point estimate from sample* |
|---|---|
| I. *Categorical variable* | |
| $\pi$ | $p$ |
| II. *Numerical variable* | |
| $\mu$ | $\bar{x}$ |
| $\sigma^2$ | $s^2 = \Sigma \frac{(x-\bar{x})^2}{(n-1)}$ |
| | or $S^2 = \Sigma \frac{(x-\bar{x})^2}{n}$ |

large sample theory. In practical terms, the large sample results are usually accurate enough if the sample size is greater than about 20.

A confidence interval has the following features:

(i) It is set out in terms of a central value (e.g. the sample mean $\bar{x}$) and a given range above and below it (e.g. $1.96(s/\sqrt{n})$). The expression for the range is known as the sampling error, since it is a measure of the extent to which the sample mean is expected to differ from the population mean. We call $s/\sqrt{n}$ the *standard error;* it is, in fact, an estimate of the standard deviation of the sample mean.
(ii) A confidence interval has attached to it a percentage such as 95%. This percentage measures the probability that the confidence

**Table 0.15** Confidence Interval Estimates: Large Sample Results

| *Population parameter* | *95% confidence interval* | *99% confidence interval* |
|---|---|---|
| I. *Categorical variable* | | |
| $\pi$ | $p \pm 1.96\sqrt{\left(\frac{p(1-p)}{n}\right)}$ | $p \pm 2.58\sqrt{\left(\frac{p(1-p)}{n}\right)}$ |
| II. *Numerical variable* | | |
| $\mu$ | $\bar{x} \pm 1.96(s/\sqrt{n})$ | $\bar{x} \pm 2.58(s/\sqrt{n})$ |

interval range correctly includes the value of the population parameter.

(iii) In large sample theory the numbers multiplying the standard error (1.96 and 2.58 in the examples) are derived from the normal distribution. 1.96 is associated with a 95% confidence interval and 2.58 with a 99% confidence interval, just as the ranges (−1.96, 1.96) and (−2.58, 2.58) are associated with central areas of 0.95 and 0.99 respectively.

(iv) The choice of confidence interval (CI) is to some extent arbitrary, though the 95% CI is the one conventionally stated. Given the central value (e.g. the sample mean) and its estimated standard deviation ($s/\sqrt{n}$), any CI can be formulated.

**EXAMPLE 0.7**

A random sample of size $n = 100$ has a sample mean $\bar{x} = 38$ and standard deviation $s = 7$. The 99% CI for the population mean is calculated thus. First, the standard error $= 7/\sqrt{100} = 0.7$. Second, a central area (for N(0,1)) is associated with the value 2.58, from the normal distribution table. Hence 99% CI $= 38 \pm 1.8$, i.e. the range (36.2, 39.8).

## 0.28 MATHEMATICAL TOPICS AND NOTATION

In this section we set out some mathematics which will be used later.

### *0.28.1 EXPONENTIAL NUMBER AND THE EXPONENTIAL FUNCTION*

The mathematical constant $\pi$ (= 3.14. . .) is well-known. Another important constant is the exponential number e (= 2.718. . .). Both these constants occur in the formula for the normal distribution curve.

Powers of $e$, such as $e^2$, are frequently used in statistics. An alternative notation for $e^2$, is exp(2) (because it is easier and clearer to print). Hence, exp(2) $= (2.718..)^2 = 7.39....$ In general, $\exp(x) = e^x$ and $\exp(x)$ is known as the exponential function. The value of $x$ need not be a positive integer. Thus, exp(2.54) has in fact the numerical value 12.7. Readers unfamiliar with this mathematics need only know that such values are calculable (a facility which is offered on scientific calculators). If the graph of $y = \exp(x)$ is plotted, it presents a smooth upward sloping curve and, for example, the value of exp(2.54) lies between the values of exp(2) and exp(3). The expression $\exp(x)$ is also defined if $x$ is negative for

$$\exp(x) = \frac{1}{\exp(-x)} \quad \text{and} \quad \exp(-x) = \frac{1}{\exp(x)}$$

For example, $\exp(-2) = 1/\exp(2) = 0.14$. For $x = 0$, $\exp(0)$ is defined as equal to unity and, of course, $\exp(1) = e$.

### *0.28.2 LOGARITHMS AND THE LOGARITHMIC FUNCTION*

We first consider 'logarithms to the base $e$'. To every *positive* number there corresponds the value of its logarithm to the base $e$. The notation is $\ln(3)$, for the logarithm of 3, and in general $\ln(x)$ for the logarithm of $x$ to the base $e$. The logarithm value is derived from the exponential function relation. Given that, for example, $\exp(2) = 7.39$, $\ln(7.39)$ is defined as equal to 2. In general, if $\exp(x) = y$, then, by definition, $\ln(y) = x$. The value of $\ln(1) = 0$ (because $\exp(0) = 1$); very small positive numbers have very large negative logarithm values, and $\ln(0)$ is not defined.

Again, the reader unfamiliar with this area need not be particularly concerned with this background mathematics, but there are two properties which are of much practical significance. The first is that, for any two (positive) numbers $x$ and $y$:

$$\ln(xy) = \ln(x) + \ln(y);$$

that is to say, the logarithm of the product of two numbers equals the sum of their logarithms. The second is that, for any positive number $x$ and any positive or negative number $k$,

$$\ln(x^k) = k\ln(x).$$

Thus, $\ln(2) = 0.693$ and $\ln(3) = 1.099$ implies that $\ln(6) = 1.792$ and $\ln(9) = 2.197$.

Logarithms are also defined for any other base and base 10 was much used in pre-electronic calculator times to calculate products. The notation $\log_{10}x$ means the logarithm of $x$ to the base 10. The two basic properties apply whatever the logarithm base. So, for example,

$$\log_{10} 2 = 0.3010,\ \log_{10} 3 = 0.4771 \text{ and } \log_{10} 6 = 0.7781.$$

A numerical property of logarithms (to whatever base) is significant in statistical and other applications. It is that the logarithms of large numbers are very much smaller than the numbers themselves: $\ln(100) = 4.61$ and $\ln(1000) = 6.91$.

### *0.28.3 FUNCTIONS AND FUNCTIONAL NOTATION*

The exponential function and the logarithmic function are particular examples of mathematical functions. A function is typically a formula or specification which depends upon a variable, often indicated by the letter $x$, and which can be evaluated for any specified value of the variable. The variable is known as the *argument* of the function. The area of a circle is a function of its radius. The area, $\pi r^2$, can be evaluated for any value of the radius, the argument, $r$.

It is useful to have a notation for a function without specifying its form. The notation is $f(x)$ where the letter $f$ preceding the bracket is a label for the form of the function and $x$ is the letter standing for the argument. Other letters can be used: $g(y)$, $h(z)$ and $f(w)$ are all possible notations for a function. If the radius of a circle is labelled $r$, then the area is a function of $r$, which could be indicated by $A(r)$, where $A(r) = \pi r^2$.

In statistics, the letters $f$ and $F$ are almost exclusively reserved for particular kinds of functions (see Chapter 5).

### *0.28.4 FACTORIALS AND COMBINATORIAL ANALYSIS*

The factorial notation 3! (spoken of as 'factorial three') stands for 3.2 = 6. Generally, $x! = x(x-1)(x-2)\ldots 3.2$, where $x$ must be a positive integer; it is known as the factorial function.

A related function, met in some frequency curve formulae, is the gamma function, written as $\Gamma(x)$. Where $x$ is a whole number, $\Gamma(x) = (x-1)!$ But $\Gamma(x)$ can also be calculated and is tabulated for non-integral values. Thus $\Gamma(6.5) = 288$ which lies between $\Gamma(6) = 5! = 120$ and $\Gamma(7) = 6! = 720$.

Combinatorial analysis is, *inter alia*, concerned with calculating the number of ways in which $r$ different objects can be chosen from $n$ different objects. The formula for this calculation is $n!/[r!(n-r)!]$ and the notation for this expression is $\binom{n}{r}$. As an example, when $n = 4$ and $r = 2$, $\binom{n}{r} = 4!/2!2! = 6$ and the reader may verify this result by writing down all possible choices of two letters from the four letters $A$, $B$, $C$, $D$. The formula for $\binom{n}{r}$ can alternatively be written as

$$\frac{n(n-1)\ldots(n-r+1)}{r!}$$

# 1

# Data Classification

## 1.1 INTRODUCTION

When one takes more than a casual glance at the various kinds of data, it is clear that a wide variety of numerical information forms the raw material for statistical analysis. We are thinking here, not so much of the many subject areas in which statistics is applied, but rather of the numbers themselves. Some are counts (frequencies) of occurrences of various kinds: of the number of people in a social survey living by themselves, or of the number of incidencies of cholera in a particular country last year. Some are measurements: measures of physical constants in scientific experiments, economic measurements such as national income and prices, and measures of crop yield in agriculture. And again, there may or may not be some inherent order in the numerical data: incidences of cholera may be an ordered sequence of figures year by year for one country, or they may be figures for one year for many countries, where the country order has no significance.

In statistics, it is essential to be clear about the nature of the basic data, if only because the choice of an appropriate analysis may crucially depend upon it. The primary aim of this chapter is to formulate a system for recording and exhibiting basic data which will cope equally with the different kinds of data normally encountered in practice. Such a system will show what are the underlying common structural features of most statistical data sets and also what are their essential differences which need to be recognised in analysis. This data recording system will also help to make clear the relation between different kinds of data and the associated different methods of analysis.

## 1.2 THE STRUCTURE OF STATISTICAL DATA

The subject matter of applied statistics is varied and so is the numerical raw material itself, but fortunately there is a common structure which underlies virtually all information of a numerical nature which is susceptible to statistical analysis. One rather obvious superficial feature of statistical data is that it typically consists of *sequences* of numbers: numbers of cholera deaths year by year, estimates of the velocity of light in repeated experiments, crop yields resulting from the applications of different fertilisers. But there are other more significant common elements which are demonstrated by the following examples.

Consider first Table 1.1 setting out an extract of the data obtained in a population census of individuals. The census is of course hypothetical; the data are real. The information for each individual is set out row by row. The same kind of information is given for each individual: his or her age, sex, marital status and occupation. In other words, there is information about four 'variables' from this census; thus the variable 'sex' assumes either the descriptive 'value' male or female, and the variable 'age' any whole number up to, say, 100. Each column of the table sets out the various 'values' of each variable for the individuals covered by the census.

**Table 1.1** Extract from International Population Census 1784

| *Name* | *Age* | *Sex* | *Marital status* | *Occupation* |
|---|---|---|---|---|
| Austen, Jane | 8 | female | single | not applicable |
| Bonaparte, Napoleon | 16 | male | single | army cadet |
| George III | 46 | male | married | king |
| Haydn, Joseph | 52 | male | married | composer |
| Johnson, Dr. Samuel | 76 | male | widowed | retired |
| Mozart, Constanze | 22 | female | married | housewife |
| Washington, George | 52 | male | married | gentleman farmer |

**Table 1.2** Deaths from Cholera in London, 1622–1627

| *Year* | *No. of deaths* |
|---|---|
| 1622 | 16 |
| 1623 | 17 |
| 1624 | 11 |
| 1625 | 35,417 |
| 1626 | 134 |
| 1627 | 4 |

*Source:* Graunt, Capt. John, F.R.S., *Natural and Political Observations on the London Bills of Mortality, 1662*

Now we turn to Table 1.2 giving deaths by cholera in London between 1622 and 1627. Although the subject matter is very different, the structure of this table has much in common with that of the previous Table 1.1. The second column, like those of Table 1.1, shows the values of a particular variable, namely, 'number of deaths' for each of a number of years. The first column names the individual years concerned, just as the first column of the previous table names the people concerned. In other words, both tables set out, row by row, information about 'individuals' (individual people in one case and individual years in the other). And for each 'individual', there is information, column by column, about the values that individual assumes for the variables concerned.

It will be apparent that we have started to use the word 'individual' in a much wider sense than its everyday usage and, indeed, will continue to do so. Specifically an individual will mean an entity relating to which there is information. A statistical data table puts together information relating to a number of such entities, or individuals, the first column giving their names or labels (personal names in the first example and year names in the second) and later columns the information about them—the variable values which they assume.

The reader should find no difficulty in fitting Tables 1.3 and 1.4 into this structure.

## 1.3 STATISTICAL VARIABLES

In the previous section, a variable has been loosely defined as information of a particular kind about individuals. The examples have

**Table 1.3** The Results of Four Experiments to Determine the Weight of a Standard Volume of Nitrogen

| *Experiment date* | *Weight* |
|---|---|
| 29.11.93 | 2.30143 |
| 5.12.93 | 2.29816 |
| 6.12.93 | 2.30182 |
| 8.12.93 | 2.29890 |

*Source:* Lord Rayleigh, *On an Anomaly Encountered in Determinations of the Density of Nitrogen Gas,* 1894

**Table 1.4** Crop Yields on Four Plots to which Different Amounts of Fertiliser have been Added

| *Plot* | *Fertiliser application (tonnes per hectare)* | *Yield (tonnes per hectare)* | *Yield: Rank Values* |
|---|---|---|---|
| A | 0.52 | 4.8 | 3 |
| B | 0.40 | 4.3 | 1 |
| C | 0.65 | 5.2 | 4 |
| D | 0.53 | 4.7 | 2 |

*Source: hypothetical*

indicated that the term is used to cover both numerical information—age in years, velocity in metres per second and so on—but also information of a descriptive nature—whether an individual is male or female and whether married or single. A classification of variables will now be set out. At the first level of categorisation this basic distinction is made between numerical and descriptive variables, but within these broad categories there are important sub-categories.

### *1.3.1 NUMERICAL VARIABLES*

Numerical variables are also synonymously known as quantitative variables. A numerical variable is a property which, for any one individual, can be specified or observed by a positive or negative number, integral or decimal.[1] All the variables in the tables above are numerical except for the descriptive variables sex, occupation and marital status in Table 1.1. Numerical variables fall into the following sub-categories:

(i) *Continuous variables* These are numerical variables which, in principle, can take any numerical value—decimal or otherwise—within the relevant range of variation of the measured property. Weight (Table 1.3) is an obvious example of a continuous variable. The numerical variable 'height of an individual' is another; in principle a height may be any number, with any number of decimal places, within the range of human height variation. Continuous variables are to be distinguished from:

(ii) *Discrete variables* These are the variables which take only a limited set of possible numerical values differing discretely from each other. The variable 'number of deaths by cholera' in Table 1.2 is a discrete variable since it can have only the values 0, 1, 2 . . . and any positive integer; it is an example of what is known as an *integer variable.* As in this example, values of the variable often arise from counting the frequencies of occurrence of some event. We shall call such a variable a *frequency variable.* It may be important in analysis to distinguish this kind of integer variable from an ordinary measurement which happens to be expressed in whole number terms, such as the age variable in Table 1.1.

The distinction between continuous and discrete variables may not be clear-cut in practice. In principle, height is a continuous variable, but in practice any measurement of height or any other entity is limited by the accuracy of the measurement process. If heights of persons are measured in metres correct to say one-tenth of a metre, then in practice this height variable is a discrete variable since it can take only values of the kind 1.1, 1.2, 1.3 and so on. However, this indeterminacy is rarely a problem in applied statistics. Although statistical theory makes a clear distinction between continuous and discrete

1. Within this category, a further distinction is often made between 'ratio scale' and 'interval scale' variables. The former are numerical variables where the zero point is absolutely defined—such as for heights—and the latter where the zero point is arbitrary—such as most temperature scales. This does not seem to us a useful distinction, partly because it is often not clear-cut and partly because it rarely matters for data analysis. It may however be relevant to the interpretation of the results of analysis.

variables, no difficulties usually arise in applying the theory of continuous variables to variables which in practice are discrete in the above sense.

### *1.3.2 CATEGORICAL VARIABLES*

These are sometimes termed qualitative variables, descriptive variables, or factors. Two examples of categorical variables were met in Table 1.1: the variable 'sex' took on, for each individual, one of the values male, female and the variable 'marital status' one of the values single, married, widowed. These are typical of categorical variables in that the values are specified, not in terms of numerical measurements or counts, but in descriptive terms. Specifically, a categorical variable is related to some concept or property and its values are specifications of the categories into which the concept or property can be classified. Categorical variables are almost inevitably discrete, since the number of categories which can be formulated, at least in practice, are limited in number. Sometimes it is difficult or tedious to set out all the values of a categorical variable and statisticians and others often fall back on the device of denominating only the more important values or categories and then defining a final value as 'not elsewhere specified'. A categorical variable which has only two values (such as the variable 'sex') is known as a *dichotomous variable.*

There are two sub-groups of categorical variables: ordinal and nominal variables.

(i) *Ordinal variables* are categorical variables whose values have an implicit or explicit ordering to them. Thus, suppose a TV company carries out a survey of reactions to a particular programme and responses are obtained under the heads: strongly dislike, dislike, indifferent, like, strongly like. The five possible responses are the five values of the variable and they have the property that in the order set out, neighbouring values have greater affinity than values separated from each other in the list. There is a natural ordering to this list of values.

(ii) *Nominal variables* on the other hand are categorical variables for which there is no ordering to the values. Thus, if the eye-colour of a set of individuals is recorded, the values of this variable might be defined as brown, blue, green, other. There is no natural order here. 'Blue' is no nearer to 'brown' than is 'green'. Any other ordering would be as satisfactory as the one given (apart from putting 'other' last for semantic reasons).

The classification of categorical variables into the sub-groups 'ordinal' and 'nominal' may appear clear-cut, but often in practice this is not so. Consider the variable 'political attitude of an individual'. The values could be defined in terms of the political party supported by the individual. The variable could be said to be a nominal variable, but on the other hand, if the parties were to be classified in order on some left-right political continuum, then the variable would be an ordinal variable. In other words, whether a categorical variable is ordinal may depend, not on some obvious inherent property of the variable, but on whether there is underlying theory from which an ordering may be inferred.

Up to this point, variables have been classified into two main groups: numerical and categorical. Overlapping these two, there is another kind of variable which may present particular problems in statistical analysis, namely the rank variable.

### *1.3.3 THE RANK VARIABLE*

Individuals are said to be put in 'rank order' when they are arranged in order according to some numerical or descriptive criterion. Conceptually, the simplest kind of ranking is one which is based upon values of a numerical variable. For example, the plots in Table 1.4 can be arranged in order of yields from, say, lowest to highest, giving the sequence B, D, A, C. Each plot can now be given a 'rank number' or rank value from 1 to 4, the plot B having the rank value 1 and the plot C the rank value 4. And if the rank values are added as another column to table 1.4, we have in order the following values of this particular rank variable: 3, 1, 4, 2. Note that a different ranking would have been obtained if the plot yields had been arranged in the reverse order from highest to lowest. It is usually an arbitrary matter which is chosen and fortunately statistical procedures for the analysis of rank variables do not depend upon it, although in this text, the *lowest* value (where appropriate) will be assigned to the rank value 1.

Individuals may also be put in rank order in relation to some descriptive criterion, that is, in relation to some categorical variable. Clearly, such a categorical variable must be ordinal since by definition values of nominal variables cannot be ordered. Consider for example the census data in Table 1.1 and particularly the categorical variable 'marital status'. This could be thought of as an ordinal variable if its values are ordered in relation to possible states in the life sequence: single, married, widowed.

A problem arises if two or more individuals have the same variable value. For descriptive purposes, it may be sufficient simply to assign successive rank values to individuals with the same variable value.

**Table 1.5** Classification of Statistical Variables

| *Name* | *Description* | *Examples* |
|---|---|---|
| 1. *Numerical or Quantitative Variables* | | |
| (a) Continuous | An infinite or 'very large' number of possible values can be taken up (within appropriate ranges): positive or negative, decimal or integral. | Measurements of various kinds: length, weight, speed. Also discrete measurements (such as money values) where the intervals between values are small. |
| (b) Discrete (i) Frequency (or other integer) | A limited number of values. Positive or negative integers or zero. | Counts of frequency of occurrences (frequency variable). Measurements rounded to the nearest whole number. |
| (ii) Other | | Otherwise rounded data. |

Suppose, for example, eight individuals have the following values for a particular numerical variable: 20, 21, 21, 24, 27, 27, 27, 30. Then ranks would be arbitrarily allocated as 1, 2, 3, 4, 5, 6, 7, 8, respectively. However, in statistical methods based upon rank variables it is sometimes required that tied ranks should be averaged in the following way: the two individuals with values 21, 21, are first allocated the ranks 2, 3 which are then averaged to 2.5, 2.5. Similarly, the individuals with values 27, 27, 27 are each given a rank equal to the average of 5, 6, 7, that is an average rank of 6. The final rank values then are 1, 2.5, 2.5, 4, 6, 6, 8.

The classification of statistical variables is summed up in Table 1.5.

| *Name* | *Description* | *Examples* |
|---|---|---|
| 2. *Categorical or Qualitative Variables* | | |
| (a) Ordinal | Descriptive categories which have an inherent or explicit ordering. | Responses specified in terms of degrees of like/dislike. Subjective assessment of intensities: cold, warm, hot; slow, fast, very fast. |
| (b) Nominal | Non-ordinal categorical variables. | Colour. Occupation. Modes of transport. Brands of detergent. |
| 3. *Rank Variables* | Each of $n$ individuals is given one of the numbers 1, 2, 3, . . . $n$ based upon the ordered values of a numerical or ordinal variable. | Ordering of: (i) Persons in order of age, height, or seniority; (ii) Cars in order of petrol consumption per mile; (iii) Countries in order of per capita national income. |

## 1.4 THE DATA MATRIX

In Section 1.2 the reader was introduced to the idea that, in general, the raw material of statistical analysis—the data—can be set out in the form of a table in which each row shows information for a given individual and each column the set of values for a particular variable. It was noted that the word 'individual' in statistics refers not only to human or living individuals, but also more widely to any entity which is a unit of record or observation. And in Section 1.3 there was a discussion of the various kinds of variables which enter into statistical analysis, that is, the various kinds of columns in the data table. We

**Table 1.6** Structure of Primary Data Matrix

| *Individual label* | *Variables* | | | | |
|---|---|---|---|---|---|
| | Var 1 | Var 2 | Var 3 | . . . . . . | Var $p$ |
| Individual 1 | | | | | |
| Individual 2 | | | | | |
| Individual 3 | | | | | |
| . . . . . . | | | | | |
| . . . . . . | | | | | |
| Individual $n$ | | | | | |

shall call such a table, containing all the information about the individual variables, and variable values, the *primary* data matrix. Schematically, the primary data matrix has the form of Table 1.6. The first column serves to identify the particular individual. It may be a name or some description which we shall call an *individual label.* In Table 1.6, the letter $n$ has been used to stand for the number of individuals and $p$ for the number of variables, a common convention particularly in multivariate analysis and one which will be preserved throughout this text. In such a table there will of course be $n \times p$ cells, that is $n \times p$ items of information.

It is always an aid to clear thinking in applied statistics if the primary data matrix is formulated so that all the variables under consideration, numerical or categorical, are each given a 'variables column' in the matrix. The first column is reserved for the individual label. One must guard against mixing up individual labels and information on variable values with a consequential confusion about the purpose and method of the statistical analysis. Table 1.2 is in danger of falling into this trap. An analysis of this data is likely to be concerned with an examination of the evolution of the number of deaths from cholera over time, that is, a study of the values of the variable 'number of deaths' in relation to the variable 'time'. Time may well be measured in terms of the year number 1622, 1623 and so on. If it is, then the years are not only 'individual labels' but also 'variable values'. Confusion is removed by making the distinction explicit and

**Table 1.7**

| *Individual label (year)* | *Variables* | |
|---|---|---|
| | *Time* | *No. of deaths* |
| 1622 | 1622 | 16 |
| 1623 | 1623 | 17 |
| 1624 | 1624 | 11 |
| 1625 | 1625 | 35,417 |
| 1626 | 1626 | 134 |
| 1627 | 1627 | 4 |

*Source:* Table 1.2

setting out the primary data matrix as in Table 1.7. This makes clear not only the relevant variables, but also shows exactly how the variable 'time' is being measured—in this case by the year numbers.

As another example, consider a data set consisting of ten experimental results: one numerical measurement obtained from each of ten experiments. The primary data matrix would have two columns: the first giving the experiment label, and the second the values of the variable 'experimental result'. But now, suppose the experimenter knows that experiments 1–5 were carried out under materially different conditions from those of experiments 6–10, and he is interested in whether the different conditions have affected the results. In other words, he is now concerned with comparing the results of two groups of experiments. He has in effect introduced an additional variable—a categorical variable—where experiments 1–5 have the value 'group 1' and experiments 6–10 have the value 'group 2'.

When the primary data matrix has been formulated in accordance with the above principles, conventional statistical analysis is in general concerned only with the information in the variables columns. For this reason, the primary data matrix excluding the column of individual labels is simply known as *the* data matrix.

## 1.5 DATA SUMMARISATION AND DATA AGGREGATION

Although the data matrix forms the raw material of statistical analysis, it is often necessary, desirable, or convenient as a first step to condense the detailed information in the data matrix and produce appropriate summary statistics before proceeding further. These, and

**Table 1.8** Aggregated Data Matrix

*Original Data Matrix (extract)*

| | *Age* | *Sex* | *Marital status* | *Place of residence* | . . . . . . |
|---|---|---|---|---|---|
| Anderson | 54 | Male | Single | Newtown | |
| Bloggs | 23 | Female | Divorced | Plainsville | |
| Carmichael | 39 | Male | Married | Sutton | |
| . . . . . . | | | | | |

*Aggregated Data Matrix (extract)*

| | *Population* | *Males* | *Females* | *Age distribution* 0–4 | 5–9 | 10–14 | . . . . . . |
|---|---|---|---|---|---|---|---|
| Newtown | 5,439 | 2,633 | 2,806 | 624 | 417 | 350 | |
| Plainsville | 682 | 357 | 325 | 57 | 48 | 41 | |
| . . . . . . | | | | | | | |

*Source:* hypothetical data

other questions concerning preliminary data analysis, will be discussed in Chapter 2.

The concept of 'data aggregation' or 'aggregated data' is best understood by way of an example. Consider a population census of, say, 50 million people in which 20 distinct questions are asked. Then the results of this census constitute a very large data matrix with 50 million rows, one for each person and 20 columns, one for the responses to each question. This kind of data matrix is never published by census authorities, partly due to its size and partly for reasons of confidentiality. Instead, other procedures are followed. One is to reduce the size of the data matrix by tabulating corresponding information for each local authority in the country rather than each individual. Table 1.8 illustrates how this aggregation process takes place. The result is an aggregated data matrix. Its 'individuals' are now the local authority areas and the variables are related to, though not identical with, the variables of the original data matrix. Note that the aggregated data matrix is formed by grouping together people with identical values for one of the census variables, in this case the variable 'place of residence'. The numerical values are all integer frequency counts in the example, though this need not be so; for example, 'number of males' could be replaced by 'percentage of males'.

**Table 1.9** Cross-Classification of Sex and Marital Status

*Thousand population*

| | Marital status | | | | |
|---|---|---|---|---|---|
| | *Single* | *Married* | *Widowed* | *Divorced* | *Total* |
| *Sex* | | | | | |
| Male | 12,688 | 8,242 | 3,214 | 1,144 | 25,288 |
| Female | 13,026 | 8,286 | 3,655 | 1,051 | 26,018 |
| Total | 25,714 | 16,528 | 6,869 | 2,195 | 51,306 |

*Source:* hypothetical data

Summary tables of various kinds may also be produced from the information in a large data matrix, usually in the form of 'cross-classifications'. Table 1.9 shows a cross-classification of two variables in the original data matrix. This table shows the 'relation' between two categorical variables recorded in the census by, in effect, allocating each member of the population to one of the cells in the table, the cells being determined by the individual values of the two categorical variables. The numbers are frequencies and tables of this kind are called 'contingency tables'. (Their analysis is substantively discussed in Chapter 15.)

The discussion so far may have suggested that there is a clear-cut difference between summarising by means of an aggregated data matrix on the one hand and by cross-classification of variables on the other. This is not so. The aggregated data matrix of Table 1.8 can itself be looked upon as a composite cross-classification of the census variable 'place of residence' with other variables such as sex, marital status and so on: at one and the same time, it has the structure of a data matrix and is also a summary table of the original data matrix. This fuzziness does not matter. What is important is that where a particular kind of statistical analysis calls for data in the form of a data matrix, it should be clearly established that the data has such a structure, that the individuals are relevant to the purpose of the analysis, and that the nature of the variables is clearly understood.

There are of course many other forms of summary tables. The tables of economic and social data to be found in government publications may at first sight seem very different from the examples of Tables 1.8 and 1.9. However, a little study usually reveals that their structure and origins are very simple indeed. Consider the two examples

**Table 1.10** Examples of Summary Tables

(a) *Universities: Courses taken by Full-Time Students in Great Britain*

| *Academic years* | | *Number of students* 1967/8 | 1968/9 | 1969/70 | ... |
|---|---|---|---|---|---|
| Education | Men | 3,731 | 4,197 | 4,440 | |
| | Women | 3,391 | 3,918 | 4,061 | |
| Medicine, dentistry & health | Men | 15,244 | 15,733 | 15,901 | |
| | Women | 5,251 | 5,477 | 5,838 | |
| Engineering & technology | Men | 31,848 | 33,535 | 34,596 | |
| | Women | 500 | 527 | 695 | |
| ... ... | | | | | |

(b) *Non-Ferrous Metals*

| | *Thousand tonnes* 1967 | 1968 | 1969 | ...... |
|---|---|---|---|---|
| Copper | | | | |
| Production of refined copper | | | | |
| Primary | 36.0 | 49.7 | 49.3 | |
| Secondary | 133.4 | 148.0 | 148.9 | |
| Home consumption | | | | |
| Primary refined | | (not reported) | | |
| Secondary refined | 514.2 | 539.1 | 546.8 | |
| Scrap metal content | 141.6 | 140.5 | 143.6 | |
| ... ... | | | | |

*Source:* Central Statistical Office, *Annual Abstract of Statistics*, 1979 Tables 5.62 and 8.41.

in Table 1.10 taken from the United Kingdom *Annual Abstract of Statistics*. Both tables offer time series data where the same data structure is repeated from one year's figures to another. In the case of Table 1.10(a), each year's figures were probably compiled from a corresponding summary table for each university. And each university table would have been based upon a data matrix in which the rows represent individual full-time students and the columns the values of the two categorical variables 'sex' and 'type of course taken'. The precise origins of the data in Table 1.10(b) are more uncertain. The units of observation for, e.g. the production figures were probably individual firms. If so, the table was in effect derived from a data matrix in which the individuals were the firms and the two numerical variables were 'production of refined copper, primary' and 'production of refined copper, secondary'.

Summarisation of a data matrix inevitably means a loss of information. Two very different cases need to be distinguished. The first is where the statistical analyst has the information in the data matrix: the loss of information in summarisation is intentional and amounts to removing the fine detail in the data matrix so that patterns can be revealed and appropriate analysis carried out. The second is where the summarisation has been made by someone else (e.g. an official statistical body); the data matrix is not published and the only available information is in the summary tables. In this case, the conclusions of any analysis will be strictly limited by the nature of the summary tables. (Thus the aggregated data matrix of Table 1.8 shows both the sex distribution and age distribution for each local authority but there is no information about the sex distribution of, e.g. the 624 children in Newtown aged 0—4.) Nevertheless, it is often enlightening to try to imagine what kind of data matrix gave rise to the summary tables: who or what were the individuals and what were the original variables.

## 1.6 COMPUTATION

Programs for the analysis of statistical data typically require the data to be input in the earlier stages of setting up the analysis on the computer. The data is almost always input in the form of a data matrix as it has been presented in this chapter with each row relating to the information for an individual and each column relating to a variable. If data are input via an on-line terminal, the variable values for each individual are typed in row by row with of course the variable values always in the same order.

Some program packages allow only the analysis of numerical variables. Others also cover categorical variable analysis in which case the variable values, which are verbal descriptions, are entered in abbreviated form with a constraint on the number of characters (letters or numbers) which may be used.

Programs do not usually cater specifically for the facility of entering the individual labels. One reason is that this is often done within the program in the sense that successive rows are automatically allocated labels such as Observation 1, Observation 2, etc. Another is that if there is a facility for categorical variable analysis, a variable column can be used to record, at least in coded form, the individual labels.

## 1.7 CONCLUSION

The taxonomy of data types set out in this chapter will underlie the organisation of later material. Chapter 2 initiates the discussion of univariate analysis, that is, where either the data matrix has a single column or the analysis proceeds separately on each of a number of variables. Bivariate analysis is introduced in Chapter 4; here the analysis proceeds simultaneously on two from possibly several variables in the data matrix. Multivariate analysis, where the data matrix has many columns and there is interest in the interactions between a number of variables, is discussed in Chapters 8 *et seq*.

The main concern in all these chapters is the analysis of numerical variables. Various methods of categorical variable analysis are presented in Chapter 15.

## EXAMPLES

**E 1.1** In terms of the classification of Table 1.5, identify the nature of the variable underlying the following data:

(i) Annual population figures of a country
(ii) Numbers on the shirts of players in a football team
(iii) Colours on paint tins in a shop
(iv) List of members of the United Nations
(v) Petrol consumption figures of various cars.

**E 1.2** In what sense, if any, can the following be said to constitute in effect a data matrix? Where appropriate, identify the individuals and the variables:

(i) Telephone directory
(ii) Classified telephone directory
(iii) Football league table
(iv) List of examination results
(v) Railway timetable
(vi) Company balance sheet.

## FURTHER WORK

**F 1.1** Examine some published data tables and, for each, consider the following questions:

(i) Is the table a primary data matrix, an aggregated data matrix or a summary table of some kind?
(ii) If it does not appear to be a primary data matrix, what individuals and what variables are embodied in the data table?

*This chapter depends upon: Sections 0.2–0.5*

# 2

# Univariate Descriptive Analysis

## 2.1 DATA ANALYSIS

As this is the first chapter in which techniques of analysis are presented, it seems desirable to begin by saying something about the actual process of statistical analysis before embarking upon an account of the techniques themselves.

The analysis of data should be thought of as proceeding in steps, each of which produces some results. In the beginning, the analyst has certain basic information which constitutes the raw material. This will include:

(i) The data itself, which may be in the form of a data matrix or, if some preliminary summarisation has been carried out, in the form of a frequency distribution.
(ii) Information about how the data were acquired.
(iii) Information about the definition of the variables and some knowledge, possibly vague, about the individuals with which the data values are associated.
(iv) Some theory or theories, again perhaps vague, concerning the possible causal associations between some or all of the variables.

There will also be one or more objectives which the analyst will hope to achieve, and we may mention in particular:

(i) Improving the theory about possible causal links between the variables, either by showing that it applies to a set of individuals which have not been the subject of previous investigation, or that it applies in circumstances which have not been previously considered.

(ii) Learning something more about the individuals themselves, perhaps by pinpointing those to whom the existing theory does not seem to apply.

(iii) Assessing the likely impact of policy changes or making better predictions about the possible consequences of leaving things as they are, this being, very often, the objective of investigations in the social/economic field.

This list is not exhaustive.

Finally, the analyst has a 'tool kit' of statistical ideas and techniques, a selection from which will be brought to bear upon the basic information, thereby producing some results. The technique selected will be directed by the basic information available and by the objectives to be achieved. The results will require interpretation and this, in turn, requires some part of the stock of basic information to be drawn upon and the exercise of some judgment on the analyst's part. From there we may proceed either to the next step in the analysis or conclude that the objectives have been fulfilled as far as possible.

To sum up, analysis is more than just the selection of an appropriate technique; in addition, the necessary background knowledge must be available and judgment must be used in interpretation if the analysis is to be effective. We will refer from time to time to the 'iterative' and 'interactive' nature of data analysis as a reminder of the need to keep in touch with whatever is known and of the need to proceed in steps, some of which may be inconclusive, rather than treating analysis as a once and for all pass through the data.

## 2.2 DESCRIPTIVE UNIVARIATE ANALYSIS

We begin with descriptive univariate analysis for three reasons. First, in any study, attention will often be drawn from the start to one (or no more than a few) of the variables, and finding out for these variables something about the way the data values vary from individual to individual may prove a useful first step towards solving the problem posed by the objectives of the study. Secondly, the interpretation of the results of more elaborate analysis may be assisted by (or may even require) knowledge of the variation present in each variable entering the analysis. Finally, the basic procedures, formulae and notation of univariate descriptive analysis underlie many of the more advanced procedures.

The scope of this chapter is limited in two ways. First, we will confine ourselves only to the case where the variable of interest is numerical. The main use of categorical variables is to identify broad groupings or common backgrounds of individuals, and they are therefore of subsidiary interest at this stage. Secondly, we assume that the order in which the values are presented is of no particular analytical significance; the values might, for example, be presented according to the alphabetical order of the individual labels. Analysis which takes explicit account of the row ordering of the data matrix (such as may be desirable, for example, when the data is a time series) is an important subject to which we shall return in Chapter 9.

Our starting point is the fact that individuals exhibit *variation*, that is to say they differ from each other to a greater or lesser extent and these similarities and differences are reflected in the corresponding values of the variables. The basic task of univariate descriptive analysis is to describe the variation in a set of values of a single variable, and we may identify two broad aspects of this. The first is a matter of producing an appropriate set of summary statistics, for example the mean and the variance, which provide a quick impression of the main features of the data, and which may form a basis for comparisons between two or more groups of individuals. The second aspect is a matter of uncovering, by visual and other means, such things as patterns in the way the values occur (for example, groups of individuals who have similar values), the exceptions to those patterns and anything which appears unusual or unexpected. At this level, no attempt is made to generalise from the set of data on hand to further data on the same variable for different individuals; we are simply making a description, noting features of possible interest and raising questions. Univariate analysis of this kind is sometimes disparaged as being 'mere description', wrongly in our opinion, since careful description and display may yield insights into the causes of the variation and may provide some guidance as to how the analysis should proceed.

The ideas of 'summarising' on the one hand and uncovering patterns on the other provide one rough classification of the techniques of univariate analysis. A second rough distinction may be made by noting that, on some occasions we will want to look at the detail in the variation and on others we will want to stand back and look at the variation as a whole. These distinctions will provide an approximate classification of the techniques and will give us a guide to their effectiveness for different purposes. Analysis which looks at the detail will be referred to as 'micro' analysis; analysis which is concerned with the overall variation will be called 'macro' analysis.

## 2.3 NOTATION

The raw material of the analysis is a univariate data set which consists, in general, of $n$ values of some numerical variable, possibly accompanied by the individual labels. The variable itself will be denoted by an upper case letter, for example $X$, and the values by $x_1$, $x_2$, ... $x_n$. In formulae, $x_r$ will be used to indicate a 'general' value of the $r$th individual. Rank values, from the smallest to the greatest, will be denoted as $x_{(1)}$, $x_{(2)}$, ... $x_{(n)}$ (cf. Section 0.2).

To avoid unnecessary repetition, we will usually refer to a univariate data set simply as a data set; later in the book, it will be necessary to distinguish between data sets which contain values of just one variable and those whose structure is more complex. The terms 'individual value' and 'data value' will be used interchangeably to denote the value associated with an individual. A data set whose values have been arranged into rank order will be called an ordered data set.

A value of $X$ is said to be *feasible* if it is a possible value in terms of the definition of the variable, even though it may not necessarily equal any of the data values; it will be denoted simply as $x$, with no subscript. For example, 15% is a feasible value for the variable referred to in Table 2.1 below, but no individual has that value. It may still be of interest, however, to ask 'which individuals have a data value less than 15% per year?'

In general, the letter $x$ will be used in the formulae given in this chapter and elsewhere. In particular instances, other letters for a variable may be convenient and in Chapter 3, for example, we use the letter $t$ (for measurements of time). There will be a need for a little more notation as the chapter proceeds, but this will be explained as it arises.

## 2.4 LOOKING FOR PATTERNS

To start the discussion, we present in Table 2.1 a simple data set which typically arises in the context of inter-country comparisons. Interest here is likely to focus particularly on the detail of the variation rather than upon the data set as a whole, since we are likely to be interested in

**Table 2.1** Average Annual Increase in Consumer Prices per cent per year, 1972–1977. Countries of the OECD.

| *Country Name* | *Label* | *Value* | *Rank* |
|---|---|---|---|
| Australia | AUST | 13.1 | 15 |
| Austria | AUS | 7.7 | 3.5 |
| Belgium | BEL | 9.7 | 10 |
| Canada | CAN | 8.9 | 7 |
| Denmark | DEN | 10.8 | 12 |
| Finland | FIN | 14.5 | 16 |
| France | FRA | 10.3 | 11 |
| Germany | GER | 5.7 | 2 |
| Greece | GRE | 16.1 | 17.5 |
| Iceland | ICE | 34.7 | 24 |
| Ireland | IRE | 16.1 | 17.5 |
| Italy | ITA | 16.5 | 20 |
| Japan | JAP | 12.9 | 13 |
| Luxembourg | LUX | 8.6 | 5 |
| Netherlands | NTH | 8.7 | 6 |
| New Zealand | NZ | 13.0 | 14 |
| Norway | NOR | 9.4 | 8 |
| Portugal | POR | 21.3 | 23 |
| Spain | SPA | 17.2 | 21 |
| Sweden | SWE | 9.6 | 9 |
| Switzerland | SWI | 5.6 | 1 |
| Turkey | TUR | 20.4 | 22 |
| United Kingdom | UK | 16.3 | 19 |
| United States | US | 7.7 | 3.5 |

*Source:* OECD Observer No. 91, March 1978.

knowing which countries are similar, which have low rates of price increases, and so on.

At this first stage of the discussion, we shall be concerned with the analysis of 'relatively small' data sets such as in Table 2.1. The final column shows the rank order of each value and, to make a start with the task of description, we will do two things:

(i) Sort the values (with their labels) into rank order, thereby producing Table 2.2. (The last two columns should be ignored for the present.) Note that Greece and Ireland have the same value (16.1) as have Austria and the United States (7.7). These pairs have,

**Table 2.2** Data of Table 2.1 Rearranged in Rank Order

| *Rank* | *Label* | *Recorded value* | *Recorded value minus median* | *Standardised values* |
|---|---|---|---|---|
| 1 | SWI | 5.6 | −6.25 | −1.7 |
| 2 | GER | 5.7 | −6.15 | −1.6 |
| 3.5 | AUS | 7.7 | −4.15 | −1.1 |
| 3.5 | US | 7.7 | −4.15 | −1.1 |
| 5 | LUX | 8.6 | −3.25 | −0.9 |
| 6 | NTH | 8.7 | −3.15 | −0.8 |
| 7 | CAN | 8.9 | −2.95 | −0.8 |
| 8 | NOR | 9.4 | −2.45 | −0.7 |
| 9 | SWE | 9.6 | −2.25 | −0.6 |
| 10 | BEL | 9.7 | −2.15 | −0.6 |
| 11 | FRA | 10.3 | −1.55 | −0.4 |
| 12 | DEN | 10.8 | −1.05 | −0.3 |
| 13 | JAP | 12.9 | 1.05 | 0.3 |
| 14 | NZ | 13.0 | 1.15 | 0.3 |
| 15 | AUST | 13.1 | 1.25 | 0.3 |
| 16 | FIN | 14.5 | 2.65 | 0.7 |
| 17.5 | GRE | 16.1 | 4.25 | 1.1 |
| 17.5 | IRE | 16.1 | 4.25 | 1.1 |
| 19 | UK | 16.3 | 4.45 | 1.2 |
| 20 | ITA | 16.5 | 4.65 | 1.2 |
| 21 | SPA | 17.2 | 5.35 | 1.4 |
| 22 | TUR | 20.4 | 8.55 | 2.3 |
| 23 | POR | 21.3 | 9.45 | 2.5 |
| 24 | ICE | 34.7 | 22.85 | 6.1 |

therefore, in accordance with the procedure described in Section 1.3.3, been assigned the ranks 17.5 and 3.5 respectively.

(ii) Plot the values as points on a line, as shown in Figure 2.1. A diagram such as this will be referred to as a line plot.

An alternative to the line plot, which requires less effort and is often good enough for a first look at a univariate data set, is provided by what Tukey (1977) has called a 'stem and leaf' plot. To illustrate the construction of such a plot we first note that several of the data values in Table 2.1 have the same integer part. For example, the values of Switzerland and Germany both have an integer part of 5. We select the integer part as a 'stem' on which there are, in this case, two 'leaves',

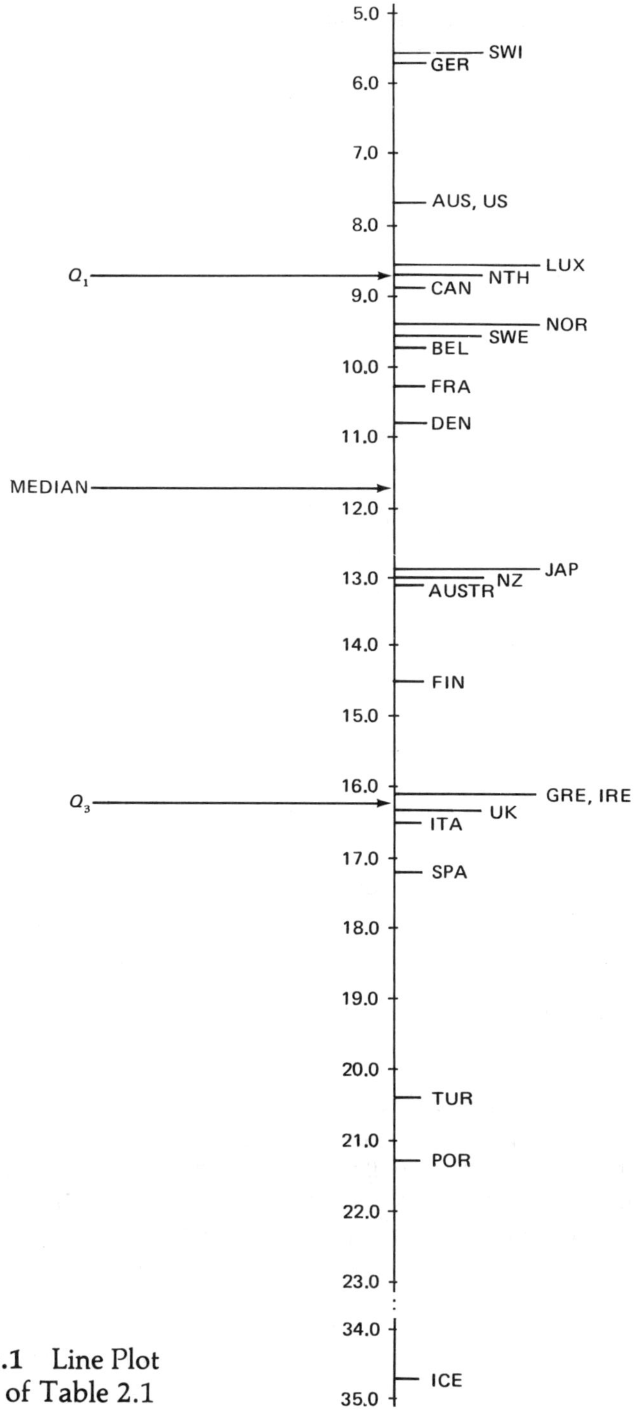


**Figure 2.1** Line Plot of Data of Table 2.1

**Table 2.3** Stem and Leaf Plot of Data in Table 2.1

| *Count* | | | |
|---|---|---|---|
| 2 | 5 | 67 | SWI, GER |
| | 6 | | |
| 4 | 7 | 77 | AUS, US |
| 7 | 8 | 679 | |
| 10 | 9 | 467 | |
| 12 | 10 | 38 | |
| | 11 | | |
| 13 | 12 | 9 | |
| 15 | 13 | 01 | |
| 16 | 14 | 5 | |
| | 15 | | |
| 20 | 16 | 1135 | |
| 21 | 17 | 2 | SPA |
| | 18 | | |
| | 19 | | |
| 22 | 20 | 4 | TUR |
| 23 | 21 | 3 | POR |
| | 22 | | |
| | ... | ... | |
| 24 | 34 | 7 | ICE |

the decimal part; these are 6 and 7 corresponding to Switzerland and Germany respectively. These two values are then written as

5 | 67

and similarly, the value for Portugal is written

21 | 3

the latter being a 'stem' (21) with a single 'leaf'. If we write all the data values in this way we obtain the 'stem and leaf' plot of Table 2.3 which may be labelled (or partly labelled, as here) with the country identifier.

The column headed 'count' is a cumulative count of the number of leaves on each stem and on the preceding stems and provides a quick indication of the rank of any value. The pattern (such as it is) is now apparent. In practice, the plot may not be quite so easy, since the choice of what is a stem (and therefore what part of the number is to be regarded as a leaf) may not be obvious. There is a clear analogy with a frequency distribution histogram. If the stem and leaf plot is turned on its side it approximates to a histogram in which the class intervals are the integers and the columns are formed from the leaf values.

The line plot (or stem and leaf plot) and the data table are complementary. The line plot shows the similarities (and dissimilarities) between the individuals with respect to the variable under consideration, and the table provides the data values themselves, in an order convenient for comparison with the diagram.

As a final preliminary we can provide some 'points of reference', such as the median and the quartile points (see Section 0.3). These are:

| | |
|---|---|
| Lower Quartile Point: | 8.75 |
| Median: | 11.85 |
| Upper Quartile Point: | 16.25 |

Their positions relative to the data values are shown on the line plot (Figure 2.1). We note also that the interquartile range is thus 7.5 and the quartile deviation 3.75.

This simple start has contributed to the description and summarisation of this data set in three ways. First, we can now see something of the pattern in the values where, by 'pattern', we mean the way in which the values are distributed[1] along the axis of the line plot. Some values are close to each other and others are detached from the rest. Secondly, we have a (provisional) standard for deciding which values are large and which are small by reference to where they fall between the quartiles. Finally, we have (or can easily derive) two useful summary statistics: the median, which has some merit as a measure of centrality, and either the interquartile range (the difference between the upper and lower quartile points) or the quartile deviation (one half of the interquartile range), both of which are measures of dispersion.

## 2.5 STANDARDISATION

A next step might be to standardise the values and this may be done in two ways:

either (i) by subtracting the median ($M$) from each and dividing by

1. The use of the word 'distributed' here is similar to the use of the word 'distribution' in the term 'frequency distribution'. The latter summarises the arrangement of data values along a line by presenting, in tabular form, the frequency with which they are distributed in each of the class intervals.

the quartile deviation ($QD$), so that the $r$th standardised value, $w_r = (x_r - M)/QD$

or (ii) by subtracting the mean ($\bar{x}$) from each and dividing by the standard deviation ($S$), so that this $r$th standardised value $z_r = (x_r - \bar{x})/S$.

For illustration we will use the first method, although the second may be more familiar. The reason for doing so will be considered more fully in Section 2.13 on outliers.

The first standardised data set $w_1, \ldots w_n$ which results, has a median of zero and a quartile deviation equal to 1 and this has two advantages for visual inspection. First, all values less than the median have a negative sign and all greater than the median are positive. Secondly, we can determine the distance from one individual to another or from one individual to one of the points of reference in units which are independent of the unit of measurement of the original data. For if the original values were expressed in a different unit of measurement, this would be equivalent to multiplying each value by some constant number, say, $k$. The median and quartile deviation would then be multiplied by $k$ and the formula for $w_r$ shows that it would remain unchanged. There are corresponding results for the standardised values $z_1, \ldots z_n$: their mean is zero, their standard deviation is 1 and they are independent of the unit of measurement.

The standardised data based upon the median is shown in the second column on the right hand side of Table 2.2. This new scale may assist us to decide, informally, whether a particular value is unusual when compared with others in the set, and thereby to raise questions concerning the individual to which it is attached. For example, it will be seen from Table 2.2 that Iceland is 6.1 quartile deviations from the median. A rule of thumb for classifying a value as apparently unusual might be based upon a rule such as 'any value more than 4 quartile deviations either side of the median', that is, any median-based standardised value numerically greater than four.

## 2.6 THE USE OF INDIVIDUAL LABELS

It will be seen that we have, so far, retained the individual labels with the variable values. This reflects the fact that there is, in practice, more information than simply 'the set of variable values'. The individual labels may enable the analyst to draw upon background knowledge of the individuals, possibly not yet quantified or not quantifiable in any satisfactory way. It is clearly necessary to have the individual labels easily accessible when the purpose of the analysis is to look for patterns in the relationship of individuals with one another, the exceptions to those patterns and the unusual or unexpected, since the objective here is to raise questions about the individuals, perhaps as a preliminary to model building. This is very much analysis in the micro mode.

On the other hand, the labels are of less importance where the objective is to compare one or more groups of individuals with one another on the basis of a set of variable values, where the individuals are assumed to form groups which are sufficiently homogeneous with respect to factors other than those on which the comparisons are to be made. We are, in other words, standing back from the detail and looking at the variation as a whole, that is, working in a macro analysis mode.

An overriding consideration in the decision to retain the individual labels may be the number of individuals (the number of variable values). As a rule, it may create problems (messy visual displays, for example) if there are more than, say, 50 individuals. In cases such as this, it may be sufficient to retain labels only for those individuals who do not form part of any pattern or which appear at one extreme or the other.

So far, we have been looking at ideas which operate very much in a 'micro' mode, concentrating upon the individual values and their relationship to one another. We now turn to look more closely at methods of analysis (including summary statistics) which move us towards the 'macro' side, that is, the broad features of a data set.

## 2.7 THE QUANTILES AND THE EMPIRICAL CUMULATIVE DISTRIBUTION

An alternative graphical representation of a data set (large or small) is provided by a plot of the empirical cumulative distribution. Table 2.4 demonstrates the construction using as an example the country data of Table 2.1. The data values are first arranged in rank order. To each value $x_{(r)}$ is associated the number $r/n$ which equals the fraction of the data set with values less than or equal to $x_{(r)}$. This number is called the cumulative relative frequency (or cumulative frequency for short). Thus, the Denmark value is $x_{(12)} = 10.8$ and its associated cumulative frequency is $12/24 = 0.50$. The cumulative frequences are shown in the third column of figures.

It is also useful to define a closely related set of figures; the *quantile orders*. To each data value, $x_{(r)}$, we associate what we shall call a quan-

**Table 2.4** Cumulative Relative Frequencies and Quantiles for Table 2.1 Data Set

| *Country* | *Rank* | *Value* | *Cumulative frequency* | *Quantile order* |
|---|---|---|---|---|
| SWI | 1 | 5.6 | 0.04 | 0.04 |
| GER | 2 | 5.7 | 0.08 | 0.08 |
| AUS | 3 | 7.7 | 0.17 | 0.16 |
| US | 4 | 7.7 | | |
| LUX | 5 | 8.6 | 0.21 | 0.20 |
| NTH | 6 | 8.7 | 0.25 | 0.24 |
| CAN | 7 | 8.9 | 0.29 | 0.28 |
| NOR | 8 | 9.4 | 0.33 | 0.32 |
| SWE | 9 | 9.6 | 0.38 | 0.36 |
| BEL | 10 | 9.7 | 0.42 | 0.40 |
| FRA | 11 | 10.3 | 0.46 | 0.44 |
| DEN | 12 | 10.8 | 0.50 | 0.48 |
| JAP | 13 | 12.9 | 0.54 | 0.52 |
| NZ | 14 | 13.0 | 0.58 | 0.56 |
| AUST | 15 | 13.1 | 0.63 | 0.64 |
| FIN | 16 | 14.5 | 0.67 | 0.64 |
| GRE | 17 | 16.1 | 0.75 | 0.72 |
| IRE | 18 | 16.1 | | |
| UK | 19 | 16.3 | 0.79 | 0.76 |
| ITA | 20 | 16.5 | 0.83 | 0.80 |
| SPA | 21 | 17.2 | 0.88 | 0.84 |
| TUR | 22 | 20.4 | 0.92 | 0.88 |
| POR | 23 | 21.3 | 0.96 | 0.92 |
| ICE | 24 | 34.7 | 1.00 | 0.96 |

tile order which is defined as $r/(n + 1)$. The data value is said to be the *quantile point* of order $r/(n + 1)$. These are shown in the final column of Table 2.4. The only difference between the cumulative relative frequencies and the quantile order is that the former embodies a divisor of $n$ and the latter a divisor of $(n + 1)$ so, when $n$ is large, the difference between the two is insignificant.

The cumulative distribution can be exhibited graphically by plotting on the horizontal axis the variable values and on the vertical axis the cumulative frequencies. Figure 2.2 shows a computer plot based upon Table 2.4. It is a basic property that, as one moves from any variable value to the next highest, the cumulative frequency increases by a con-

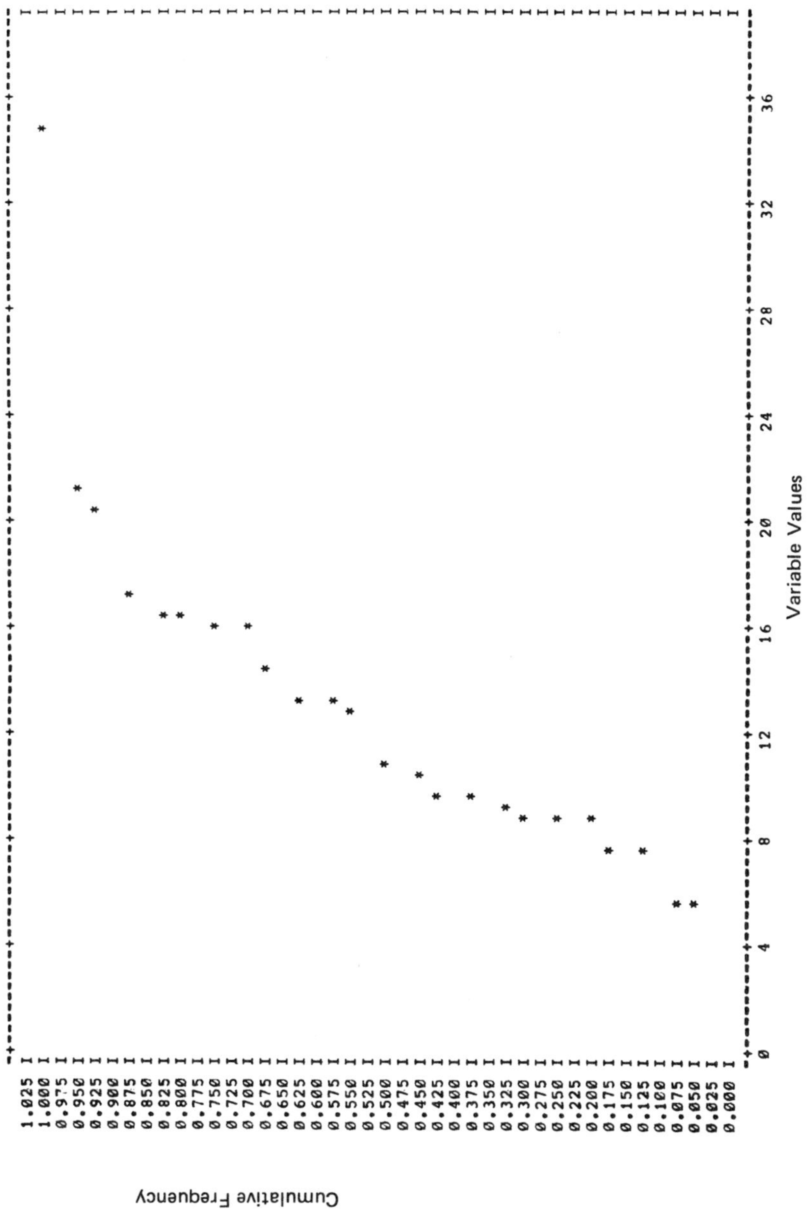


**Figure 2.2** Empirical Cumulative Distribution: Computer Plot Based upon Table 2.4

stant amount $1/n$, or 1/24 in the example. We note however that this feature is not exactly represented on the computer output because the plotted points must lie on a line of print so that, due to rounding, the increase is not always the same constant value. This is typical of the imperfections which one must sometimes face for the facility of computational graphing.

The cumulative distribution plot gives a useful visual appreciation of the data; thus the extreme value for Iceland in the top right hand corner is immediately apparent. However, the line plot or the stem and leaf plot and the frequency distribution (to be discussed below) are usually more immediately helpful.

We now return to the quantile orders shown in the final column of Table 2.4 and demonstrate how they may be used to determine the values of the median and the quartile points. First, the median must be some value between that of Denmark ($x_{(12)} = 10.8$) and Japan ($x_{(13)} = 12.9$) since then one-half of the values would be below and one-half above the median. The same result would have followed if we had defined the median as that (feasible) data value having a quantile order equal to 0.50, for 0.50 lies between the quantile for Denmark (0.48) and Japan (0.52). Similarly, defining the lower quartile point as the feasible value with a quantile order of 0.25, since 0.25 lies between the quantiles of Netherlands and Canada, the lower quartile point must lie between the values 8.7 and 8.9.

The concept of quartiles, dividing the data set into four parts, can be generalised. For example *percentiles* divide the data set into one hundred parts. The twentieth percentile corresponds to a value having a quantile order of 20/100 = 0.20. The data value for Luxembourg has exactly a quantile order of 0.20 and so 8.6 is the value of the twentieth percentile.

As is clear from the above examples, the median, quartile or percentile points may not be precisely determined. However, they may be estimated by proportionate interpolation. Thus, for the median of the Table 2.4 data, 0.50 is half way between the quantile orders in the final column of 0.48 and 0.52 and so the median is estimated as half way between the corresponding data values of 10.8 and 12.9, that is, as 11.85.

## 2.8 SUMMARY STATISTICS

Summary statistics are single numbers calculated from the data set, each of which measures some collective property of that data. Their purpose is to describe some of the essential features of the data set, no matter what its size is, in terms of a limited number of informative values which may, in turn, form the basis for comparison of one or more predefined groups with each other. Alternatively, as we have seen in Section 2.4, some of them have uses in the analysis of a single set of data by providing points of reference.

It is convenient and useful to classify univariate summary statistics under the following heads: measures of centrality, dispersion, skewness and kurtosis and this classification will be used in the discussion which follows. There are alternative measures available under each of these heads. Which are the more relevant or useful depends upon the nature of the data set and the purposes of the analysis. The virtues and drawbacks of different measures will be considered in Section 2.13.

There are three basic measures of centrality: the mean, the median and the mode (see Section 0.3). There are two other measures of centrality, little used in practice but sometimes defined in textbooks. The geometric mean is defined as the $n$th root of the product of the data values, i.e.

$$\text{geometric mean} = {}^{n}\sqrt{(x_1 \times x_2 \times \ldots x_n)}$$

The harmonic mean is defined as the reciprocal of the mean of the reciprocals of the data values, i.e.

$$\text{harmonic mean} = 1 \Big/ \frac{1}{n}\left(\frac{1}{x_1} + \ldots \frac{1}{x_n}\right)$$

The various measures of dispersion have been defined in Section 0.3 (Section 0.4 in the case of grouped data). Note that there are two alternative formulae for the variance (and correspondingly for the standard deviation) depending upon whether a divisor of $n$ or $(n - 1)$ is used. For descriptive statistics purposes, it does not much matter which is used (so long as one is consistent), though the former is usually more convenient. We shall use the notation $S^2$ for the variance with a divisor of $n$ and $s^2$ with a divisor of $(n - 1)$. The latter is more convenient in the analysis of statistical models and its use will be substantively discussed in Chapter 4. When $n$ is large, the difference between the two is negligible.

## 2.9 THE FREQUENCY DISTRIBUTION AND THE HISTOGRAM

The frequency distribution is derived from a set of data by appropriate grouping of values into classes and counting the number of values in each class. The counts may be left as such, and are then refered to as absolute frequencies, or each may be divided by the total of the counts (the 'total frequency'); this produces a set of relative frequencies. The construction of the empirical frequency distribution and the histogram is reviewed in Section 0.4. Sometimes only the frequency distribution of the data set is available and any data analysis is then bound to be in the macro mode.

The frequency distribution is therefore a partial summarisation and, as such, achieves some of the objectives of univariate descriptive analysis and it is likely to be of use when interest is focused on the overall nature of the variation. In particular, the frequency distribution, together with one of its visual counterparts, the histogram, leads to a consideration of the 'shape' of any pattern present in a set of data values, and we now look more closely at this.

It is convenient to begin by identifying certain idealised shapes as reference for empirical frequency distributions encountered in practice. These idealised shapes are usually presented as frequency curves (see Section 0.17) and Figure 2.2 shows some of the more common. These 'ideal' shapes may be classified according to:

(i) The presence or absence of one or more 'humps' or modes (each mode corresponding to a class whose relative frequency is larger than those on either side of it). Shapes which have only one mode are called uni-modal; those with two or more modes are referred to as being bi- or multi-modal, respectively, although it may be convenient to disregard the presence of modes which are minor by comparison with the classes on either side.

(ii) If there is only one mode, then it is useful to distinguish between symmetry and skewness. Symmetry is illustrated by Figure 2.3(a) and is best regarded as a standard of reference rather than something encountered in practice, although, in the interests of parsimony of description, it may be worth disregarding minor deviations from symmetry and describing a distribution as being 'practically symmetrical'. As a matter of terminology, the value around which the frequency curve (or the histogram) is symmetrical is called the 'point of symmetry'.

   Skewness (or asymmetry) manifests itself by the frequency curve (or the underlying histogram) having an elongated 'tail' which may be to the right (an elongated upper tail), as in Figure 2.3(b), or to the left (an elongated lower tail), as in Figure 2.3(c). These shapes may also be referred to as positive and negative skewness, respectively. In Section 2.10, we discuss the measurement of skewness.

It may be that two or more distinct groups of individuals exist

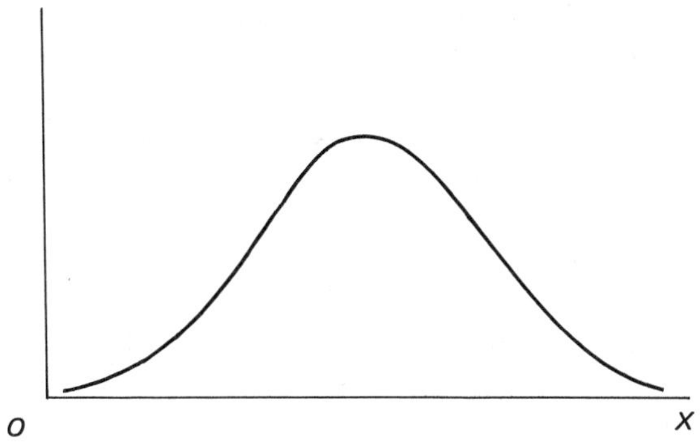


(a) Symmetric frequency curve

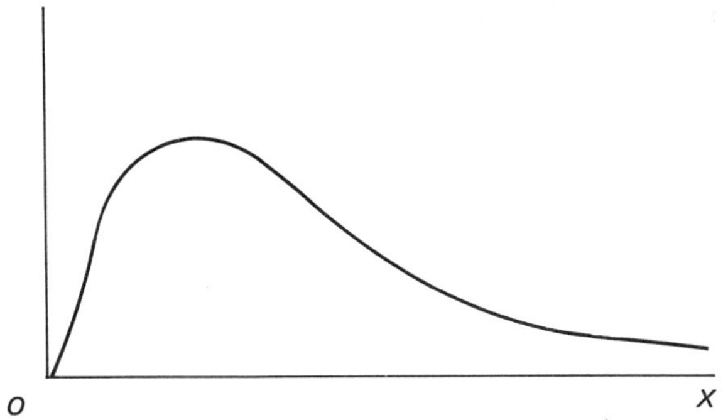


(b) Positively skewed frequency curve

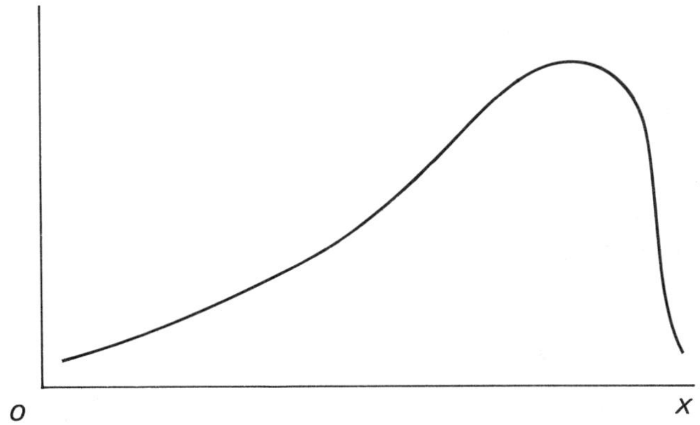


(c) Negatively skewed frequency curve

**Figure 2.3** Shapes of Frequency Curves

within the data and these give rise to the multi-modal shape. If such groups can be identified, it would be clearly sensible to proceed with analysis on each separately.

With these ideas in mind, we may add to the description of a data set for which the histogram has been drawn, a qualitative statement concerning its shape. It needs to be stressed, however, that the frequency distribution is very much a macro analytic device since, by its nature, it obscures the fine detail. For this reason, it may be of only limited practical use as a means of examining the kind of data with which this book is concerned, particularly (although by no means exclusively) if the data set contains only a small number of values. Two reasons for this may be mentioned. First, these data sets often contain values which are detached from the others in the set, possibly (but not necessarily) at one extreme or the other, thereby making it difficult to define satisfactory class intervals, since it is not clear where the detached values should be placed. Even if this problem is overcome, a histogram drawn from the frequency distribution may have a ragged appearance, with no useful interpretation in terms of shape. Secondly, grouping the individuals together will usually mean that their identities are lost and we have already argued that these may be of importance; it is, however, possible to overcome this second objection by the use of labelled histograms. These problems may be less severe with large data sets, since detached values are less likely to occur and we have more to gain than to lose by some partial summarisation.

The more general difficulty is that the appearance of the histogram may be affected by the choice of class interval width (or the number of classes). Data whose histogram possesses an almost symmetric appearance under one choice can appear to be skewed or multi-modal under another. In Section 2.11, we offer some suggestions for choosing the number of class intervals to be used in constructing the frequency distribution; meanwhile, we remark that while a histogram is better than no visual display, there are some other graphical methods which are more effective as techniques for detecting departures from symmetry, such as line plots or stem and leaf plots which have already been described.

## 2.10 THE MEASUREMENT OF SKEWNESS

If a set of data values has a uni-modal distribution (disregarding, if necessary, any minor modes), the measurement of skewness may be approached from two directions. The first approach depends upon one or other of the following consequences of skewness: (i) the difference between the median and the lower quartile point is NOT the same as the difference between the upper quartile point and the median; (ii) the median does not equal the mean. The first of these consequences leads to the interquartile coefficient of skewness (*IQCS*), namely:

$$\frac{(Q_3 - M) - (M - Q_1)}{(Q_3 - M) + (M - Q_1)}$$

where $Q_3$ denotes the upper quartile point
$Q_1$ denotes the lower quartile point, and
$M$ denotes the median,

and the second leads us to examine the standardised difference between the mean and the median, i.e.

$$\left(\frac{M - \bar{x}}{QD}\right)$$

where $QD$ is the quartile deviation.

The IQCS may vary between −1 and +1, with 0 implying that $(Q_3 - M)$ equals $(M - Q_1)$, i.e. the distribution is practically symmetrical between the upper and lower quartile points. The sign of the IQCS provides information about the direction of the skewness, a negative sign implying a long 'tail' to the left and a positive sign implying a long 'tail' to the right. It should be borne in mind that the calculation of the IQCS ignores values which are less than the lower quartile point or greater than the upper quartile point and 'symmetry' in terms of this measure implies no more than symmetry between these two values. Little can be said about the measure based on the standardised difference between the mean and the median, except that a value close to zero suggests symmetry and that its sign has the same interpretation as with the IQCS.

A second approach depends upon a measure usually denoted as $g_1$ and defined as:

$$g_1 = \frac{(\sqrt{n})\Sigma(x - \bar{x})^3}{[\Sigma(x - \bar{x})^2]^{3/2}} \qquad \textbf{(2.10.1)}$$

with values close to zero implying approximate symmetry and the sign having the same interpretation as it does with the IQCS.

Interpretation of these skewness measures requires care. If a visual display suggests skewness is present, one of the measures above will assist in determining its severity; without a visual representation, any measure is of doubtful practical use. If the data set has only a few values in it, then the interpretation of ANY visual representation is likely to be ambiguous.

## 2.11 THE CONSTRUCTION OF FREQUENCY DISTRIBUTIONS

We can now tackle the question, outstanding from Section 2.7, concerning the number of classes to be used in constructing the frequency distribution.

Following a proposal due to Doane (1976), the number of classes is calculated, for an approximately symmetrical distribution, by the integer nearest to

$$K = 1.44 \times \log_e (n) = 3.3 \times \log_{10} (n)$$

Where skewness is present, Doane suggests a formula for adding extra classes. This usually results in the calculated value of $K$ being increased by 2 or 3. $K$, of course, will not necessarily be a whole number and Doane discusses this point together with the question of the choice of the lower limit of the first interval and the choice of the interval width, the issue being that these should be chosen in such a way that the class limits are aesthetically pleasing 'round' numbers. These latter choices can require a small change in the value for $K$ produced by the formula above, and Doane presents some methods for reconciling any conflict which may arise between the choice of the number of class intervals and the need for 'pleasing' values for the class limits; the reader is referred to the article for details. As a rule of thumb, it will generally be satisfactory to take the number of classes as being the nearest integer to $K$; if the grouping is being done by pencil and paper, inspection and intuition will then supply suitable values for the class limits. If the frequency distribution is being constructed by computer, the choices may, in any case, be pre-empted.

## 2.12 KURTOSIS

We now come to the last heading under the classification of summary statistics set out in Section 2.6. Kurtosis is a property of symmetric (or practically symmetric) distributions, and, in principle at least, provides a means for distinguishing one symmetric distribution from another on the basis of (i) the length or 'thickness' of their tails and (ii) the extent to which they are 'peaked', with the normal frequency curve providing the standard of reference. Distributions which are 'thicker' in the tails and more 'peaked' than the normal distribution are said to be *leptokurtic;* those which are 'flat' rather than 'peaked' (relative to the normal distribution) and 'short-tailed' are said to be *platykurtic.* The normal distribution itself is said to be *mesokurtic.*

The degree of kurtosis can be measured by the coefficient of kurtosis.

This is conventionally denoted by $g_2$ and its formula is:

$$g_2 = \frac{n\Sigma(x-\bar{x})^4}{(\Sigma(x-\bar{x})^2)^2} - 3$$

The normal distribution itself has a coefficient of kurtosis of 0; values less than 0 are indicative of platykurtosis and those greater than 0 are indicative of leptokurtosis. These guidelines refer to data whose frequency distributions are EXACTLY, not approximately, described by a smooth frequency curve, however. The computed measure therefore, needs to be interpreted with some caution. If the data set is large, and the histogram has a reasonable 'smooth' appearance and is symmetric and uni-modal, then the computed kurtosis may be useful in deciding whether the data may be 'modelled' by means of a normal distribution. Exploration of this possibility must await the further discussion of statistical modelling.

In principle, of course, the kurtosis of any set of variable values may be computed, but the results are likely to be difficult to interpret if the distribution is not uni-modal and practically symmetric.

## 2.13 OUTLIERS AND 'ROBUST' SUMMARY STATISTICS

We have mentioned from time to time throughout this chapter the possibility that a data set may contain values which appear to be inconsistent with the others, often because they are unusually small or unusually large by comparison with the remainder. Such values are called outliers, and the possibility that they may be present in a data set raises four related questions which may be broadly stated as follows:

(i) How do we know whether to treat one or more values as outliers?
(ii) Can any use be made of outliers?
(iii) What effect will outliers have on the process of data description and summarisation, with which this chapter has so far been concerned?
(iv) If the effects noted as possibilities at (iii) are considered undesirable, what alternatives are there?

We can provide only a partial answer to the first of these questions at present, since any rule for classifying a value (and its associated individual) as an outlier is based, ultimately, upon a statistical model for the data. We must defer a fuller discussion therefore until later, but we can draw upon the simple techniques of this chapter in order to make one or two suggestions for detecting the apparently ususual. In particular, the line plot will draw attention to values which are detached at one extreme or the other and standardisation will show how far these are from the selected central value. One simple rule is to look more closely at values which are more than four quartile deviations from the median and check on any background knowledge of these individuals (having checked, as far as possible, that the value has not been incorrectly recorded). If the data are approximately symmetrically distributed, this is roughly equivalent to identifying as an outlier any value more than about $2\frac{1}{2}$ standard deviations from the mean.

In selecting values, and therefore individuals, for closer inspection, one question which will be in the analyst's mind is this: have we any background knowledge which may explain why this value is so small (or so large) by comparison with the others? Having checked that the recorded value is not simply in error, one might then look more closely at the other variables and see whether any unusual behaviour is exhibited there also. For example, in a survey enquiry into expenditure patterns, it might be found that the unusually small values are recorded for households consisting of the elderly or for single person households and this may lead to the question, 'to what extent is household expenditure systematically associated with household size or with the age of the occupants of the household (or with both of these factors)?'. To take another example, we might find that mortality from a particular cause is unusually high in a certain locality. This finding may lead the investigator to ask what else is known about this area. Is it heavily industrialised? Is there anything unusual about the diet of the people who live there? Are medical services poor by comparison with other localities in the study? Each of these questions suggests a possible systematic relationship between mortality from the specified cause and certain other factors, some of which will be measured by variables in the data matrix, others will not. Questions of this kind therefore push the analysis in two directions: towards some tentative model building on the one hand and towards some further data collection on the other.

There are occasions on which outliers are a nuisance and should be discarded. To take an example, suppose it is desired to compare the incomes of people in two different occupations. This may be done by means of a suitably chosen set of summary statistics, but before any calculations are performed it will be necessary to identify and exclude any individuals whose incomes and whose backgrounds give reason for them to be considered unusual. Before discarding anything, we still need to make a decision as to what is to be considered unusual, and this requires us to call once more upon any other knowledge available to help to decide whether an unusual value is an inherent feature of the data, or whether it arises through factors which are peculiar to the individuals and which thereby make them typical of the group to which they have been assigned. In comparing, for example, the incomes of medical practitioners and lawyers, it seems reasonable to discard those whose incomes are low because they are partially retired or whose incomes are large because they pursue a difficult or unusual speciality. Judgment is necessary and should be made explicit; discarding is not the same as suppressing.

The informal means for detecting outliers suggested above used the median rather than the mean and the quartile deviation rather than the standard deviation as the basis for standardising the data values. This brings us to the third point in the introduction to this section, namely, the effects of outliers on the standard methods for summarisation. To illustrate some of these effects, let us consider the calculation of two measures of centrality—the mean and the median—and two measures of dispersion—the standard deviation and the quartile deviation—in the context of the data set in Table 2.5 below. There is one obvious outlier—the United States—and the values of the summary statistics, with and without this value, are as follows (figures rounded to nearest hundred):

**Table 2.5** Defence Expenditures 1977, NATO Countries (Millions of US dollars)

| | |
|---|---:|
| Belgium | 2,476 |
| Britain | 12,103 |
| Canada | 3,348 |
| Denmark | 1,085 |
| France | 13,666 |
| Germany | 17,130 |
| Greece | 1,328 |
| Italy | 4,730 |
| Luxembourg | 29 |
| Netherlands | 3,716 |
| Norway | 1,130 |
| Portugal | 545 |
| Turkey | 2,652 |
| United States | 104,250 |

*Source:* International Institute for Strategic Studies, *The Military Balance 1978/79* London, 1978.

| | *Mean* | *Median* | *SD* | *QD* |
|---|---|---|---|---|
| All values included | 12,000 | 3,000 | 26,100 | 5,700 |
| US value excluded | 4,900 | 2,700 | 400 | 3,700 |

Adding the US value to the data set more than doubles the mean and increases the standard deviation by more than five times, making both potentially misleading as summary statistics for the main part of the data set; by contrast, the median and the quartile deviation are increased relatively little. Of course, with a small data set such as this, one's attention is drawn to the value for the United States from the outset, and its inclusion would at least be questioned before any calculations were made. It should be borne in mind, however, that opportunities for visual inspection of a large data set may be limited or non-existent, particularly if the analysis is being done by computer and the data itself is only available in computer readable form,for example on magnetic tape. In cases such as this, the measures of centrality and dispersion chosen for the descriptive stage of the analysis should include those like the median and quartile deviation which are relatively unaffected by—and resistant to—the presence of outliers. Summary statistics of this kind are said to be *robust* in the sense that they are resistant to the effect of unusual extreme values.

Although the median and quartile deviation have the desirable property of robustness, they nevertheless have a less important role than the mean and standard deviation (or variance) in later work. The main reason is that the latter are more tractable measures in theoretical analysis.

## 2.14 TRIMMING AND WINSORISATION

This section presents, briefly, some other methods of calculating robust summary statistics. They will be less familiar to the reader and, at this stage, are best regarded as reference material.

To begin, we note that the formula for the mean may be written, albeit in a somewhat cumbersome fashion, as:

$$(w_1x_1 + w_2x_2 + \ldots + w_{n-1}x_{n-1} + w_nx_n)/\Sigma w_r$$

where each $w_r = 1/n$ (for $r = 1, \ldots, n$)

This formula expresses the mean as a weighted average of the data values with equal weights of $1/n$. This suggests that one approach to modifying the effect of outliers on the calculated value of the measure of centrality would be to give a smaller weight to some of the extreme values. A proposal of this kind has already been encountered, in a different guise, since the median may be thought of as a mean calculated as above by giving a weight of one to the value with rank $\frac{1}{2}(n+1)$ (if $n$ is odd) or weights of one-half to the value with rank $(n/2)$ and to the value with rank $(n/2)+1$ (if $n$ is even) and weights of zero to all the other values in the data set. The values to which a weight of zero have been given may be said to have been 'trimmed off' and, in effect, the median results from calculating, in the usual way, the mean of the values which remain.

Less drastic trimming is clearly possible, and this leads to what are called the '$\alpha$% trimmed means'. The usual mean is calculated by trimming nothing off—it is the zero % trimmed mean—and the median trims off all but one (or two) of the data values, and, as we will see in a moment, is the same as the 50% trimmed mean.

Let us first present a formula for calculating a trimmed mean:

$$m(\alpha) = \frac{1}{h} \sum_{r=g+1}^{n-g} x_{(r)}$$

where $m(\alpha)$ denotes the '$\alpha$% trimmed mean'

$g$ is largest integer less than $n\alpha$, and

$h$ is $n-2g$

Thus, using the following (hypothetical) data set (with $n = 10$)

0.9 1.3 2.5 3.6 4.9 5.3 6.2 8.3 10.1 20.6

we calculate the trimmed mean for $\alpha = 0.0$, 0.25 and 0.5:

| $\alpha$ | $n\alpha$ | $g$ | $h$ | $n-g$ | *Values Used* | *Trimmed Mean* |
|---|---|---|---|---|---|---|
| 0.0 | 0.0 | 0 | 10 | 10 | ALL | 6.4 |
| 0.25 | 2.5 | 2 | 6 | 8 | 2.5 3.6 4.9 5.3 6.2 8.3 | 3.8 |
| 0.5 | 5.0 | 4 | 2 | 6 | 4.9 5.3 | 5.1 |

The 50% trimmed mean, it will be observed, is simply the average of the values with rank 5 and rank 6, and is thus the same as the median. The 25% trimmed mean is sometimes called the 'midmean'.

Winsorisation is a related procedure and consists of replacing the extreme values at both ends by those adjacent in rank, and the extent to which this replacement is done defines the level of Winsorisation. A fuller discussion of this procedure may be found in Dixon and Massey (1969).

There are a number of other procedures for protecting summary statistics (in particular, measures of centrality) from the effect of outliers, and the interested reader is referred to the article by Smith, Michie and Pope (1977) and the references cited there for further information.

## 2.15 TRANSFORMATIONS

We have, so far, taken the data in the units in which it was measured (or at least reported). This may not always be the most effective thing to do, however, and further insights and more useful interpretation can sometimes be obtained by transforming the values mathematically.

To the inexperienced user of statistical methods this may seem an odd (and possibly invalid) thing to do. In this chapter, we will deal only with the logarithmic transformation. There are many other transformations possible, but this one yields values which have, in certain circumstances, both useful practical interpretations and useful statistical consequences.

Suppose the data are measures of the sizes of individuals, for example firms in an industry or particles in a sample of soil. Then, in making comparisons between one individual and another, we are often more interested in knowing that individual '*B*' is twice the size of individual '*A*' than in knowing that the difference between the two values is, say, 100 units. Comparisons such as this are facilitated if the data values are first transformed by taking their logarithms. To illustrate this point, consider the following (hypothetical) data set:

| *Individual* | *A* | *B* | *C* | *D* | *E* |
|---|---|---|---|---|---|
| Size | 100 | 200 | 400 | 1000 | 2000 |

(arbitrary units)

In this case, comparison in relative terms is easy to make (after a little mental arithmetic), because of the way the values have been chosen. A line plot would not make it clear, however, that *E* is twice the size of *D* and that the same relative relationship holds between *B* and *A*. If the data values are transformed by taking logarithms, we obtain the following:

| *Individual* | *Size (Original Units)* | *Size (Logarithmic Units)* |
|---|---|---|
| *A* | 100 | 4.6 |
| *B* | 200 | 5.3 |
| *C* | 400 | 6.0 |
| *D* | 1000 | 6.9 |
| *E* | 2000 | 7.6 |

In this table, the logarithms are natural logarithms to the base *e* since these are what are produced by most computer programs and pocket calculators. In the present context, the conclusions are unaffected whether natural logarithms or logarithms to the base 10 are used.

It is now apparent that if the difference between one pair of logarithmic values is the same as the difference between another, then the original pairs of values stand in the same proportionate relationship to one another. The differences between values are one of the features of the data to which the line plot draws attention, and if the logarithmic values are displayed in this way then individuals which are the same (or similar) distances apart may be interpreted as having the same proportionate relationship to each other. If the line plot is augmented by marking on it a central value, for example the median, then equal distances from this central point have, of course, similar proportionate interpretation.

There is another consequence of the transformation which is of some interest. If we calculate a measure of skewness for the original and for the transformed data sets, we will find that the transformation has reduced skewness. For the record, the IQCS for the original data is 0.63 and for the transformed data is 0.09. A further reason for transforming a set of values, therefore, is that to do so will sometimes reduce or eliminate the skewness, and it is necessary to consider why this might be desirable.

The first reason is that of parsimony of description in terms of summary statistics. If there is a transformation which achieves approximate symmetry in the distribution, then the only summary statistics which need to be displayed (for comparative purposes, for example) are a measure of centrality and a measure of dispersion (and, perhaps, kurtosis, if the data set is large enough).

The second reason returns us once more to the question of outliers. We have already suggested that an informal classification of a value as an outlier may be based upon a rule which states that an outlier is a value which is more than a certain number of standard deviations from the mean, on both sides of the mean. Suppose we use this rule in the context of a data set whose distribution has a long right-hand tail. Having plotted the values as points on a line, we mark off the two standard deviation limits on both sides of the mean; it will now be clear that there are more values classified as outliers in the right-hand tail (the relatively high values) than in the left-hand tail. This is a simple consequence of skewness, but it does mean that decisions concerning what might be an outlier depend upon whether the value is above or below whatever has been chosen as the measure of centrality.

The third reason is concerned with statistical model building. If the values in a data set have a symmetrical distribution, then it might be possible to make use of the normal distribution as a model for them. This, in turn, may enable statements with a predictive content to be made and will, in addition, make available a number of useful statistical procedures which are theoretically based upon the data being normally distributed. To use these procedures when it is obvious that the normal distribution is an inappropriate model (as, for example, when the data possess a skewed distribution) may weaken or possibly destroy any conclusions which may be drawn from them. If the data can be transformed so that their distribution is reasonably symmetrical, these methods may then be of more use.

## FURTHER WORK

**F 2.1** Select a univariate data set and consider:

(i) What would be the most appropriate and effective form of visual display.
(ii) Whether the retention of individual labels is practicable and useful.
(iii) What are its main features, in terms of summary statistics and, if relevant, its frequency distribution.
(iv) The effect of standardising in two ways, the first using the median and quartile deviation and the second using the mean and standard deviation.
(v) Whether there are any individuals which appear to be outliers.
(vi) Whether any skewness is present: if so, examine the effect of a logarithmic transformation.

**F 2.2** The logarithmic transformation discussed in Section 2.15 is only one possible device to remove skewness. Another is to take the square root of each of the data values. Compare the results (by means of the IQCS) of using this transformation with those obtained by using the logarithmic transformation on a univariate data set of your choice.

**F 2.3** On the basis of the data of Table 2.1, make a comparison of the EEC countries with the non-EEC countries.

**F 2.4** Examine the effect of trimming on the measure of centrality of a data set containing:

(i) An outlier at one extreme
(ii) An outlier at both extremes.

Does there appear to be, for the data set you have chosen, any case for applying these techniques in a non-symmetric fashion?

*This chapter depends upon: Sections 0.18, 0.20–0.21, 0.27*

# 3

# Statistical Models

## 3.1 INTRODUCTION

The methods of analysis discussed in Chapter 2 are usually known as those of descriptive statistics. These methods amount to taking the data as it stands, describing its numerical properties and summarising its structure. In the main, the descriptive statistics approach to analysis is not usually determined by the way the data originated, although the results of the analysis may well throw light on the background to the data. The statistical model approach to data analysis takes a wider perspective. It seeks to relate the data to some knowledge or theory about the way the data was generated. In consequence, it offers a means of using data to test the validity of some background theory or to put quantitative substance to a theory previously formulated only in qualitative terms.

The central purpose of this chapter is to introduce the reader to the methods and aims of statistical modelling in relation to a set of hypothetical data. Although the data set and its analysis are relatively simple, the procedure described in this chapter is fundamental to the more complex analytical methods studied in later chapters. Section 3.2 will introduce the data set and consider the possible aims of analysis. It will also apply the methods of descriptive statistics to the data set and will demonstrate the essential limitations of these methods. The following sections will then show what additionally is achieved by statistical modelling and the procedures which it adopts.

## 3.2 THE DATA SET AND ITS DESCRIPTIVE ANALYSIS

The data set comprises the observed journey times by car (in minutes) between home and work. There are two alternative routes, *A* and *B*, and the journey times were observed over routes *A* and *B* on ten and twelve occasions respectively. The data are shown in Table 3.1. Note that in terms of the data matrix structure described in Chapter 1, this is essentially a two variable data matrix. Each 'individual', that is each journey, has allocated to it the values of two variables: the numerical variable 'journey time' and the categorical variable 'route taken'—a data matrix with 22 rows and two columns. Table 3.1 has put this data matrix in a more convenient form.

**Table 3.1** Journey Time Data (minutes)

| *Route A* | *Route B* |
|---|---|
| 29.2 | 29.6 |
| 34.2 | 31.4 |
| 31.1 | 32.1 |
| 32.4 | 31.0 |
| 28.8 | 28.9 |
| 25.5 | 30.4 |
| 30.2 | 30.6 |
| 29.1 | 31.6 |
| 28.1 | 34.9 |
| 27.7 | 31.5 |
| | 30.2 |
| | 31.8 |

*Source:* hypothetical data

We presume the question to be resolved is: which is, on the whole, the quicker of the two routes? Before considering how the statistical modelling approach would be applied to this situation, it is instructive to see how far the descriptive statistical methods of Chapter 2 will take us: what questions they answer and, more importantly, what questions they cannot answer. Figure 3.1 makes a start by setting out a line plot of the two groups of data and Table 3.2 shows the various summary statistics. There are too few data values to formulate useful frequency distributions.

The line plots and the summary statistics all paint the same picture.

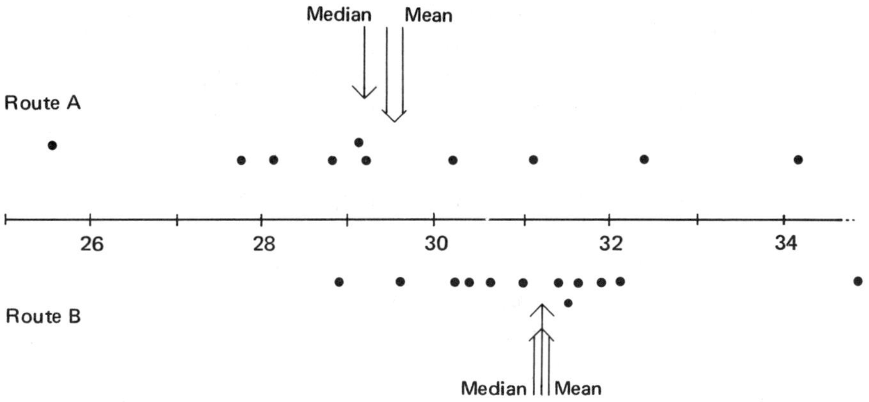


**Figure 3.1** Line Plots of Journey Times (minutes)

On average, Route *A* was quicker than Route *B*. But this was merely the experience of 22 journeys. The journey times obviously vary from one occasion to another and it would be rash to assume that future experience will *exactly* repeat the past experience. And equally it would be foolish to assume on the basis of this limited experience that Route *A* will tend to be the quicker. The same point can be made slightly differently by observing that if, by chance, the single quickest journey on Route *A* and the single slowest journey on Route *B* had not been taken, the average times for the two routes would have been much closer at 30.1 and 30.8 minutes respectively.

Clearly, any conclusions about the significance of this data are bound to be tentative, but the methods of descriptive statistics offer no means of quantifying their degree of reliability or unreliability. Intuitively, it seems reasonable to suppose that the average journey time recorded on Route *B* is the more reliable indicator of what the

**Table 3.2** Journey Time Data (minutes): Summary Statistics

| | *Route A* | *Route B* |
|---|---|---|
| Mean | 29.6 | 31.2 |
| Median | 29.15 | 31.2 |
| Mode | none | none |
| Range | 25.5–34.2 | 28.9–34.9 |
| | (8.7) | (6.0) |
| SD (s) | 2.5 | 1.5 |
| IQR | 3.0 | 1.4 |

long-run average would be because there is less variability of journey time over Route *B*. But again the descriptive statistics approach has nothing to say on this point. Statistical model building resolves questions of this kind, though at the expense of introducing others.

## 3.3 THE NATURE OF A STATISTICAL MODEL

It is fairly typical of statistical data that it exhibits two features: a regularity of some kind coupled with some irregular variation. Thus, the individual journey times appear to fluctuate in a random fashion around some central value which can be regarded as the 'norm' for the particular route. The data suggest that the norm for Route *A* is less than that for Route *B*. The norm is the regularity, the fluctuation above or below the norm is the irregular variation. Other examples of this kind are easy to imagine. The figures of a smoker's consumption of cigarettes would change irregularly from week to week, but the general level could be expected to be fairly stable at a level typical of the smoker. A slightly more complicated example would be data of weekly shop sales of some vegetable such as potatoes. The 'normal' level of sales would itself vary during the year because of the seasonal nature of the product but, on top of this, the sales in any one week would inevitably show some additional irregular movement.

A statistical model is a mathematical-statistical specification which embodies these two features of regularity and irregularity. In the context of the journey time data, we would think of the regularity as being the normal journey time if there were no unusual traffic hold-ups, if the traffic lights did not all happen to be green (or all red), if the density of traffic were not unusually light (or heavy) and so on. The irregularity represents the net influence of all these perturbing factors, sometimes combining to give a longer and sometimes a shorter journey time than the norm. That is, the observed journey time is looked upon as the sum of two components: the norm which depends upon the fixed features of each route and an irregular effect which varies in an unpredictable fashion from one particular journey to another.

In general, the regular and irregular components of this model will be usually known as the *systematic* and *random components* respectively.

This modest theory about the generation of the journey time data can be expressed algebraically as follows:

$$\underset{\substack{(r\text{th journey time}\\ \text{on Route } A)}}{t_{rA}} = \underset{\substack{(\text{'normal' time}\\ \text{on Route } A)}}{T_A} + \underset{\substack{(\text{random component for}\\ r\text{th journey on Route } A)}}{u_{rA}} \qquad \textbf{(3.3.1)}$$

and

| $t_{sB}$ | = | $T_B$ | + | $u_{sB}$ | (3.3.2) |
|---|---|---|---|---|---|
| (*s*th journey time on Route *B*) | | ('normal' time on Route *B*) | | (random component for *s*th journey on Route *B*) | |

Setting this model in relation to the data of Table 3.1, it postulates that, for the Route *A* times,

$$29.2 = T_A + u_{1A}, \quad 34.2 = T_A + u_{2A} \quad \text{and so on}$$

and for the Route *B* times,

$$29.6 = T_B + u_{1B}, \quad 31.4 = T_B + u_{2B} \quad \text{and so on.}$$

The model, as specified so far, is unproductive. We cannot put any values to $T_A$ and $T_B$ without knowing the values of the random components, and conversely. Moreover, the reader may have noted that we have refrained from any precise statement of what is meant by the 'normal' time of journey. These problems will be resolved by adding to the model a statistical specification of the nature of the random component. To this we now turn.

## 3.4 SPECIFICATION OF THE RANDOM COMPONENT

In the model, the random component reflects the net influence of a miscellany of perturbing factors. Since there is presumed to be no hard information about them, it would be impossible to try to offer any mathematical formulation which would account for or would predict the individual values of the random components. Instead, we add to the model by specifying that the random component will have some particular frequency distribution *in the long run*. The individual value of the random component on the occasion of a particular journey is looked upon as if it were a 'random observation' or a 'random drawing' from the long-run frequency distribution.

As regards the choice of an appropriate long-run frequency distribution for the random component, we shall simply assume that it has a long-run normal distribution, say the distribution $N(\mu_A, \sigma_A^2)$ for Route *A* and $N(\mu_B, \sigma_B^2)$ for Route *B* (see Section 0.20). It would be premature to assume that the distributions for the two routes are identical. Indeed, the line plots of Figure 3.1 and the dispersion statistics of Table 3.2 suggest that the variance of the random component for Route *A* is greater than that for Route *B*. However, one simplification is reasonable to make: we assume that the two means $\mu_A$ and $\mu_B$ are each equal to zero. This is equivalent to assuming that the average long-run effect of the perturbing factors is zero.

Having specified the long-run frequency distribution of the values of the random component, we have in effect specified the *probability distribution* of any one value of this component for any particular journey. For each such value is looked upon as a random selection from a normal distribution $N(0, \sigma_A^2)$ or $N(0, \sigma_B^2)$ and so its probability distribution is similarly $N(0, \sigma_A^2)$ or $N(0, \sigma_B^2)$ (see Section 0.21). To put it in another way, the long-run frequency distribution implies that, for example, 95% of the Route $A$ random component values lie within the range $\pm 1.96\sigma_A$ and, correspondingly, the probability distribution implies that there is a probability of 0.95 that any one value lies within the range $\pm 1.96\sigma_A$.

## 3.5 THE COMPLETE SPECIFICATION OF THE MODEL

The complete model can now be set out in formal terms. For the data of journey times on Route $A$:

$$t_{rA} = T_A + u_{rA}, \text{ where } u_{rA} \sim N(0, \sigma_A^2) \qquad \textbf{(3.5.1)}$$

and for Route $B$:

$$t_{sB} = T_B + u_{sB}, \text{ where } u_{sB} \sim N(0, \sigma_B^2) \qquad \textbf{(3.5.2)}$$

The notation $u_{rA} \sim N(0, \sigma_A^2)$ is conventional; it is simply a concise way of stating that a random value, such as $u_{rA}$, has a particular probability distribution, here $N(0, \sigma_A^2)$.

At this point, it is useful to stand back and take a look at the different kind of statements and assumptions made in the model specifications of expressions (3.5.1) and (3.5.2). Taking (3.5.1) as an example, $t_{rA}$ is an observed data value and the suffix $r$ takes the successive values 1, 2, . . . 10. Each $t_{rA}$ can be said to be 'modelled' as the sum of an unknown fixed value $T_A$ and a value $u_{rA}$, where $u_{rA}$ is presumed to be a random observation from a normal distribution having zero mean and an unknown variance $\sigma_A^2$. One may sum this up by saying that (i) the model has a particular structure—a constant systematic component plus a random component with a normal distribution and (ii) the model is formulated in terms of two unknown constants or *parameters* as we shall call them—$T_A$ (the systematic parameter) and $\sigma_A^2$ (the variance of the random component).

## 3.6 MODEL ESTIMATION

The next stage of the analysis is to estimate values or ranges of values for the model parameters $T_A$ and $\sigma_A^2$ and $T_B$ and $\sigma_B^2$. Of course, in the present context, the main interest is in the values of $T_A$ and $T_B$: what are their best estimates, how much do they differ, what is the range of uncertainty? In general, the method of estimation depends upon the structure of the model and the specification of the distribution of the random component.

In the case of the present example, it is a convenient consequence of modelling the random component in terms of a normal probability distribution that the results of statistical theory of the normal distribution can be used to arrive at estimates of the parameters $T_A$ and $T_B$. The same estimation procedure will apply both to the Route $A$ data and the Route $B$ data so, simplifying the notation, we are essentially concerned with estimating the model:

$$t_r = T + u_r,$$

where $u_r \sim \mathrm{N}(0, \sigma^2)$ and $r$ successively takes each of the values 1, 2, 3, . . . $n$ (corresponding to each of the $n$ individuals). In future this will be written: $r = 1, \ldots, n$.

Now the $t_r$ values equal a constant ($T$) added to a corresponding random value $u_r$ having the distribution $\mathrm{N}(0, \sigma^2)$. It follows that $t_r$ must also have a normal distribution, the same as that of $u_r$, but with the constant $T$ added to the mean (see Section 0.20). That is, $t_r \sim \mathrm{N}(T, \sigma^2)$.

Thus, in the language of statistical theory, the model postulates that the journey times are a random sample from a normal distribution with mean $T$. Therefore the estimation problem is that of estimating the 'population mean' from the 'sample data'. The observed journey times are the 'sample', the long-run distribution of journey times is the 'population'. A confidence interval for $T$, centred on the data mean, can be calculated (see Section 0.27). Denoting the data mean by $\bar{t}$ and the data standard deviation by $s$, the 95% confidence interval estimate of $T$ is:

$$\bar{t} \pm 1.96 \frac{s}{\sqrt{n}} \tag{3.6.1}$$

Applying this theory separately to the data for Routes $A$ and $B$ gives the following results:[1]

$$\begin{aligned} &95\% \text{ confidence interval for } T_A \text{ is } 29.6 \pm 1.5 \\ &\text{and for } T_B \text{ is } 31.2 \pm 0.8 \end{aligned} \tag{3.6.2}$$

1. More accurately, small sample theory should have been used for this confidence interval calculation since the numbers of data values are relatively small. Small sample theory will be introduced in Chapter 4.

Finally, we return to the original question: which of the two routes tends to be the quicker? In terms of the model, this question has been reinterpreted as: which of the two parameters $T_A$ and $T_B$ is the greater? The two 95% CI's do not give a simple answer. Their respective ranges are (28.1, 31.1) and (30.4, 32.0) which overlap each other. The implication of the analysis is clearer if instead we estimate the difference between the two paramaters. The 95% CI for this difference $T_B - T_A$ can be shown to be:[2]

$$(31.2 - 29.6) \pm 1.96\sqrt{\left(\frac{2.5^2}{10} + \frac{1.5^2}{12}\right)}$$

$$= 1.6 \pm 1.8, \text{ that is, the range } (-0.2, 3.4) \qquad \textbf{(3.6.3)}$$

This range just includes the value zero (which would imply no difference between the two routes), but on balance the data does suggest that Route $A$ is likely to be the quicker one in the long run *if the model is a valid one.*

## 3.7 DIAGNOSTIC CHECK OF THE MODEL

A statistical model, and indeed any theory about a real-world situation, is almost inevitably a simplification and an abstraction. To some extent or other, it is bound to be imperfect and the important questions are: does it capture the essence of the data and its context? are its assumptions plausible? are the analytical results of any utility for subsequent decisions or actions?

In the present example, it would certainly appear plausible to look upon the time of journey as comprising two elements: the systematic component dependent upon the particular route, and the random component dependent upon all other factors affecting the journey time. A stronger and more questionable assumption was made about the long-run frequency distribution of the random component, namely, that it has a normal distribution. Two comments may be made on this point. Firstly, the assumption has an *a priori* plausibility based upon the Central Limit Theorem. The Central Limit Theorem states that a variable which is the net outcome of a large number of small independent factors will tend to have a normal distribution. Secondly, and more importantly, it is possible when the model has been estimated (and sometimes before it is estimated) to apply

2. See Footnote 1.

statistical checks of its validity. We shall call this final stage of the analysis the *diagnostic check.*

The diagnostic check is all too often omitted from statistical analysis. This is unfortunate. For one thing, the diagnostic check may reveal that the model was inappropriate for the data and therefore the analytical results ill-founded. And for another, even where the model is a good one, the diagnostic check may well give additional insight into the data and possibly suggest further, hitherto unforeseen, lines of inquiry.

The technical procedures for diagnostic checking depend upon the nature of the model and they will be discussed in Chapter 7. Diagnostic procedures often depend upon a study of the *residual values* and it is possible to say something useful about this aspect at this stage, in the context of the simple journey time model.

First recall that the model for the journey time data was

$$t_{rA} = T_A + u_{rA} \qquad \text{(from } \mathbf{(3.5.1)}\text{)}$$

for Route $A$ and a similar expression for the Route $B$ data. In this expression, the $t_{rA}$ values are the known data values, but the value of $T_A$ and the values of the random component $u_{rA}$ are unknown. As a result of the analysis, we arrived at a central estimate of the parameter $T_A$, namely 29.6. This estimate can be used to calculate corresponding estimates of the random component values by substituting the estimate of $T_A$ in (3.5.1). Thus we have

$$t_{1A} = 29.2 = 29.6 + e_{1A}$$ (where $e_{1A}$ denotes the estimated random component value)

and so $e_{1A} = 29.2 - 29.6 = -0.4$

In general, we have

$$e_{rA} = t_{rA} - 29.6 \qquad \mathbf{(3.7.1)}$$

and similarly

$$e_{sB} = t_{sB} - 31.2 \qquad \mathbf{(3.7.2)}$$

Table 3.3 sets out the estimated random component values for the data on each route. Henceforth, these will be called the *residuals.*

The residuals can be examined at two levels. At a simple descriptive level, one can identify any residuals with apparently relatively extreme values. The 3.7 under Route $B$ is an example. This in turn would lead to an examination of the corresponding data value (34.9 in Table 3.1) and a search for some explanation. Of course, in the case of this present simple model, an extreme residual will correspond to a

**Table 3.3** Journey Time Data: Residuals

| *Route A* | *Route B* |
| --- | --- |
| −0.4 | −1.6 |
| 4.6 | 0.2 |
| 1.5 | 0.9 |
| 1.8 | −0.2 |
| −0.8 | −2.3 |
| −4.1 | −0.8 |
| 0.6 | −0.6 |
| −0.5 | 0.4 |
| −1.5 | 3.7 |
| −1.9 | 0.3 |
| | −1.0 |
| | 0.6 |

clearly extreme data value and so Table 3.3 does not offer much more information than the original data Table 3.1, but this is not usually true of more complicated data and models.

At a more technical level, the residuals themselves can be subjected to further diagnostic analysis. The theory and practice will be considered in later chapters, but a general idea of what is possible can be suggested at this stage. It can be shown that *if the model is valid* the residuals will have a normal distribution with zero mean. That is, the long-run frequency distribution of the residuals would exhibit a normal distribution. Theory tells us that the estimated standard deviation of this normal distribution is the standard deviation of the data values. In the case of Route *A*, calculations lead to an estimated standard deviation of 2.5 and of 1.5 for Route *B*. If we now divide the residuals by this standard deviation, we get a set of standardised residuals which should approximately have a standard normal distribution.

The results of this standardisation are shown in Table 3.4. The results are more readily evaluated in the corresponding line plots of Figure 3.2. Visually, the standardised residual values for Route *A* appear to be a plausible set of values from a standard normal distribution. They lie comfortably within the ± 1.96 SD range. This is not however true of the Route *B* values. The outlier value of 2.5 (which originates from the original outlier in the data of 34.9) is an improbably large value to be observed for a standard normal distribution. It suggests one of two things. Either the original model assump-

**Table 3.4** Journey Time Data: Standardised Residuals

| *Route A* | *Route B* |
|---|---|
| −0.2 | −1.1 |
| 1.8 | 0.1 |
| 0.6 | 0.6 |
| 0.7 | −0.1 |
| −0.3 | −1.6 |
| −1.6 | −0.5 |
| 0.2 | −0.4 |
| −0.2 | 0.3 |
| −0.6 | 2.5 |
| −0.8 | 0.2 |
| | −0.7 |
| | 0.4 |

tion, that the random component had a normal distribution, was invalid. Or it was generally valid but did not apply to the outlier data value to which some special circumstances applied. If the second alternative were to be thought the more likely, then the sensible thing might be to reject the outlier value from the data set and recompute the analysis, or—if the information is available—to see if there were any really extraordinary circumstances which applied to the particular journey.

## 3.8 AREAS AND QUANTILES OF THE NORMAL DISTRIBUTION

In the present chapter we have used the normal distribution in two ways: first, as a model for the random component and second, in making large sample confidence interval estimates of the systematic

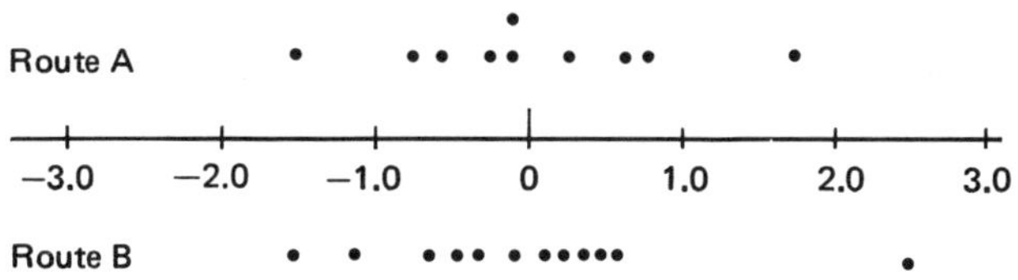


**Figure 3.2** Line Plot of Standardised Residuals

parameters $T_A$ and $T_B$. Confidence interval estimation (and for that matter other procedures such as hypothesis testing) depend upon area properties. For example, the range for a 95% confidence interval involves the value 1.96 because, for the distribution N(0, 1), the range of values (−1.96, 1.96) determines a central area of 0.95.

Other theoretical distributions will be introduced in later chapters and again we shall be concerned with questions such as, 'what range of values determines a central area of 0.95?' or, 'what value determines a right-hand tail of 0.01?'. It will therefore be helpful to have a standard notation and terminology for expressing the connection between values and associated areas. The discussion at this point will be in terms of the standard normal distribution but it will prove of wider applicability.

Published tables of the normal distribution vary in the way they specify the relation between area and value. Thus, for the variable value 1.96, a table may give 0.95 (the central area), 0.975 (the total area to the left of 1.96) or 0.025 (the tail area to the right of 1.96). This variability need not cause any fundamental difficulty since it is arithmetically easy to calculate one kind of area from another. Nevertheless, it can cause confusion in describing the properties of distributions and the way they are used in practice. We therefore introduce a terminology and notation which focuses attention on one kind of area, namely the total area under the curve lying to the left of a specified value; we shall call this the *cumulative area.* (The central area is not chosen, since it is useful only for symmetrical distributions.) This area measures the relative frequency of values less than or equal to the specified value, or, in the context of a probability distribution, the probability that a random observation of the variable is less than or equal to the specified value.

The letter $p$ will be used to indicate the cumulative area corresponding to a variable value $x$. Thus, for the standard normal distribution, when $x = 1.96$, $p = 0.975$ and when $x = 1.0$, $p = 0.84$. These connections are illustrated in Figure 3.3. An alternative terminology is to call a variable value a *quantile point*[3] and the corresponding cumulative area the *order of the quantile point* (or, simply, the *quantile order*). Sometimes we are given the quantile order, $p$, numerically and need to determine the corresponding value of $X$ (the quantile point); sometimes we know the value of $X$ and need to determine the corresponding value of $p$ (the quantile order).

3. The single word 'quantile' is often used for what we call the 'quantile point'. We prefer the latter because it emphasises that the quantile point is a single value, not a range of values.

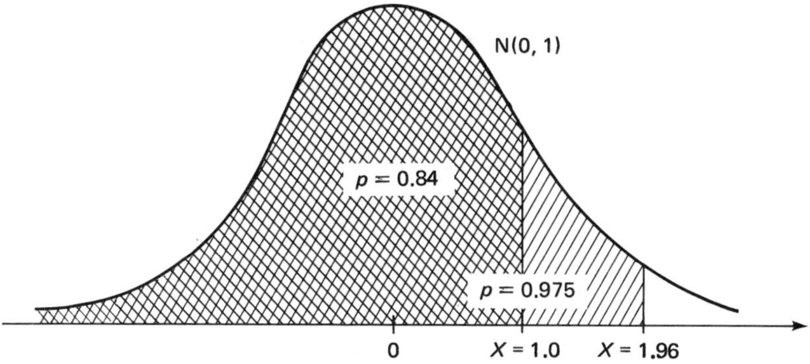


**Figure 3.3** Standard Normal Distribution Areas and Quantiles

Table 3.5 shows some selected pairs of values of $x$ and $p$ derived from the normal distribution Table A.1. It also shows central areas corresponding to positive variable values, that is, the area between the values $\pm\, x$. We shall use the Greek letter $\gamma$ to indicate a central area. For any symmetrical distribution such as the normal, there is a simple arithmetical relation between the cumulative area, $p$, and the corresponding central area, $\gamma$, namely:

$$p = \tfrac{1}{2}(1 + \gamma) \text{ and } \gamma = (2p - 1) \qquad \textbf{(3.8.1)}$$

Hence we can say that a central area of $\gamma$ is determined by the quantile point of order $\frac{1}{2}(1 + \gamma)$.

**Table 3.5** Quantiles and Central Areas of the Standard Normal Distribution

| | | | | | | |
|---|---|---|---|---|---|---|
| Variable value (quantile point) | $x$ | −1.96 | 0 | 1.64 | 1.96 | 2.58 |
| Quantile order | $p$ | 0.025 | 0.500 | 0.950 | 0.975 | 0.995 |
| Central area | $\gamma$ | . . . | 0 | 0.90 | 0.95 | 0.99 |

## 3.9 REVIEW

The methods introduced in this chapter comprise three distinct stages: the model *specification*, embodying unknown parameters; the model *estimation*, in which the model is fitted by putting numerical values to the parameters; and the *diagnostic check* in which the validity of the model is examined. We use the terms 'fitting (or fitted)' and 'estimating (or estimated)' almost interchangeably but, insofar as a distinction is made, the process of estimation is the calculation of numerical estimates of the model parameters and, when these are substituted into the model specification, we speak of a 'fitted' model.

The following Chapter 4 will extend the analysis in two ways. We shall still be concerned with models based upon the sum of a systematic component and a random component modelled by the normal distribution, but the systematic component will be extended to bivariate rather than univariate analysis. At the same time, a deficiency in the estimation procedure of this chapter will be removed in that we shall show how to estimate the model when the data values are not sufficiently numerous for the application of large sample theory.

## EXAMPLES

**E 3.1** Given that $X \sim N(0, 1)$ determine
a) the value of $X$ corresponding to
(i) $p = 0.975$ (ii) $p = 0.99$ (iii) $p = 0.05$
b) the value of $p$ corresponding to
(i) $x = 1.64$ (ii) $x = -2.0$ (iii) $x = 2.58$

**E 3.2** Given that $X \sim N(4, 9)$ determine
a) the value of $X$ corresponding to
(i) $p = 0.975$ (ii) $p = 0.99$ (iii) $p = 0.05$
b) the value of $p$ corresponding to
(i) $x = 9.88$ (ii) $x = 1.64$ (iii) $x = 1.0$
(iv) $x = 2.0$

**E 3.3** For the distribution N(4, 9) find the two values of $X$ which determine a central area, $\gamma$, equal to
(i) 0.90 (ii) 0.95 (iii) 0.99

**E 3.4** Reformulate Table 3.5 so that it applies to the distribution N(4, 9).

## FURTHER WORK

**F 3.1** Apply the model and methods of this chapter to a suitable univariate data set and: (i) consider whether model assumptions *a priori* appear appropriate, (ii) estimate the systematic parameter and its 95% confidence interval, (iii) calculate the residuals, and (iv) examine them for consistency with the model assumptions. A suggestion for data: meteorological series. (If monthly rainfall figures were used, how would the residuals indicate the inappropriateness of an assumed constant systematic component? If average annual temperature figures were used, how could one examine whether or not recent years have tended to get colder (or warmer)?

*This chapter depends upon: Sections 0.8 and 0.9 and Ch*

# 4

# Linear Models

## 4.1 INTRODUCTION

We may write the model introduced in the last chapter in the following general form:

variable value = systematic component + random component

and in this chapter we will consider three possible specifications of the systematic component, namely:

(i) models in which the systematic component is a constant;
(ii) models for a data set which relates to two predefined groups of individuals with a constant systematic component for each group;
(iii) models in which the systematic component depends upon the value of another variable.

The random component, which differs for each individual, will be assumed to be a random drawing from a normal distribution with a mean of zero and a variance, $\sigma^2$, which is invariably unknown but which will be assumed to be constant from one individual to another. These three models will be referred to as Models A, B and C respectively.

Let us now be a little more precise and introduce some terminology and notation. The starting point is a set of $n$ values of a variable which we will denote by $Y$ and refer to as the *dependent variable*. The values themselves will be denoted by $y_1, y_2, \ldots y_n$ with $y_r$ denoting a typical value. In the case of Model C, the variable which enters into the specification of the systematic component will be denoted by $X$, with individual values $x_1, x_2, \ldots x_n$, and is called an *explanatory variable*. The random component values will be denoted by $u_1, u_2, \ldots u_n$. With this notation, we may write the three models sketched above as follows, introducing, as we do so, the notation for the parameters:

Model A assumes that:

$$y_r = \alpha + u_r, \text{ where } u_r \sim \mathrm{N}(0, \sigma^2)\ (r = 1, \ldots, n) \qquad \textbf{(4.1.1)}$$

This is the typical expression for the $r$th individual; putting $r$ successively equal to the integers from 1 to $n$ gives the model specification for each variable value. The parameters are $\alpha$ (a fixed constant) and $\sigma^2$, the variance of the random component.

Model B is a little more complex. First let us assume that there are $n_1$ individuals in the first group, $n_2$ individuals in the second group and $n_1 + n_2 = n$. We then write the model as:

$$y_r = \alpha_1 + u_{1r}, \text{ where } u_r \sim \mathrm{N}(0, \sigma_1^2)\ (r = 1, \ldots, n_1)$$
if the individual is in the first predefined group, and
$$y_s = \alpha_2 + u_{2s}, \text{ where } u_2 \sim \mathrm{N}(0, \sigma_2^2)\ (s = 1, \ldots, n_2)$$
if the individual is in the second predefined group. **(4.1.2)**

For clarity we are using two subscripts on the variables, the first denoting the group to which the individual belongs and the second denoting the individual within that group. The parameters are $\alpha_1$, $\alpha_2$, $\sigma_1^2$ and $\sigma_2^2$, and one implication is that if there is a distinction between the groups, then $\alpha_1$ will differ from $\alpha_2$. Initially we shall proceed by assuming that $\sigma_1^2 = \sigma_2^2 = \sigma^2$, a common value, and this defines what will be referred to as Model B (I). The case of $\sigma_1^2$ not equal to $\sigma_2^2$ defines what we will call Model B (II).

Model C can be formally stated as:

$$y_r = \alpha + \beta x_r + u_r, \text{ where } u_r \sim \mathrm{N}(0, \sigma^2)\ (r = 1, \ldots, n) \qquad \textbf{(4.1.3)}$$

where $x_r$ is explained above. Here there are three parameters: $\alpha$, $\beta$, and $\sigma^2$.

These three models are all examples of linear models. That is to say, the systematic component is a linear function of its parameters and therefore does not involve expressions such as $\alpha^2$, or $\log \beta$ or $\alpha\beta$. Model C also happens to be a linear function of the explanatory variable, but this is not a crucial feature for the definition of a linear model.

The discussion in this chapter will centre around the following questions:

(i) Under what circumstances might each model be considered appropriate?
(ii) How can the data be used to estimate values for the parameters which define the systematic component of the model and also the variance of the random component?
(iii) Having obtained estimated values for the parameters, how can we proceed to construct confidence intervals for them?

Question (ii) defines the process which we will call 'fitting'. It will be recalled that the parameters are initially unknown. When their values have been estimated numerically and are substituted into the model specification, we have what is known as a fitted model. Question (iii) concerns the calculation of confidence intervals which will give an impression of the range of parameter values which are consistent with the data on hand and with the proposed model. This will require a brief digression to present (or review) the properties and use of the $t$-distribution and the idea of the standard error.

One important assumption which will not be questioned in this chapter is that the random component behaves as a random drawing from a normal distribution; methods for investigating the plausibility of this assumption will be presented in Chapter 7.

Two data sets will be used to illustrate the ideas and methods of this chapter, and these are set out in Tables 4.1 and 4.2. The first sets out the results of the determination of the aluminium content of certain archaeological specimens and the second sets out the mortality rate from heart disease in several countries together with information on their consumption of dairy products.

## 4.2 MODEL A

The assumption of a constant systematic component added to the random component is practically the simplest possible assumption which we can make and realistic examples of models of this kind are not easily found. Two possibilities may be mentioned. First, the model may be appropriate either when the individual values are measurements repeated under some controlled, identical conditions as in, for example, the determination of some physical constant. Secondly, it may be of use in the case where the individuals have different identities but are considered, provisionally at least, to form a well-defined homogeneous group with the interest being upon the

**Table 4.1** Percentage Aluminium Content of Theban and Chania Stirrup Jars

| | | | | | | | | | |
|---|---|---|---|---|---|---|---|---|---|
| Theban | 11.4 | 13.4 | 13.5 | 13.8 | 13.9 | 14.4 | 14.5 | 15.0 | 15.1 |
| ($n$ = 18) | 15.8 | 16.0 | 16.3 | 16.5 | 16.9 | 17.0 | 17.2 | 17.5 | 19.0 |
| Chania | 9.0 | 10.5 | 10.8 | 12.0 | 12.7 | 12.8 | 13.4 | 14.2 | |
| ($n$ = 16) | 14.3 | 14.7 | 15.0 | 15.7 | 16.0 | 18.2 | 18.2 | 19.1 | |

*Source:* Catling, H.W. and Jones R.E., *Archaeometry*, 19, 2 (1977)
*Summary statistics:*
Theban: Mean = 15.40, median = 15.45, variance ($s^2$) = 3.44
Chania: Mean = 14.16, median = 14.25, variance ($s^2$) = 8.30
Theban and Chania combined: Mean = 14.8, median = 14.85

**Table 4.2** Country Data on Mortality Rate from Heart Disease (*IHDMR*) and Per Capita Consumption of Dairy Products (*TDP*)

| | *IHDMR* | *TDP* | | *IHDMR* | *TDP* |
|---|---|---|---|---|---|
| Australia | 263.2 | 422.7 | Japan | 36.4 | 67.1 |
| Austria | 237.5 | 447.6 | Netherlands | 182.1 | 477.8 |
| Bel/Lux | 181.4 | 292.5 | New Zealand | 243.0 | 529.4 |
| Canada | 226.8 | 472.1 | Norway | 269.9 | 559.1 |
| Denmark | 297.1 | 526.7 | Portugal | 77.2 | 121.4 |
| Finland | 240.9 | 860.6 | Spain | 46.6 | 220.3 |
| France | 80.9 | 332.7 | Sweden | 333.9 | 501.8 |
| Germany | 171.5 | 317.0 | Switzerland | 105.3 | 445.9 |
| Greece | 66.5 | 240.1 | UK | 293.0 | 454.5 |
| Ireland | 263.9 | 706.9 | USA | 326.1 | 490.4 |
| Italy | 134.0 | 211.0 | | | |

*Source:* Table 8.3

central value in that group. For illustration, we shall use the data in Table 4.1 and consider the Theban and the Chania jars as forming two such groups.

If a normally distributed variable is standardised by subtracting its immediate implication of the model specification (4.1.1) is that

$$y_r \sim N(\alpha, \sigma^2) \qquad (r = 1, \ldots, n)$$

which is a simple consequence of the properties of the normal distribution (see Section 0.20). The parameter $\alpha$ has the interpretation, therefore, that it is the mean of the normal distribution model which generates the data; the variation around this mean, measured by $\sigma^2$, is simply a reflection of the differences from one individual to another (or of measurement error). To fit the model, therefore, we may use the data mean as an estimated value of $\alpha$ and the data variance as an estimated value of $\sigma^2$. We denote these values by $a$ and $s^2$ (based upon a divisor of $(n - 1)$) respectively; that is, we use the corresponding letter from the Roman alphabet to denote the estimates. For the archaeological data, they are shown at the foot of Table 4.1. Thus, for the Theban jars, $a = 15.4$ and $s^2 = 3.44$.

The proposal to use these quantities derived from the data as estimates of $\alpha$ and $\sigma^2$ can be justified more formally but we defer doing so until Chapter 6; for the moment, we trust that it seems, intuitively, a reasonable procedure to follow.

Our final task is that of constructing a confidence interval for $\alpha$. (A confidence interval for $\sigma^2$ can be constructed but it requires some theoretical results which will be presented later; interest, in any case, usually centres upon $\alpha$.) The construction requires, first the idea of a standard error and, secondly, the $t$-distribution. We digress momentarily, therefore, to discuss these and return to the confidence interval construction in Section 4.5.

## 4.3 STANDARD ERRORS AND LARGE SAMPLE ESTIMATION

We first consider the theoretical foundation behind the distribution of the sample mean and the large sample confidence interval estimate of the population mean which is set out in Tables 0.13 and 0.15. Relating what is set out there to the notation of Model A, we here have a population mean $\alpha$, a population variance $\sigma^2$ and we draw upon the result that the distribution of the sample mean, $a$, is $N(\alpha, \sigma^2/n)$.

If a normally distributed variable is standardised by subtracting its mean and dividing by its standard deviation, the resulting variable has a standard normal distribution N(0, 1) (see Section 0.20). Hence

$$\frac{(a - \alpha)}{\sigma/\sqrt{n}} \sim N(0, 1) \qquad \textbf{(4.3.1)}$$

If $n$ is large, this result still stands (approximately) if $\sigma$ is replaced by the standard deviation of the data values, $s$, so:

$$\frac{(a - \alpha)}{s/\sqrt{n}} \sim N(0, 1) \text{ if } n \text{ is large} \qquad \textbf{(4.3.2)}$$

We shall call $s/\sqrt{n}$ the *standard error* of the data mean. The standard error, $s/\sqrt{n}$, of the data mean is an estimate (based upon the data values) of the standard deviation, $\sigma/\sqrt{n}$ of the distribution of the data mean. Verbally, one may say that:

> the difference between the data mean (the estimate of the model parameter) and the population mean (the model parameter) divided by the standard error has a standard normal distribution.

More generally, the standard error (SE) of a data estimate (such as the data mean) will mean an *estimate* of its standard deviation.[1]

The large sample confidence interval for $\alpha$ is derived from result (4.3.2). Taking 95% as an example, it first follows that the probability is 0.95 that:

$$-1.96 < \frac{(a-\alpha)}{s/\sqrt{n}} < 1.96$$

and therefore, multiplying by $s/\sqrt{n}$, that:

$$-1.96\, s/\sqrt{n} < (a-\alpha) < 1.96\, s/\sqrt{n} \qquad \textbf{(4.3.3)}$$

The expression $1.96\, s/\sqrt{n}$ measures the range of values around the model parameter $\alpha$ within which the data mean $a$ is likely to be. Expression (4.3.3) leads directly to the 95% confidence interval for $\alpha$, namely:

$$a \pm 1.96\, s/\sqrt{n} \qquad \textbf{(4.3.4)}$$

Verbally, this can be stated:

> Confidence interval for model parameter = data mean ± (normal distribution value) × (SE of data mean) **(4.3.4a)**

## 4.4 t-DISTRIBUTION AND 'SMALL' SAMPLE ESTIMATION

The large sample theory of the last section depends upon result (4.3.2) which is valid only if $n$ is large. Fortunately, a very similar theory can be developed when $n$ is any value, large or small, *but only if the data values can be regarded as independent random observations from some normal distribution.* This is indeed the assumption made about

1. The reader should be warned that the term 'standard error' is not used consistently in the literature. It is often applied synonymously to the standard deviation itself: the $\sigma/\sqrt{n}$ in the present context. This does not seem to us a helpful terminology since it leaves no convenient term for the *estimated* standard deviation.

the data values in Model A and it is a necessary assumption for what follows. The theory is founded upon a theoretical result that, for any value of $n$, the expression in (4.3.2) has, not the standard normal, but another theoretical distribution known as the *t-distribution*. This is sometimes referred to as 'small sample' theory, but this is a misnomer since the theory is valid for all values of $n$, large or small, given the assumptions of Model A.

The $t$-distribution is in fact a family of distribution which is based on just one parameter conventionally given the Greek letter $\nu$. The parameter $\nu$ can take only positive integral values, so when $\nu$ equals, say, 9, a particular $t$-distribution is thereby identified. As a notation for the distribution, we shall use $t(\nu)$, so that, for example, $t(9)$ will indicate that particular $t$-distribution with $\nu = 9$. Figure 4.1 shows the shape of different members of the $t$-distribution family in comparison to that of the standard normal to which the $t$-distribution is closely related.

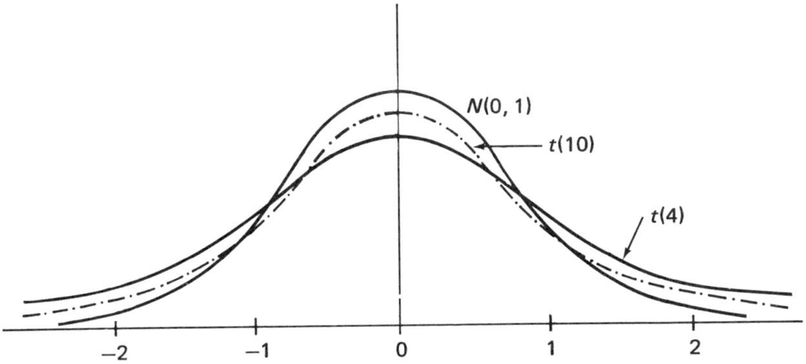


**Figure 4.1** t-Distribution and Standard Normal Curves

Figure 4.1 demonstrates three properties of the $t$-distribution:

(i) like the standard normal, it is symmetrical about the value zero and bell-shaped;
(ii) it is 'flatter' than the standard normal. This has the practical consequence that, for any member of the family, a given central area corresponds to a greater range of values than that for the standard normal distribution;
(iii) as the value of the parameter $\nu$ increases, the shape of the corresponding $t$-distribution curve gets closer to that of the standard normal. In fact, for $\nu$ greater than about 30, the $t$-distribution is practically indistinguishable from that of the standard normal.

For this reason, the notation $t(\infty)$ stands for the standard normal distribution.

In making use of the standard normal distribution in confidence interval estimation (and in other applications such as hypothesis testing), attention is often focused on variable values which correspond to particular central areas and tail areas. This is also true of the $t$-distribution and its published tables allow the relation between areas and variable values to be evaluated for any particular value of the $t$-parameter. The appendix Table A.2 shows selected $t$-distribution values and areas, each row of the table relating to a particular value of $\nu$. Table 4.3 is an extract drawn up on the same lines as Table 3.5 for the standard normal distribution (see Section 3.8). Table 4.3 shows, for the three distributions $t(10)$, $t(30)$ and $t(\infty)$, the variable values corresponding to the quantile orders 0.50, 0.95, 0.975 and 0.995 (together with the associated central areas). Note that the final row, $\nu = \infty$, sets out values for the standard normal distribution.

**Table 4.3** Quantile Orders and Variable Values for Particular t-distributions

| | | | | | |
|---|---|---|---|---|---|
| Quantile order, $p$ | | 0.500 | 0.950 | 0.975 | 0.995 |
| Central area, $\gamma$ | | 0 | 0.90 | 0.95 | 0.99 |
| Variable values, $x$ | | | | | |
| ($\nu = 10$) | $t_p(10)$ | 0 | 1.81 | 2.23 | 3.17 |
| ($\nu = 30$) | $t_p(30)$ | 0 | 1.70 | 2.04 | 2.75 |
| ($\nu = \infty$) | $t_p(\infty)$ | 0 | 1.64 | 1.96 | 2.58 |

Table 4.3 introduces an important notation which will be much used subsequently. In general, the expression $t_p(\nu)$ means the quantile point of order $p$ for the $t$-distribution with parameter $\nu$. Thus, for the distribution $t(10)$, $t_{0.95}(10)$ is the quantile point (that is, variable value) of order 0.95 and equals 1.81. In other words, for the distribution $t(10)$, it is the value 1.81 which determines a cumulative area of 0.95 (and a central area of 0.90). Similarly $t_{0.975}(\infty) = 1.96$.

The notation can also be used if central areas (rather than cumulative areas) are of primary concern. Recalling that $p = \frac{1}{2}(1 + \gamma)$ (see result (3.8.1)), the variable value determining a central area of $\gamma$ is $t_{(1+\gamma)/2}(\nu)$. As an example, the 95% large sample confidence interval for the parameter $\alpha$, set out in result (4.3.4), can be restated as

$$a \pm t_{0.975}(\infty)s/\sqrt{n} \tag{4.4.1}$$

where $0.975 = \frac{1}{2}(1 + 0.95)$.

## 4.5 CONFIDENCE INTERVAL ESTIMATE OF PARAMETER $\alpha$ IN MODEL A

We start by stating formally the theoretical result referred to at the beginning of Section 4.4, namely:

If a set of data values is correctly modelled by a normal distribution with mean $\alpha$, then

$$\frac{(a - \alpha)}{s/\sqrt{n}} \sim t(n-1) \qquad \textbf{(4.5.1)}$$

for any value of $n$.

Note that the result says that a particular member of the $t$-distribution family is specified: that with $\nu$ equal to the number of data values less one. Furthermore, it is necessary that the data standard deviation should be calculated with a divisor of $(n - 1)$.

Result (4.5.1) is similar to the large sample result (4.3.2), the only difference being that where the latter refers to the standard normal distribution, the former refers to the $t(n - 1)$ distribution. A confidence interval can be derived exactly as in Section 4.3, the only difference being that $t$-distribution values (with $\nu = (n - 1)$) are substituted for standard normal distribution values. If we make this change to the verbal statement of the large sample result at (4.3.4a), we have:

Confidence interval for model parameter
= data mean ± ($t$-distribution value) × (SE of the data mean)
**(4.5.2)**

More formally, and on similar lines to the large sample result (4.4.1), we have:

For Model A, the 95% CI for $\alpha$
$= a \pm t_{0.975}(n-1) \times s/\sqrt{n}$ **(4.5.3)**

Other confidence intervals easily follow. In general, a $100\gamma\%$ CI is given by replacing the suffix 0.975 by $\frac{1}{2}(1 + \gamma)$.

It is perhaps worth restating that this simple equivalence between large sample and 'small sample' confidence intervals is only true if the data standard deviation is measured by $s$ (with a divisor of $(n - 1)$) rather than by $S$ (with a divisor of $n$). It explains why we have used $s$ consistently in the present chapter.

The above theory can now be readily applied to the Stirrup jar data of Table 4.1. For the Theban jars we have:

$$a = 15.4 \qquad s = 1.85 \qquad n = 18$$

and so the 95% CI for $\alpha$ is:

$$15.4 \pm 2.11 \times 1.85/\sqrt{18} = 15.4 \pm 0.9$$

## 4.6 FUTURE APPLICATION OF CONFIDENCE INTERVAL ESTIMATION BASED UPON t-DISTRIBUTION

In the following sections of the present chapter, and in Chapter 8 which considers multiple regression analysis, we shall be developing rather more complicated models than Model A, namely models based upon more than one systematic parameter. However, the procedure for setting up confidence intervals of these parameters is little different from the procedure for the estimation of the single parameter $\alpha$ in Model A. Indeed, the verbal specification of the confidence interval (result (4.5.2)) will still apply and the only substantive change will be that the *t*-distribution value will be derived from a *t*-distribution with its parameter set equal to some number other than $(n - 1)$. Specifically:

The 95% CI of a systematic parameter of a linear model is equal to:

data estimate $\pm\ t_{0.975}(\nu) \times$ (SE of data estimate)

where $\nu$ is calculated according to the following rule:

$\nu =$ (total number of individuals in data set)
− (total number of systematic parameters in model) **(4.6.1)**

For Model A, $\nu$ was set equal to $(n - 1)$, there being only one systematic parameter, $\alpha$, in the model.

The *t*-distribution parameter, $\nu$, is often called the *degrees of freedom* and one speaks of a *t*-distribution with, for example, $(n - 1)$ degrees of freedom.[2] The degrees of freedom are thus reduced by one for each additional systematic parameter in the model.

## 4.7 MODEL B

Model B brings us into the realm of comparison of two groups of individuals such as the journey time data for the two routes *A* and *B* of Chapter 3 and the measurements on the two groups of archaeological jars from Thebes and Chania of Table 4.1.

2. A satisfactory explanation of this terminology cannot be given without appeal to the background mathematical statistics.

We start by setting up a Model A type specification for each group separately. Thus, for the aluminium content of the Theban jars, we have:

$$y_{1r} = \alpha_1 + u_{1r}, \text{ where } u_{1r} \sim \mathrm{N}(0, \sigma_1^2)\ (r = 1, \ldots, n_1)$$

and for the aluminium content of the Chania jars:

$$y_{2r} = \alpha_2 + u_{2r}, \text{ where } u_{2r} \sim \mathrm{N}(0, \alpha_2^2)\ (r = 1, \ldots, n_2)$$

In Section 4.5, we have in effect already set out 95% confidence interval estimates for $\alpha_1$, and that for $\alpha_2$ may similarly be calculated. However, we are now more concerned with the magnitude of the *difference* between $\alpha_1$ and $\alpha_2$. Is the difference $(\alpha_1 - \alpha_2)$ large, which would indicate that Theban and Chania jars are substantively different as regards their aluminium content? Or are the data consistent with a model in which $\alpha_1$ and $\alpha_2$ are the same, that is $(\alpha_1 - \alpha_2) = 0$? To answer these questions we proceed to calculate a 95% CI for the single value $(\alpha_1 - \alpha_2)$. At this stage we consider only Model B(I) as defined in Section 4.1, that is, it is assumed that $\sigma_1^2 = \sigma_2^2 = \sigma^2$.

The general results set out in Section 4.6 can be applied here. Suppose that the data means of the two groups are $a_1$ and $a_2$, and the data variances are $s_1^2$ and $s_2^2$. Then the data estimate of $(\alpha_1 - \alpha_2)$ is $(a_1 - a_2)$. To calculate the SE of this data estimate, we first need to calculate $[(n_1 - 1)s_1^2 + (n_2 - 1)s_2^2]/(n_1 + n_2 - 2) = s^2$, say. $s^2$ is known as the 'pooled estimate' of the common model variance, $\sigma^2$, of the two groups. Then the SE of $(a_1 - a_2)$ is $s\sqrt{[(1/n_1) + (1/n_2)]}$. By result (4.6.2), the $t$-parameter $\nu = (n_1 + n_2 - 2)$, since there are $(n_1 + n_2)$ individuals and two systematic parameters in the model, $\alpha_1$ and $\alpha_2$. Hence:

$$\text{the 95\% CI for } (\alpha_1 - \alpha_2) = (a_1 - a_2) \pm t_{0.975}(n_1 + n_2 - 2)\, s\sqrt{[(1/n_1) + (1/n_2)]} \qquad \textbf{(4.7.1)}$$

For the archaeological data: $a_1 = 15.40$, $a_2 = 14.16$, $s_1^2 = 3.44$, $s_2^2 = 8.30$, $s^2 = 6.02$ and $n_1 + n_2 - 2 = 32$. Hence the 95% CI for $(\alpha_1 - \alpha_2) = 1.24 \pm t_{0.975}(32)\, 2.45\sqrt{(\frac{1}{16} + \frac{1}{18})} = 1.2 \pm 1.7$

This confidence interval includes zero and so the data is consistent with a model in which $\alpha_1 - \alpha_2 = 0$ although the central estimate is 1.2.

However, the fact that $s_2^2$ is about 3 times as large as $s_1^2$ suggests that the assumption that $\sigma_1^2 = \sigma_2^2$ is untenable. Model B (II) appears more appropriate.

The data estimate of $(\alpha_1 - \alpha_2)$ is the same as that for model B (I), namely $(a_1 - a_2)$. Unfortunately it is not possible to set up an exact CI formula for $(\alpha_1 - \alpha_2)$ and even an approximate solution is technically complex. A practical, albeit approximate, procedure is first to calculate the data estimate of $(\alpha_1 - \alpha_2)$, namely $(a_1 - a_2)$ and use the standard error

$$\sqrt{\left(\frac{s_1^2}{n_1} + \frac{s_2^2}{n_2}\right)}$$

Then, as before, we may calculate a CI for $(\alpha_1 - \alpha_2)$ based upon (4.6.1). The calculation of $\nu$ is, however, more complicated; if $n_1$ is roughly equal to $n_2$, $\nu$ is given by the integer nearest to

$$n_1 \frac{(\lambda + 1)^2}{\lambda^2 + 1} - 2$$

where $\lambda = s_1^2/s_2^2$.[3]

Reworking the Stirrup jars data in terms of Model B (II) we have:

$(a_1 - a_2) = 1.2$, with standard error 0.84
$s_1^2/s_2^2 = 0.41$ and so $\nu \doteq 28$

Hence the 95% CI for $(\alpha_1 - \alpha_2)$ is $1.2 \pm 2.05 \times 0.84 = 1.2 \pm 1.7$

It is worth noting that the more complicated calculations based upon Model B(II) have led to the same result as before.

## 4.8 A SUMMING UP

To sum up so far, we have seen how two simple models may be fitted. Model B underlies the procedures often presented under the heading of estimating a difference between two means. Our approach, however, draws particular attention to the assumptions which are being made, assumptions which tend to get overlooked in the more usual approach to this problem.

It is often a common misunderstanding that the relevant theory for constructing confidence intervals is that provided by the normal distribution unless the number of data values is 'small'. *Providing the assumption that the random component behaves as a random drawing from a normal distribution can be sustained,* the crucial distinction turns not upon whether the data set is 'small' or not, but upon whether the variance parameter $\sigma^2$ is 'known' or has to be estimated from the data. If, as seems realistic, the variance has to be estimated, the $t$-distribution theory is all that is needed. Formal methods for checking the assumptions will be considered in Chapter 7. Meanwhile one could check informally, on the lines suggested in Chapter 2, that at

3. This result is derived from a more general formula given in Hoel (1962).

least the distribution of the data values seems reasonably symmetrical. If not, we might, as a start, investigate whether a transformation might help (see Section 2.15).

## 4.9 MODEL C

The Model C specification is

$$y_r = \alpha + \beta x_r + u_r, \quad \text{where } u_r \sim \mathrm{N}(0, \sigma^2) \qquad (r = 1, \ldots, n) \qquad \textbf{(4.9.1)}$$

Model C allows us to specify the dependence of $Y$ upon the value taken by another variable $X$. The variable $X$ is known as the explanatory variable (or sometimes the independent variable, a usage which we will, however, avoid).

This model is known as the bivariate regression model. In Chapter 8 we will investigate the use and consequences of allowing the systematic component to depend upon several different explanatory variables, but many of the ideas which occur in this extension have their counterpart in the simple case of Model C. It is therefore worth studying not only for the simple uses to which it may be put, but also as an *entrée* into the multiple regression model.

The starting point for considering a model of this type is often provided by a simple theory of relationship between two variables; for example:

(i) economic theory predicts an association between the price of a good and the quantity sold, other things being equal;
(ii) a theory of the variation in mortality from some specified cause suggests that diet may be an important factor: in particular, a country in which the inhabitants have a diet which is high in animal fats will exhibit a high mortality rate from heart disease compared with a country whose inhabitants' diet is less rich in these products;
(iii) in examining variation in crime rates, a sociologist may have a tentative view that they vary from one area to another according to the density of population in each area.

Observe that the underlying theory may be fairly precise or rather vague; all that is needed to begin with is a notion that two variables may have an association which can be defended by some theory of causation. The data requirements are a data matrix which contains, for each individual, values of two numerical variables. Which

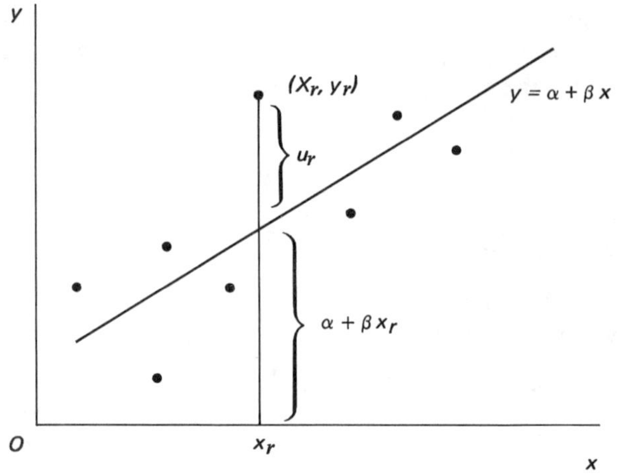


**Figure 4.2** Diagram of Bivariate Regression Model

variable is dependent and which is the explanatory variable depends uon the theory underlying the investigation.

The implication of the specification is that a plot of the pairs of values $(x_r, y_r)$ (the scatter diagram) should appear to lie around some straight line $y = \alpha + \beta x$ which reflects the systematic dependence of $Y$ on $X$. This is sometimes known as the 'underlying' relation between $Y$ and $X$. The deviations from this straight line reflect the effect on $Y$ of the random component. Figure 4.2 is a diagrammatic representation of the model for the case of a positive relation between the two variables, i.e. $\beta > 0$. It should be recalled that the scatter could exhibit a downward slope (when viewed from left to right) rather than an upward slope as shown here; the implication then would be that higher values of $X$ were associated with lower values of $Y$ and $\beta$ is then negative.

To fit Model C, we first need to calculate the data estimates, $a$ and $b$, of the systematic parameters $\alpha$ and $\beta$ respectively. The appropriate estimates are identical to those given by calculating the least squares linear regression line. In other words, the regression line $y = a + bx$ can be looked upon as an estimate of the underlying model relation. (The descriptive linear regression procedure is summarised in Section 0.8 and the formulae for $a$ and $b$ are set out in Section 0.9.)

Confidence intervals for $\alpha$ and $\beta$ can be calculated by applying the general result (4.6.1) described in Section 4.6. Thus:

the 95% CI for $\beta$ is

$$b \pm t_{0.975}(n-2) \times (\text{SE of } b) \qquad \textbf{(4.9.2)}$$

A similar statement applies for $\alpha$. Note that the $t$-parameter is put equal to $(n-2)$ because there are two systematic parameters in Model C. We do not present the calculations of the standard errors of $a$ and $b$; their values are shown in standard computer printout (see the following Section 4.10).

It is also possible to calculate from the data an estimate of the variance of the random component, $\sigma^2$, just as $s^2$ was calculated as an estimate of the $\sigma^2$ in Model A. Again, this is part of standard computer output and it will be discussed in Section 4.11.

## 4.10 A PRACTICAL EXAMPLE

A standard computer program was used to fit Model C to the data of Table 4.2

where $Y$ is the mortality rate from ischaemic heart disease (labelled in the computer output as *IHDMR*)

and $X$ is the per capita consumption of all dairy products (*TDP*)

The use of Model C may be justified in a similar way to that for the journey time data of Chapter 3. The $Y$ variable, mortality rate, is viewed as being determined by (i) the level of consumption of dairy products, $X$ and (ii) by many other factors which are unspecified and whose net effect is covered by the random component. A computer scatterplot is given in Figure 4.3. Part of the output of a typical program is as follows:

Regression Coefficients
Dependent Variable: *IHDMR*

| | *Estimate* | *SE* | *T* |
|---|---|---|---|
| Constant | 40.54438 | 36.06082 | 1.12 |
| TDP | 0.37008 | 0.07964 | 4.65 |

The first column of figures gives the value of the regression coefficients $a$ and $b$.

The construction of a confidence interval for e.g. $\beta$ is straightforward using result (4.9.2). The standard errors are shown in the second column of figures; $(n-2)$ here equals $(21 - 2) = 19$, and $t_{0.975}(19) = 2.09$. Hence the 95% confidence interval for $\beta$ is

$$0.37 \pm (2.09) \times (0.08) = 0.37 \pm 0.17.$$

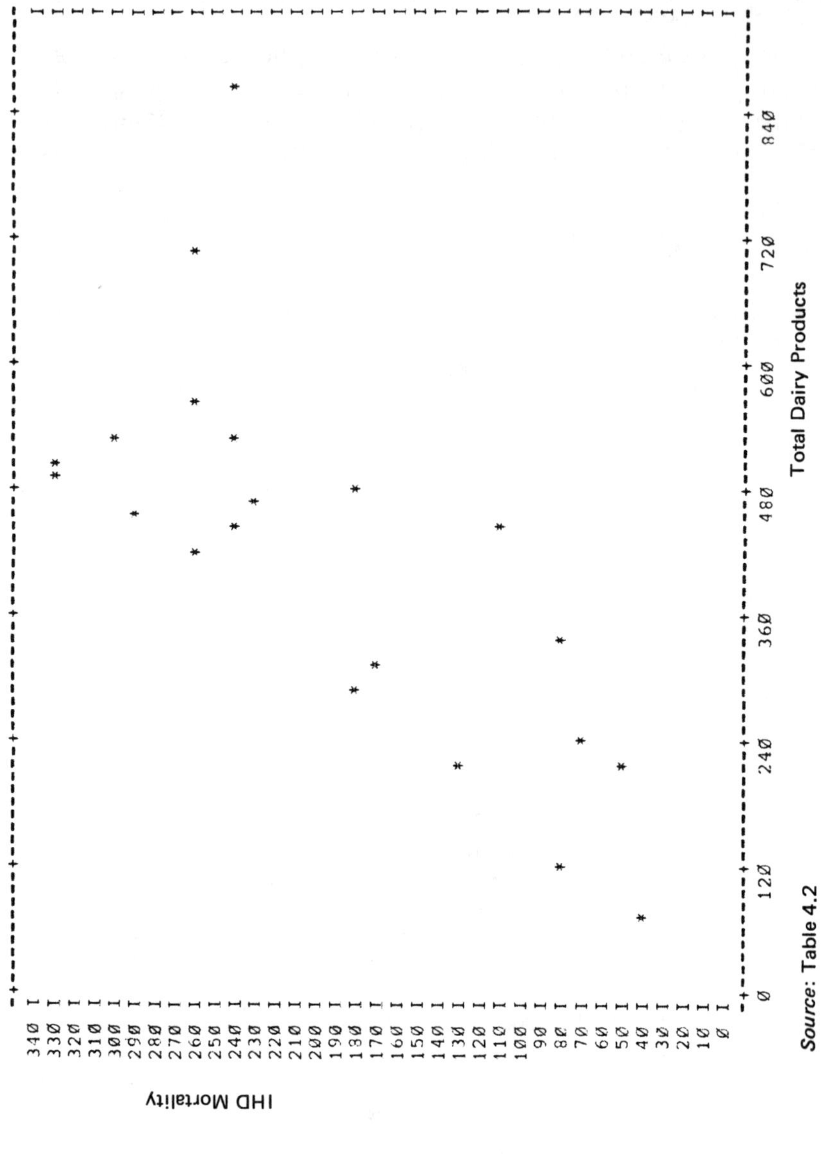


*Source*: Table 4.2

**Figure 4.3** Scatterplot of IHDMR and TDP

The column headed $T$ simply sets out the ratio of the figures in the first two columns; its utility will be considered in Section 7.10.

This completes the first part of the fitting process. There is, however, some further information provided by the computer output and we now turn to consider this.

## 4.11 FITTED VALUES AND RESIDUALS

Once the estimates $a$ and $b$ of $\alpha$ and $\beta$ respectively have been found, we may write the regression equation as follows:

$$\hat{y} = a + bx \qquad \textbf{(4.11.1)}$$

where $\hat{y}$ denotes what is called the *fitted* value of $Y$, corresponding to some value $x$ of the explanatory variable. For a *particular* value of $X$ in the data, the corresponding fitted value of $Y$ is:

$$\hat{y}_r = a + bx_r \qquad r = 1, \ldots, n \qquad \textbf{(4.11.2)}$$

In the context of the data used above, we have for example, the fitted value of *IHDMR* for Australia equals $40.54 + (0.37 \times 422.7) = 196.9$.

In general, these fitted values will differ from the corresponding values of $Y$ in the data set and the difference

$$y_r - \hat{y}_r = e_r \qquad r = 1, \ldots, n \qquad \textbf{(4.11.3)}$$

defines the residuals. Thus the residual for Australia is $(263.2 - 196.9) = 66.3$. Making use of (4.11.2), we have:

$$y_r = a + bx_r + e_r \qquad \textbf{(4.11.4)}$$

The set of residuals is important in two respects: first, the residuals allow us to measure the goodness of fit of the regression line to the data points and secondly, as we will see in Chapter 8, they have a part to play in checking some of the underlying assumptions of the model.

Before taking up this first point more fully, we note, without proof, three algebraic properties:

(i) the mean of the fitted values of $Y$ is equal to the mean of the data values of $Y$, that is

$$\bar{\hat{y}} = \bar{y} \qquad \textbf{(4.11.5a)}$$

(ii) the mean of the residuals is zero, that is

$$\bar{e} = 0 \qquad \textbf{(4.11.5b)}$$

(iii) the fitted values and the residuals are uncorrelated. **(4.11.5c)**

It can also be shown that there is an algebraic identity connecting the fitted values of $Y$, the data values of $Y$ and the residuals, namely:

$$\Sigma(y_r - \bar{y})^2 = \Sigma(\hat{y}_r - \bar{y})^2 + \Sigma e_r^2 \quad \textbf{(4.11.6)}$$

It is this relation which allows us to proceed to a measure of goodness of fit. The term $\Sigma(y_r - \bar{y})^2$ defines what is called the *total sum of squares* (TSS) and the term $\Sigma e_r^2$ is the *residual sum of squares* (RSS). The term $\Sigma(\hat{y}_r - \bar{y})^2$ which equals the sum of squares of the *fitted* values of $Y$ around their mean is called the *explained* (or, sometimes, the regression) *sum of squares* (ESS). Result (4.11.6) may be restated therefore as

$$\text{TSS} = \text{ESS} + \text{RSS} \quad \textbf{(4.11.7a)}$$

or, dividing through by TSS

$$1 = \text{ESS/TSS} + \text{RSS/TSS} \quad \textbf{(4.11.7b)}$$

An intuitive understanding of the use of this identity may be obtained from the three hypothetical scatter diagrams, with their regression lines, in Figure 4.9. Case 1 relates to the situation in which there is a close linear fit, Case 3 shows a very poor fit and case 2 lies in-between. In Case 1, RSS is very small and so TSS is practically equal to ESS, that is ESS/TSS is close to 1. The ratio ESS/TSS equals 1 if the data points lie exactly on the regression line. As we move from Case 2 to Case 3, RSS increases and the ratio ESS/TSS decreases; in fact, in the limit, when the correlation between the two variables is zero, ESS/TSS is equal to zero. Thus it can be seen that ESS/TSS is a measure of goodness of fit and it can be shown that this ratio is equal to the square of the usual correlation coefficient $r$, that is:

$$r^2 = \frac{\text{ESS}}{\text{TSS}} \quad \textbf{(4.11.8)}$$

which ratio is also called the *coefficient of determination.*

This analysis is exemplified in the computer program printout which, for the data and model considered in the last section, is as follows:

| | *DF* | *SS* | *MS* |
|---|---|---|---|
| Regression | 1 | 96263 | 96263 |
| Residual | 19 | 84709 | 4458 |
| Total | 20 | 180972 | 9049 |

The column of figures headed *SS* (sum of squares) sets out, respectively, ESS, RSS and TSS and, as in (4.9.4a), 96262 + 84709 = 180972 and, dividing through by TSS (180972): 0.53 + 0.47 = 1. Thus, for this data set, the coefficient of determination equals 0.53. It

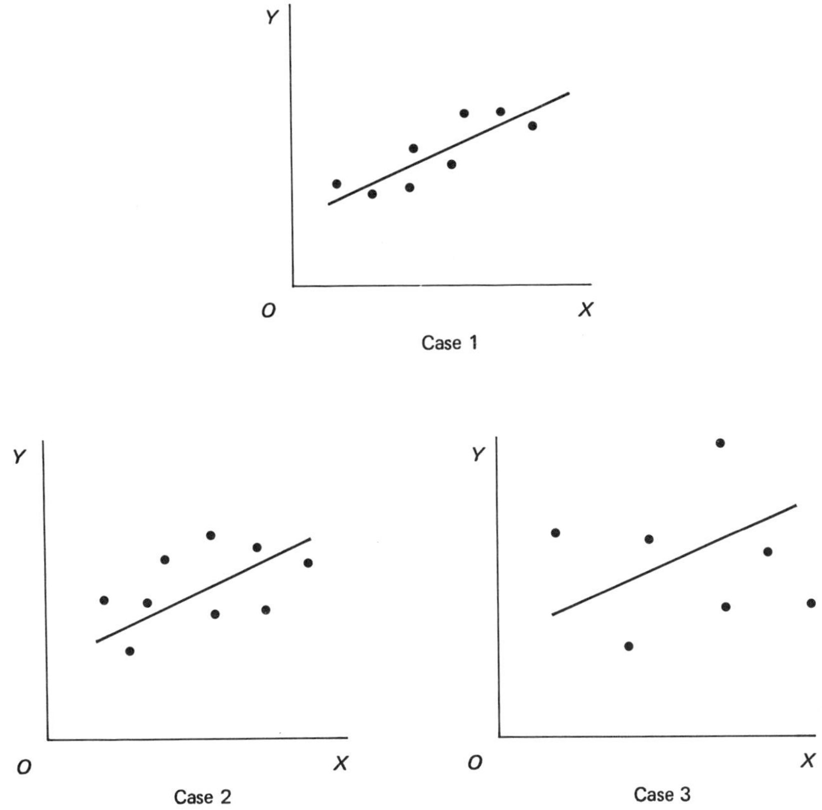


**Figure 4.4** Allocation of Sums of Squares

is often useful to express this as a percentage and one speaks of the explanatory variable in the regression explaining 53% of the variability of the data values of $Y$. The remaining 47% reflects that part of the variability of the dependent variable attributable to the residuals.

An estimate, $\hat{\sigma}^2$, of the variance $\sigma^2$ of the random component can be made from the residual sum of squares ($\Sigma e_r^2$). Specifically:

$$\hat{\sigma}^2 = \Sigma e_r^2/(n-2) \tag{4.11.9}$$

In terms of our numerical example, $\hat{\sigma}^2 = 84{,}709/19 = 4{,}458$. It will be seen that the divisor is shown opposite Residual in the *DF* column and the ratio $\hat{\sigma}^2$ in the column headed *MS*. A full explanation of the columns headed *DF* (degrees of freedom) and *MS* (mean square) will be given in Chapter 8.

A final word of warning: the value of $r^2$ expresses the goodness of fit of a straight line to the data. It could well be the case that the relationship between $Y$ and $X$ is curvilinear, thereby implying that a different specification of the systematic part is required, or that the relationship is linear only after transformation, a point which we now take up.

## 4.12 TRANSFORMATION OF VARIABLES

It may be desirable to make a transformation of the data in order to render more appropriate the basic assumptions used to formulate the models of this chapter. For example, it would be inappropriate to apply Models A or B to data whose distribution is skewed. As suggested in Section 2.15, a logarithmic transformation may lead to a more symmetrically distributed data set which could then be more reasonably represented by models based upon the normal distribution.

Similarly, the assumption of Model C, that $Y$ depends upon $X$ in a linear fashion, may not be supported, initially, by the data, in that the scatter diagram may suggest a *curvilinear* systematic component. We can sometimes get around this problem by fitting Model C to transformed values of either or both of the dependent and explanatory variables. For example, if the explanatory variable is first transformed by taking the logarithm of each value, we may then write model C as

$$y_r = \alpha + \beta \log x_r + u_r$$

and the method of least squares, using values of $Y$ and $\log X$, will yield estimates, $a$ and $b$, of $\alpha$ and $\beta$. Note that the fitted equation is

$$\hat{y} = a + b \log x$$

which is a *curvilinear* relation between $X$ and $Y$.

## 4.13 CONCLUSION

This chapter has been almost entirely concerned with the specification and fitting of three simple linear models. The final step in analysis—checking the validity of a proposed model—will be deferred until Chapter 7. The use of, say, Model A rather than Model C may depend upon the perspective that one wishes to take. Given a set of values of a variable $Y$ it may, at times, be desirable to treat the individuals as a homogeneous group and information on other variables as being of secondary importance. At other times it may be helpful to examine the extent to which the values of a dependent variable can be explained by a single explanatory variable as in Model C. In practice, realistic causal modelling of a dependent variable usually requires the use of several explanatory variables and, as we shall see in Chapter 8 which deals with the multiple regression model, bivariate modelling can sometimes be positively misleading.

These simple models are therefore of limited interest. Their main importance is that they introduce, in a simple setting, methodology and ideas which pave the way for future work.

## EXAMPLES

**E 4.1** a) Determine $t_p(\nu)$ for $\nu = 6, 10, 20$ and $\infty$ and
(i) $p = 0.975$ (ii) $p = 0.99$
b) What is $p$ such that
(i) $t_p(8) = 1.86$, (ii) $t_p(14) = 2.98$, (iii) $t_p(20) = 1.72$
(iv) $t_p(\infty) = 1.64$

**E 4.2** For the following values of $\nu$ determine $t_p(\nu)$ such that $\gamma$ (the central area) is 0.95
$\nu = 12, 16, 20$

**E 4.3** Given the following from fitting Model A to four different data sets, determine the 95% and 99% confidence intervals for $\alpha$:

| | *Set 1* | *Set 2* | *Set 3* | *Set 4* |
|---|---|---|---|---|
| $a$ | 3.0 | 2.5 | 2.5 | 2.5 |
| $s^2$ | 9 | 9 | 9 | 9 |
| $n$ | 11 | 11 | 15 | 31 |

**E 4.4** Fitting Model B to a data set in which $n_1 = 15$ and $n_2 = 17$ gives estimates as follows:
$a_1 = 3.0$, $a_2 = 2.5$, $s_1^2 = 9.0$, $s_2^2 = 9.6$
Calculate a 95% confidence interval for $(\alpha_1 - \alpha_2)$.

**E 4.5** Fitting Model C to a data set in which $n = 20$ gives
$a = 4.0$, $b = 0.8$, $\mathrm{SE}(b) = 0.4$
Calculate a 95% confidence interval for $\beta$.

**E 4.6** On fitting Model C to a bivariate data set ($n = 10$) it is found that the variance of $Y$ (based upon a divisor of $n$) is 8.3 and the residual sum of squares is 15.0. Calculate TSS, ESS, the coefficient of determination and the correlation coefficient.

## FURTHER WORK

**F 4.1** Using the data in Table 8.3, fit the bivariate relationship between *IHDMR* (the dependent variable) and *SUG* (the explanatory variable). Consider by graphical means the plausibility of a linear systematic component. Are there any reasons for preferring this model to that considered in Section 4.10 (with *TDP* as the explanatory variable)?

*This chapter depends upon: Sections 0.13–0.15, 0.17–0.21, 0.28*

# 5

# Theoretical Frequency and Probability Distributions

## 5.1 INTRODUCTION

Statistical theory has led to the development of a large number of theoretical distributions which are used in the analysis of data. We have already met the normal distribution and the *t*-distribution. In the first part of this chapter (Sections 5.1–5.4) we discuss in general terms the nature of theoretical distributions and their role in data analysis. The second part (Sections 5.5 *et seq.*) concentrates upon the properties of some of the more important distributions. The information about individual distributions will be drawn upon freely in the rest of the text. However, an understanding of all the material in Sections 5.5 *et seq.* will not be immediately required for subsequent chapters and readers may find it convenient to regard it as reference material to be consulted as the need arises.

Essentially, a theoretical distribution is one whose properties are specified exactly in mathematical terms. Thus, if the distribution relates to a continuous variable, as does the normal distribution, it can be specified by the mathematical expression which will determine the height of its frequency curve for any value of the variable. This mathematical expression is known as the *density function.* Again, if the distribution relates to a discrete variable, then the distribution will be defined by a mathematical expression or density function which specifies the relative frequency for each discrete value of the variable. The details of these mathematical procedures need not detain us at the moment. The important point is that, because a theoretical distribution is specified mathematically, its statistical properties are predetermined and known.

The simplest way in which theoretical distributions are used is in 'fitting' them to a univariate data set. By fitting is meant the identification of some theoretical distribution which closely matches the relative frequency distribution of the data. Suppose, for example, that it has been found that a particular set of data is closely approximated by the normal distribution having mean of 50 and standard deviation of 3. Then two things have been achieved. First the data can be more economically described: 'approximately a normal distribution with mean 50 and standard deviation 3'. Secondly, the known properties of the normal distribution can be used to make more detailed statements about the properties of the data. Thus, because, for all normal distributions, 68% of the values lie within one standard deviation of the mean, for this data set approximately 68% of the values will lie within the range $50 \pm 3$. Or again, because all normal distributions have a kurtosis of 0, then the data set will have a kurtosis approximately equal to 0.

The process of fitting a theoretical distribution to an empirical distribution is a simple example of statistical modelling. The theoretical distribution constitutes a statistical model of the data. In chapter 4, other models were discussed of the type in which the observed values in the data set are represented as the sum of a 'systematic component' and a 'random component'. There it was the random component which was modelled by a theoretical distribution, in fact the normal distribution.

Apart from this use of the theoretical distributions in setting up a statistical model of data, some distributions enter into data analysis by the back-door as it were. Thus, in Chapter 4, the *t*-distribution was used in setting up confidence intervals for the parameters of models based upon the normal distribution. Two other distributions—the $\chi^2$ and the *F*-distribution, which *inter alia* are used in analysing the effectiveness and validity of a model, will be introduced in Chapters 7 and 8 respectively. These three distributions are rarely used as direct models of data and will receive no further mention in the present chapter.

## 5.2 FREQUENCY DISTRIBUTIONS AND PROBABILITY DISTRIBUTIONS

There is a basic duality between frequency distributions and probability distributions in the sense that statistical models can usually be described either in terms of a frequency distribution or a probability distribution at choice. An example has been met in Chapter 3 where the random component was modelled by a normal distribution. We can either say that, in the long-run, the values of the random component are presumed in the model to have a normal frequency distribution. Alternatively, values of the random component can be said to be modelled by the corresponding normal probability distribution. These two statements are equivalent. This is a consequence of interpreting probability in the sense of long-run relative frequency. This duality makes it usually unnecessary to specify whether a theoretical distribution is a probability or a frequency distribution: it will depend upon the context.

For simplicity, the term 'frequency curve' will be used for continuous distributions such as the normal, whether the context be that of a frequency distribution or a probability distribution.

It is to be understood that, where a set of data values is modelled by a probability distribution, it will be implicitly assumed that any one data value is statistically independent of any other value. In the context of the journey time data of Chapter 3, this means that there is no dependence between the times of different journeys, e.g. that an unusually long journey time does not tend to be followed by another unusually long time. This assumption therefore forms part of the initial specification of the model and the methods of analysis and estimation we shall use depend upon the assumption being valid. If the assumption of independence seems to be an implausible one—either by reason of prior background knowledge or later analytical validation analysis—then a more complex model would have to be formulated. Time series data typically exhibit dependence and the particular analytical problems to which this gives rise are examined in Chapter 9.

## 5.3 THE CHARACTERISTICS OF THEORETICAL DISTRIBUTIONS

Since the theoretical distributions are used in modelling statistical data and in the subsequent analysis, the applied statistician needs to have a knowledge of the more important distributions as part of his or her tool-kit.

Typically, any theoretical distribution is not just one distribution but a family of distributions. Thus, the normal distribution comprises a family of different distributions with different means and variances, but all having the typical normal distribution symmetrical bell shape. For the normal distribution, the mean and variance are said to be the *parameters* of the distribution. If the two parameters take up specific numerical values, then a particular normal distribution is thereby specified. The mathematical specification of the density function includes in its formula these two parameters, conventionally labelled $\mu$ and $\sigma^2$.

The two parameters of the normal distribution themselves measure two simple properties: the mean and the variance. This is not generally the case; the binomial distribution has two parameters $n$ and $\pi$, but the mean is $n\pi$ and the variance $n\pi(1 - \pi)$. In general, the summary statistics—the mean, the variance etc.—of any theoretical distribution are mathematical functions of the parameters, that is they can be calculated numerically for given numerical values of the parameters.

When making use of any theoretical distribution in modelling and analysis, a number of prior considerations need to be borne in mind:

(i) *What is the nature of the variable described by the distribution?* A distinction has already been made in Chapter 1 between categorical variables, whose values are specified in descriptive terms, and numerical variables. Most theoretical distributions in fact describe numerical variables. There are two reasons. The primary one is that statistical data mostly come in numerical form. But even where the data are based upon observations of a categorical variable, frequency counts of numbers of occurrences of the various categories yield values which can be modelled by distributions of numerical variables. The binomial and Poisson distributions fall into this category.

(ii) *If the variable is numerical, is it discrete or continuous?* The mathematics of theoretical frequency distributions makes a clear distinction between discrete and continuous distributions. In the discrete case, the density function formula specifies the relative frequency (or probability) of occurrence of each of a discrete set of values, such as the positive integers. It has no relevance to any other values in-between, such as 1.5 or 3.7. In the continuous case, the density function formula specifies the height of the frequency curve and the density function can be calculated (in principle) for every possible value of the variable over the relevant range.

(iii) *What is the mathematical specification of the density function?* This is basic and defines the distribution. Many density functions have complicated mathematical forms—the normal distribution is an example. However, for many purposes, a technical knowledge of this mathematics is relatively unimportant for an informed use of a theoretical distribution in practice. It is more important to have an understanding of the answers to the following questions:

(iv) *What is the theoretical context of the distribution?* A theoretical distribution often describes some idealised situation. In applications, one may need to consider how far the data context matches this idealised situation. In other words, enough needs to be known about the conceptual background of the theoretical distribution to ensure that it is used appropriately in application.

(v) *What is the shape of the distribution and how does it vary as the parameter values vary?* It has already been pointed out that a theoretical distribution typically comprises a family of distributions generated by the different possible values of the parameters of the distribution. In the case of the normal distribution, they all have the uni-modal bell shape; the centre of the frequency curve is above the value of $\mu$ and the bigger the value of $\sigma^2$ the wider the curve. Usually individual members of a family of distributions have similar shapes, sometimes they vary significantly.

## 5.4 CONTINUOUS THEORETICAL DISTRIBUTIONS

As we have already remarked, a continuous theoretical distribution is defined by its density function, $f(x)$, which equals the height of the frequency curve at the value $x$. The cumulative area to the left of $x$ is known as the distribution function and is denoted by $F(x)$. For example, for the standard normal distribution $F(1.96) = 0.975$. In other words, $F(x) = p$, the quantile order of $x$ which was introduced in Section 3.8.

The relation between the distribution function and the density function is shown in the first diagram of Figure 5.1. The second diagram shows, in general terms, a graph of the distribution function. The distribution function (that is, the cumulative area) increases from 0 to its maximum value of 1 as $x$ increases from its smallest feasible value to its largest.

Table 5.1 sets out the main properties of four distributions which will be covered below: the normal, lognormal, Pareto and uniform

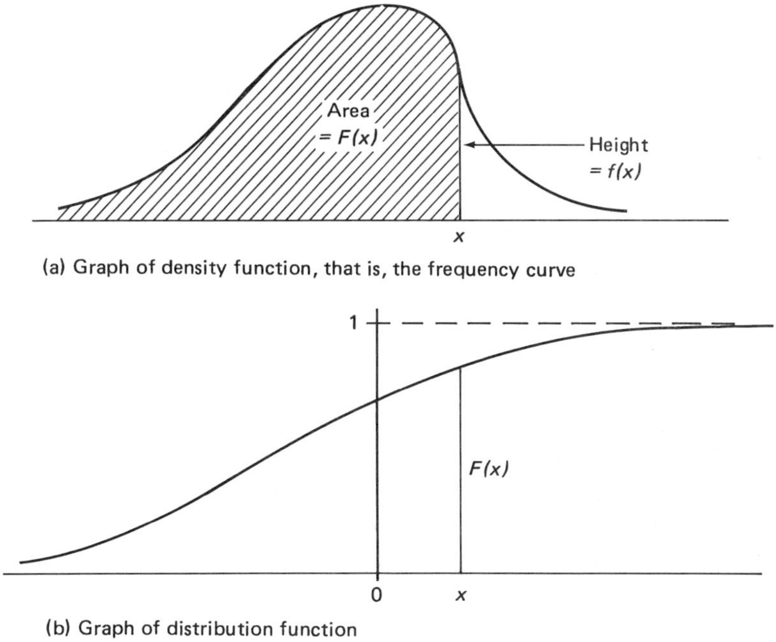


**Figure 5.1** Density Function and Distribution Function for a Continuous Distribution

distributions. The information in the table will be drawn upon in what follows. For reference purposes, the table also includes information on another four distributions: the exponential, gamma, beta and logistic distributions; the Bibliography includes texts containing extended accounts.

There is one simple but important distinguishing feature of distributions which is crucial in their use as models for univariate data sets. It concerns the possible range of values of the variable, that is, the feasible values. In the first column of Table 5.1, under the heading 'density function', we note that for the normal distribution, and also for the logistic distribution, the variable can take any positive or negative value. However, for all the other distributions, there are limitations on the variable values and these are shown in the table. These theoretical restrictions on the feasible values of the variables can be important in applications. For example, a data set containing negative values cannot be modelled by the lognormal distribution.

Although the distributions of this section relate to continuous variables, they can nevertheless be used to model situations where the

**Table 5.1** Properties of Continuous Distributions

| *Density Function* | *Parameter values* | *Mean* | *Variance* | *Shape* |
|---|---|---|---|---|
| *Normal* | | | | |
| $\frac{1}{\sqrt{(2\pi)}\sigma} \, exp\left[-\frac{1}{2}\frac{(x-\mu)^2}{\sigma^2}\right]$ ($x$ any value) | $\mu$ any value $\sigma > 0$ | $\mu$ | $\sigma^2$ | 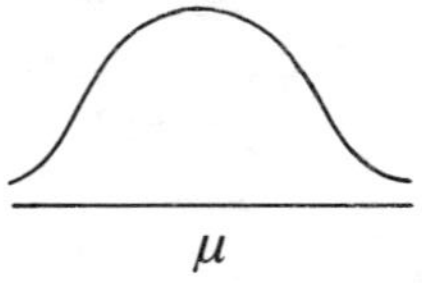 |
| *Lognormal* | | | | |
| $\frac{1}{x\sqrt{(2\pi)}\sigma} \, exp\left[-\frac{(\ln x-\mu)^2}{2\sigma^2}\right]$ ($x > 0$) | $\mu$ any value $\sigma > 0$ | $\exp(\mu + \frac{1}{2}\sigma^2)$ | see below | 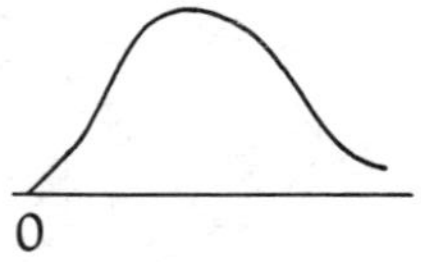 |
| *Pareto* | | | | |
| $\frac{\alpha k^\alpha}{x^{\alpha+1}}$ ($x > k$) | $k > 0$ $\alpha > 0$ | $\frac{\alpha k}{\alpha - 1}$ ($\alpha > 1$) | $\frac{\alpha k^2}{(\alpha-1)^2(\alpha-2)^2}$ ($\alpha > 2$) | 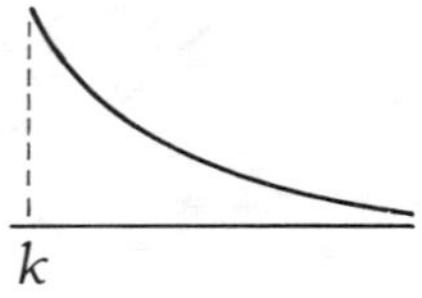 |
| *Uniform* | | | | |
| $\frac{1}{b-a}$ ($a \leqslant x \leqslant b$) | $a$ and $b$ any value ($b > a$) | $\frac{1}{2}(a+b)$ | $\frac{(b-a)^2}{12}$ | 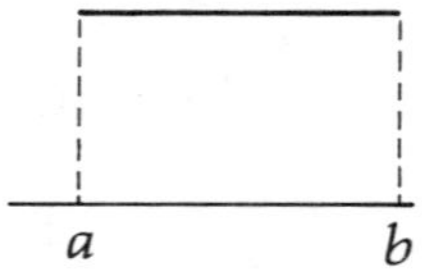 |

| | | | | |
|---|---|---|---|---|
| *Exponential* | | | | |
| $\lambda \exp(-\lambda x)$<br>$(x \geqslant 0)$ | $\lambda > 0$ | $\frac{1}{\lambda}$ | $\frac{1}{\lambda^2}$ | 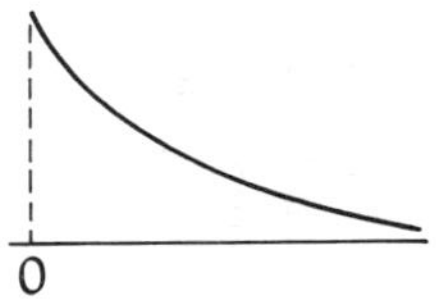 |
| *Gamma* | | | | |
| $\frac{\theta r}{\Gamma(r)} x^{r-1} \exp(-\theta x)$<br>$(x \geqslant 0)$ | $\theta > 0$<br>$r > 0$ | $\frac{r}{\theta}$ | $\frac{r}{\theta^2}$ | 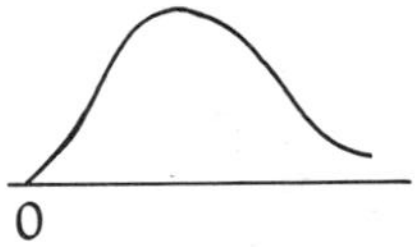 |
| *Beta* | | | | |
| $\frac{\Gamma(a)\Gamma(b)}{\Gamma(a+b)} x^{a-1}(1-x)^{b-1}$<br>$(0 \leqslant x \leqslant 1)$ | $a > 0$<br>$b > 0$ | $\frac{a}{a+b}$ | see below | 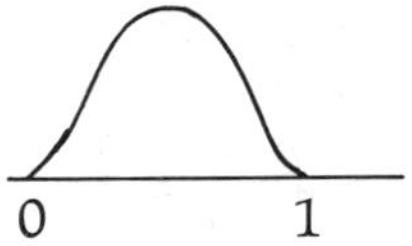 |
| *Logistic* | | | | |
| $\frac{\frac{1}{b} exp\left(-\frac{(x-a)}{b}\right)}{\left[1+\exp\left(-\frac{(x-a)}{b}\right)\right]^2}$<br>$(x$ any value$)$ | $a$ any value<br>$b > 0$ | $a$ | $\frac{b^2\pi^2}{6}$ | 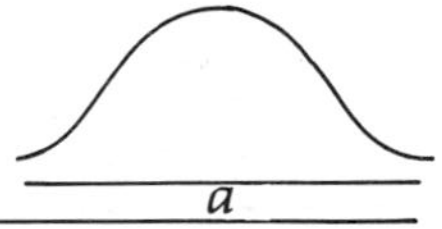 |

Variance of lognormal = $[\exp(\mu + \frac{1}{2}\sigma^2)]^2 \, [\exp \sigma^2 - 1]$
Variance of beta: $ab/(a+b+1)(a+b)^2$

observed data are not strictly continuous. As has been pointed out in Section 1.3.1, some variables are in principle continuous, but observed data may be discrete due to limitations of measurement or rounding of data values. There would be no objection to modelling such data by an appropriate continuous distribution so long as the gaps between successive feasible discrete observations are relatively small. Conversely, a continuous distribution may also be used as a model where the variable is discrete but the feasible discrete values are close together and there are very many of them: 'population values of different countries' is an example of such a variable.

## 5.5 NORMAL DISTRIBUTION

The basic properties of the normal distribution have been set out in Section 0.20. It is symmetric; its two parameters $\mu$ and $\sigma^2$ respectively measure the mean and variance; and the coefficient of kurtosis of any normal distribution equals 3. The notation for the distribution, $N(\mu, \sigma^2)$, was introduced in Section 0.20.

The normal distribution plays a central role in statistics in two ways. It is useful for setting up data models, particularly models of the kind considered in Chapter 4 where a random component is modelled by the normal distribution. Also, it is often important in data analysis: an example is large sample estimation as used in Chapter 3.

There are empirical and theoretical justifications for its use in data modelling. Empirically, it is one of the few distributions available for modelling data having an approximately symmetrical distribution. It is flexible in that the range of feasible variable values is unrestricted. It is also simple in the sense that its two parameters correspond to two fundamental summary statistics, the mean and the variance.

The theoretical justification stems from the Central Limit Theorem which can be explained in terms of the use of the normal distribution for modelling the random component of the journey time data of Chapter 3. There, the random component reflected the net effect of a very large number of factors: the effect on the journey time of each set of traffic lights, the situation at each junction, the density of traffic on each section of the route, and so on. Now the effect on the journey time of each of these factors will have its own frequency distribution in the long run so that the random component effectively is the sum of a large number of variables having small numerical values. The Central Limit Theorem states that the sum of a large number of variables of this kind has approximately a normal distribution, and this is the justification for the use of the normal distribution for the random component. The theorem applies whatever the frequency distribution of the variables corresponding to each factor. Two conditions however must be fulfilled. The factors must be very many, with no dominant ones, and they must be statistically independent, one from another. It seems plausible to expect both these conditions to hold, approximately at least, for the journey time situation.

Textbooks typically lay much stress upon the natural occurrence of data having a normal distribution and refer to the Central Limit Theorem as a reason. In fact, outside the experimental sciences, normally distributed data is not easily come by, the main reason being that the source of variability in the data values is not normally restricted to minor factors, as is required by the theorem. Thus, the distribution of weights of a varied collection of people of all ages and both sexes is unlikely to be normally distributed. But if the data were 'controlled' by restricting the coverage to, say, boys aged 10 from the same social background, then a normal distribution might well emerge. The source of variation would be confined to the 'uncontrollable'. Such a procedure really amounts to setting up a statistical model for the weight data on similar lines to those discussed in Chapter 4. The systematic effects due to age, sex and social class have been separated out and the residual variation is that of a random component. The importance of the normal distribution in data modelling lies more in its use as a model for the random component rather than as a model for a univariate variable itself.

## 5.6 LOGNORMAL DISTRIBUTION

The lognormal is a skew distribution with an extended right-hand tail and the variable being restricted to positive values (see frequency curve diagram in Table 5.1). There are two parameters, conventionally labelled $\mu$ and $\sigma^2$; these parameters are *not* respectively equal to the mean and variance of the distribution. The mean and standard deviation or variance are in fact rather complicated functions of $\mu$ and $\sigma$ (see Table 5.1); the mode equals $\exp(\mu - \sigma^2)$ and the median $\exp(\mu)$. Our notation for the distribution is $LN(\mu, \sigma^2)$.

Empirically, the lognormal can be simply viewed as a convenient model for positively skewed data. It suffers from the disadvantage that only positive variable values are permitted, though extended versions of the distribution (which introduce additional parameters into the density function) do not have this restriction (see references in Bibliography).

As its name suggests, the lognormal is closely related to the normal distribution. Indeed, the lognormal can be defined by the following property: if a set of values has a lognormal distribution, then their logarithms have a normal distribution. In other words, if a set of data values has a skewed distribution and if it is found that a logarithmic transformation results in a set of transformed values having approximately a normal distribution, then the data values would have approximately a lognormal distribution. This logarithmic transformation property accounts for a number of features of the lognormal. It explains why a lognormal variable has values over the range $(0, \infty)$ since taking logarithms leads to values over the range $(-\infty, \infty)$, that of the normal distribution. This transformation property does not depend upon the base of the logarithms, but if they are calculated to the base *e*, then if the original data have a particular distribution $\mathrm{LN}(\mu_0, \sigma_0^2)$ say, the transformed data will have the distribution $\mathrm{N}(\mu_0, \sigma_0^2)$. This explains the presence of the parameters $\mu$ and $\sigma^2$ in the lognormal density function.

The lognormal distribution can also be approached through its own form of the Central Limit Theorem. It will be recalled that if a variable is formed by adding together the effects of a large number of independent random factors, each having their own distribution, the variable itself will have a normal distribution. If however, the variable is built up by a *multiplicative* rather than an additive process, the resulting variable will have a lognormal distribution. A generation process of this kind can be looked upon in terms of a temporal sequence of proportionate effects. One can imagine a variable standing initially at a particular value and its value changed successively by multiplication by random amounts. At each stage the effect in absolute terms will be proportionate to the value so far attained. The final value will have a probability distribution of the lognormal.

The lognormal distribution has been proposed and fitted for a wide variety of empirical variables including personal income, size of firms in a particular industry, size of particles in geological deposits and so on. The latter example gives an illustration of the practical meaning of a random sequence of proportionate effects. One imagines a deposit of particles of identical size at some distant point in geological time. This is followed by a sequence of geological events of diminution or accretion in size; at each event the size of each particle is, as it were, multiplied by a random value from some distribution, that is to say, at each event the effect on the size is proportionate to its existing size. Those particles which happened to have a larger proportionate effect will tend to grow relatively the bigger thereafter and the distribution of size values will stretch out with the long right-hand tail typical of the lognormal distribution.

## 5.7 PARETO DISTRIBUTION

The Pareto distribution was originally developed as a model for income distributions. It has a 'reversed J-shape' and has the unusual feature that the variable values are restricted to positive values greater than some constant, $k$, which is a parameter of the distribution (see Table 5.1). The second parameter, $\alpha$, is restricted to values greater than zero. Our notation for the distribution is Pa($k$, $\alpha$). There is a mathematical complication in that the mean of this distribution does not exist unless $\alpha > 1$, nor the variance unless $\alpha > 2$.

The distribution function rather than the density function is often used in practice. For the Pareto:

$$F(x) = 1 - \left(\frac{k}{x}\right)^{\alpha} \qquad (5.7.1)$$

Since $F(x)$ measures the relative frequency of values less than or equal to $x$, $1 - F(x)$ equals the relative frequency greater than $x$, which, for the Pareto, has therefore the simple form

$$1 - F(x) = \left(\frac{k}{x}\right)^{\alpha} \qquad (5.7.2)$$

As an example the authors have found that the distribution of length of text allocated to words in the Collins-Robert French dictionary is well fitted by the Pareto: at the lower end of the distribution there are very many short textual entries for infrequently used words and at the upper end relatively few very long entries for common words.

The Pareto is important in that it has been found that it can give a good fit to the right-hand tail of some empirical distributions. Income

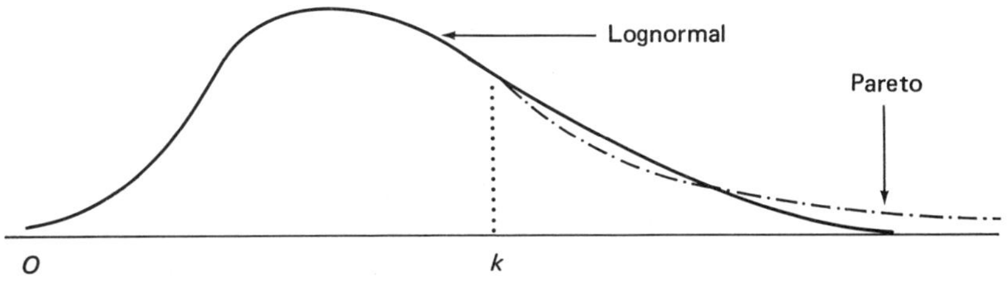


**Figure 5.2** Lognormal and Pareto Distributions Compared in Right-Hand Tail (Diagrammatic)

distributions are an example. These are typically skew with a minimum value of zero and a long right-hand tail. The lognormal is a potentially suitable model on empirical and theoretical grounds. However, it is sometimes found that the empirical distribution has a 'thicker' tail than is provided by the lognormal and, above a certain value of the variable, the Pareto gives a better representation. Figure 5.2 illustrates this diagrammatically.

## 5.8 CONTINUOUS UNIFORM DISTRIBUTION

By reason of its shape (see Table 5.1), the uniform is also known as the rectangular distribution. It has two parameters $a$ and $b$ which determine the range of variation of the variable and we shall use the notation U($a$, $b$). It may be looked upon as a neutral distribution in the sense that ranges of values of the same width all have the same relative frequency. Thus, for the distribution U(2, 10), the relative frequencies of values in the ranges (2, 3), (3.4, 4.5), (7.1, 8.1) all equal 1/8.

The distribution of errors in rounding data is one example of an empirical uniform distribution. Suppose that a set of numbers expressed to many decimal places is rounded to the nearest integer and consider the variable 'rounding error' equal to the difference between the unrounded and rounded number. Then the rounding error will lie within the range (−0.5, 0.5), no particular values can be expected to be preferred and the rounding error distribution would be closely approximated by U(−0.5, 0.5).

There is also a uniform distribution of a discrete variable and this is reviewed in Section 5.16 below.

## 5.9 DISCRETE DISTRIBUTIONS AND THEIR DISTRIBUTION FUNCTION

A discrete distribution is distinguishable from a continuous distribution in that one can at least in principle list the feasible variable values. If, for example, a variable has a Poisson distribution, the only feasible values are 0, 1, 2, 3, . . . and any positive integer. Note that the feasible values may nevertheless be infinite in number, but of course the relative frequencies of extremely large values become negligibly small.

The density function of a discrete distribution, indicated by $f(x)$, determines the relative frequency (or probability) of each feasible value. The distribution function, $F(x)$, is defined as for a continuous distribution; it equals the cumulative relative frequency of all values

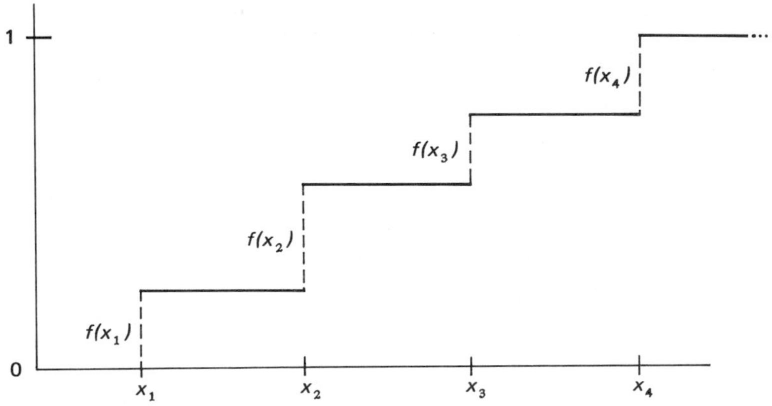


**Figure 5.3** Distribution Function of a Discrete Distribution

less than or equal to $x$. The diagrammatic representation of $F(x)$ for a discrete distribution is given in Figure 5.3. This diagram presents a stepwise appearance. For variable values below the smallest, $x_1$, the cumulative relative frequencies are all necessarily zero. Then, at the value $x_1$, the cumulative relative frequency jumps to $f(x_1)$ where it remains until the variable value $x_2$ when the cumulative relative frequency jumps to $f(x_1) + f(x_2)$, and so on.

## 5.10 THE FREQUENCY VARIABLE AND BERNOULLI TRIALS

The most common kind of discrete variable, and the one which will chiefly engage our attention is that which arises from a process of counting. The variable values are, therefore, integers and we call such a variable a *frequency variable.*

Data which consist of counts can arise in various ways and it will be useful to sketch out some of these. To make a classification we will ask:

(i) what *primary* data matrix underlies the data, and
(ii) what process of aggregation, if any, has been applied to this primary data matrix.

Consideration of these possibilities leads us to three types of data consisting of counts.

First, suppose the primary data matrix consists of values of a single

**Table 5.2** Examples of Frequency Variable Data

| | | | | | | | |
|---|---|---|---|---|---|---|---|
| (i) Sentence length (no. of words) | 20 | 21 | 26 | 22 | 22 | 46 | |
| (ii) No. of rainy days in September (1973–79) | 14 | 20 | 17 | 18 | 9 | 8 | 6 |
| (iii) Proportion of food shops in 6 towns | 0.20 | 0.17 | 0.12 | 0.26 | 0.14 | 0.19 | |

*Sources:* (i) Jane Austen, *Northanger Abbey*, first six sentences.
(ii) *Whitaker's Almanack*, Observations at Kew, London.
(iii) Hypothetical data.

variable which, by definition, takes only integer values. This is the simplest case to envisage. No aggregation is called for and analysis will be based directly upon this data matrix. Example (i) in Table 5.2 is of this form.

Consider now a primary data matrix whose individuals consist of days of a month. For each day we record the value of the categorical variable: rainfall/no rainfall. At the end of a month we may summarise the rainfall pattern by counting the number of days in which rainfall was recorded. This gives us a single line of a new data matrix in which the individual is a given month in a particular year and the variable value is a count of rainfall occurrences. A new data matrix may now be built up by repeating this process for the same month across several different years. Example (ii) in Table 5.2 is of this form.

Consider finally the case where the individuals are, say, shops in a town and we record, for each shop, the value of the categorical variable food shop/non-food shop. Our interest may be in the proportion of food shops relative to the total number of shops. The primary data matrix would be aggregated to produce a new data matrix which shows, for each town, the values of two frequency variables: the number of food shops and the total number of shops. *Both* these figures will (usually) be different from one town to another.

Data in the first category may be directly modelled by a variety of distributions, including continuous distributions if the feasible values are closely spaced. In fact, the lognormal distribution has been applied to sentence length data (see Williams (1970)). Data in the third category, involving the recording and analysis of two frequency variables, will be considered in Chapter 15.

The distributions to be considered below are mainly relevant to the second category where the frequency variable value results from counting the number of occurrences of a particular categorical variable value in an underlying primary data matrix. We shall call an observation of a particular categorical variable value (such as 'rainy day') the occurrence of a *basic event*. With this terminology, the frequency variable counts the total number of basic events in the primary data matrix.

Models of frequency variables are usually founded upon a probability model of the occurrence of a basic event. An individual observation where a basic event has a constant probability of occurrence, say $\pi$, is known as a Bernoulli trial. Rolling a dice when the basic event is 'two is uppermost' is a Bernoulli trial with $\pi = 1/6$. Observing whether a randomly chosen September day is rainy is a Bernoulli trial with $\pi$ equal to the long-run proportion of rainy September days. Bernoulli trials are said to be *independent* if the outcome of one trial is not affected by the outcome of any other trial. Rolling a dice 60 times could be expected to constitute a sequence of 60 independent Bernoulli trials. On the other hand, observing 30 successive days in September would not, because rainy days tend to be followed by rainy days: the probability of rain on 3 September would depend upon whether or not it had rained on 2 September.

In what follows, $\pi$ will always stand for the probability of the basic event and $n_B$ will stand for the number of Bernoulli trials (independent or otherwise) which generate the frequency variable. That is, $n_B$ stands for the number of individuals in the primary data matrix—the number of days in September.

Finally, we note one simple property of a sequence of Bernoulli trials. The frequency of basic events observed in $n_B$ trials is a random value in the sense that it depends upon the random outcome of each trial. Thus, the number of twos observed in 60 rolls will vary from one set of rolls to another. But the long-run average number of twos in repeated sets of 60 rolls will equal $1/6 \times 60 = 10$ and in general the long-run average number of basic events (each having probability $\pi$) in $n_B$ independent Bernoulli trials equals $n_B\pi$. In the language of Section 0.15, the number of basic events in $n_B$ trials is a random variable with expected value $n_B\pi$.

## 5.11 THEORETICAL DISCRETE DISTRIBUTIONS

In Sections 5.12—5.16, we discuss four discrete distributions: the Poisson, binomial, negative binomial and the discrete uniform. Their properties are summarised in Table 5.3 which also includes, for reference, the geometric and Pascal distributions. With the exception of the discrete uniform, all these are distributions of frequency

**Table 5.3** Properties of Discrete Distributions

| *Density function* | *Parameter values* | *Mean* | *Variance* | *Shape* |
|---|---|---|---|---|
| *Poisson* | | | | |
| $\frac{\lambda^x}{x!} \exp(-\lambda)$<br>$x = 0, 1, 2, \ldots$ | $\lambda > 0$ | $\lambda$ | $\lambda$ | 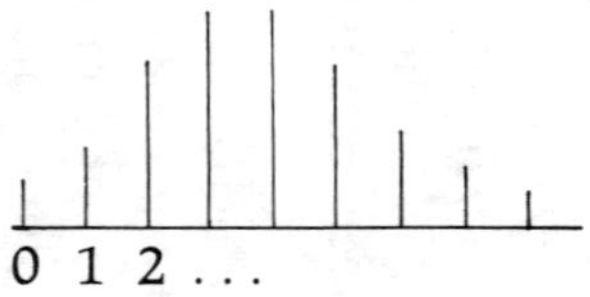 |
| *Binomial* | | | | |
| $\binom{n}{x} \pi^x (1-\pi)^{n-x}$<br>$x = 0, 1, 2, \ldots n$ | $0 < \pi < 1$<br>$n$, any positive integer | $n\pi$ | $n\pi(1-\pi)$ | 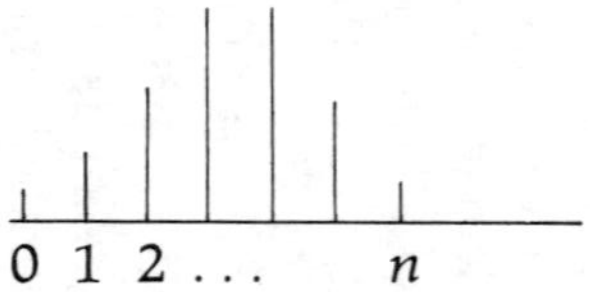 |
| *Negative binomial* | | | | |
| See Section 5.14 for density function etc.<br>$x = 0, 1, 2, \ldots$ | | | | 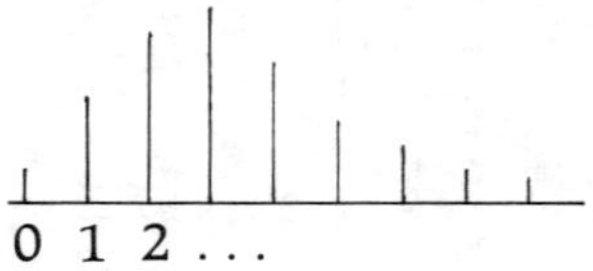 |

| | | | | |
|---|---|---|---|---|
| *Geometric* | | | | |
| $\pi(1-\pi)^x$<br>$x = 0, 1, 2, \ldots$ | $0 < \pi < 1$ | $(1-\pi)/\pi$ | $\dfrac{(1-\pi)}{\pi^2}$ | 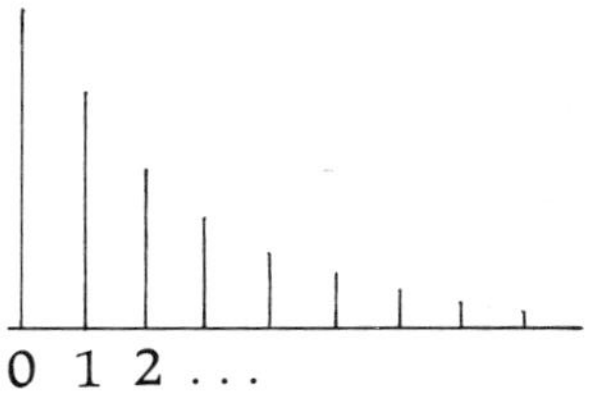<br> |
| *Pascal* | | | | |
| $\binom{x-1}{x-k}\pi^k(1-\pi)^{x-k}$<br>$x = 1, 2, \ldots$ | $0 < \pi < 1$<br>$k$, any positive integer | $k/\pi$ | $\dfrac{k(1-\pi)}{\pi^2}$ | 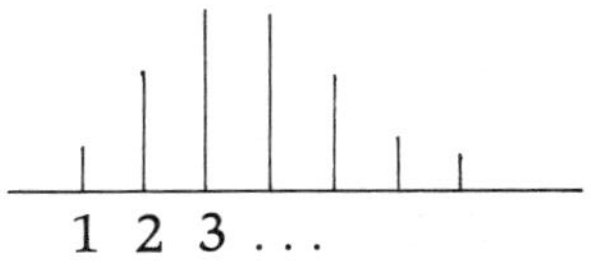<br> |
| *Discrete uniform* | | | | |
| $1/n$<br>$n$ values of $x$ equally spaced between $a$ and $b$ | $n$, any positive integer<br>$a$ and $b$ any values where $a < b$ | $\frac{1}{2}(a+b)$ | $\frac{1}{12}d^2(n^2-1)$<br>where<br>$d = \dfrac{b-a}{n-1}$ | 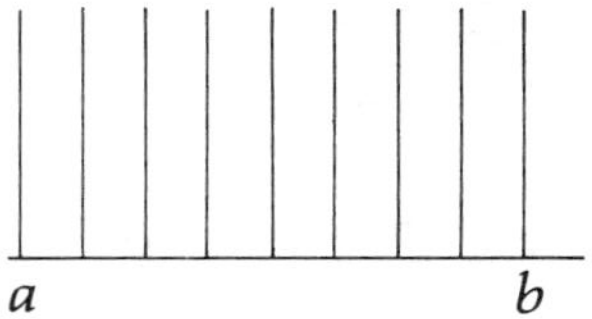<br> |

*Note:* The notation $x = 0, 1, 2 \ldots$ or $x = 1, 2, \ldots$ indicates that the variable can take any positive integral value without limit.

variables. The parameter $\pi$ which appears in many of the density functions stands for the probability of the basic event generating the particular frequency variable which is why this parameter invariably lies between 0 and 1. (It should not be confused with the mathematical constant $\pi = 3.14..$)

## 5.12 POISSON DISTRIBUTION

The Poisson distribution has a single parameter denoted by $\lambda$. Our notation for the distribution is Po($\lambda$). As shown in Table 5.3, the mean and the variance both equal $\lambda$. If $\lambda$ is not an integer, the mode is at the whole number between $(\lambda - 1)$ and $\lambda$; if $\lambda$ is an integer, the mode is at $(\lambda - 1)$. The general shape is skew with an infinite right-hand tail.

The Poisson distribution can be a model for the distribution of the total number of basic events in a fixed number, $n_B$, of Bernoulli trials with constant probability $\pi$. There are however two important conditions: $n_B$ must be large, and $\pi$ must be small. The Poisson can thus be said to describe the distribution of rare events. (The reader can think of the dice example with $n_B = 60$ and $\pi = 1/6$ as satisfying these conditions.) Given these conditions, it can be shown that the distribution of the number of basic events depends only upon $n_B\pi$, the expected number of basic events. Writing $n_B\pi = \lambda$ leads to the density function based upon the single parameter $\lambda$ as shown in Table 5.3. Thus the distribution Po(10) would describe the distribution of the number of twos thrown in 60 rolls of a dice; in this case, $n_B = 60$, $\pi = 1/6$ and $\lambda = 60(1/6) = 10$. If sets of 60 rolls of the dice were repeatedly carried out, the number of twos in each set would behave as a random sample from the Poisson distribution and in the long-run would build-up to a Poisson relative frequency distribution.

Outside the natural sciences, empirical data sets rarely match exactly the context of a theoretical distribution, which is why textbooks rely for illustrative purposes on rather artificial examples such as penny tossing, dice rolling and the like. Nevertheless, it is important to consider carefully the structure of an empirical data set to see where it might not match a theoretical context. For applying the Poisson distribution, this means considering the nature of the underlying Bernoulli trials and the underlying parameters $n_B$ and $\pi$. Table 5.4 sets out some examples. In each case $n_B$ is large and $\pi$ is small, but each has some defect or other.

In the first example, if one day is frosty it is more likely that the following day will be frosty: this would violate the assumption for the Poisson distribution of *independent* Bernoulli trials. In the second example the number $n_B$ of individuals (place names) observed will

**Table 5.4** Data Sets which might be Modelled by the Poisson Distribution

| *Data* | *Individuals observed for each data value* | *Basic event* | $\pi$ | $x$ *frequency variable)* |
|---|---|---|---|---|
| 1. No of frosty days in September, over a period of years | Each day in September ($n_B = 30$) | Individual day is frosty | Probability that any one day is frosty | Total no. of frosty days in September |
| 2. No. of place names ending in '-field' in each grid square on a map | Each place name in a grid square ($n_B$ varying) | Individual place name ending in '-field' | Probability that any one place name ends in '-field' | Total no. of such place names in a grid square |
| 3. No. of occurrences of the word 'but' on successive pages of a book | Each word on a page ($n_B$ varying) | Individual word is the word 'but' | Probability that a word is 'but' | Total no. of 'buts' on a page |

vary from one grid square to another; the primary data matrices are not of constant size and hence the count for each grid square would relate, not to one constant Poisson distribution, but to different Poisson distributions with varying parameter values $\lambda$. In the third example, it may be objected that words are most certainly not used 'randomly' in literature and a Poisson distribution (or any other) is inappropriate. This however is not the point if the Poisson distribution is found to provide an informative description of the data.

In fact, the restrictive conditions of the Poisson distribution—a single parameter and the mean being equal to the variance—prevent it being a widely applicable model for frequency variable data. As we shall see, the negative binomial distribution (Section 5.14 below) is more flexible and more useful.

However, the Poisson distribution is often a good model for the occurrence of some rare event over successive periods of time. Suppose, for example, that frequency variable data arise from counting the number of explosions in mines year by year. Then the Poisson distribution can be shown to be an appropriate model if the following conditions are satisfied: (i) there is a constant and very small probability that an explosion will occur in a short time interval (e.g. one day), (ii) the probability is negligible that two explosions happen within the same short time interval, and (iii) the occurrence of an explosion on any one day is independent of the occurrence or non-occurrence of an explosion on any other day.

## 5.13 BINOMIAL DISTRIBUTION

The binomial distribution, like the Poisson, describes the number of basic events in a fixed number $n_B$ of independent Bernoulli trials but, unlike the Poisson, there are not the constraints on the values of the parameters $\pi$ and $n_B$ and these appear in the formula for the density function. Whereas the Poisson distribution provides only an approximation (though a very close one) for the dice rolling example in Section 5.10 (with $\pi = 1/6$, $n_B = 60$ and $\lambda = 10$), the binomial density function is exact and is so for any values of $\pi$ and $n_B$.

Our notation for the distribution is $\mathrm{Bi}(n, \pi)$, where, to accord with standard practice, $n_B$ is written simply as $n$. (We prefer $\pi$ to the more usual notation of $p$ in line with the practice of using Greek letters for model parameters, so far as possible).

The binomial distribution plays an important role in the analysis and interpretation of some practical sampling situations. The outcome of a social survey carried out by taking a simple random sample of a human population, is well modelled by the binomial distribution. The selection of a sample size $n_0$ is equivalent to $n_0$ Bernoulli trials where the basic event is that an individual is in one of the categories covered by the survey: being male, being an old-age pensioner and so on. The probability $\pi$ that a basic event occurs equals the proportion possessing the category in the population. And the number, $x$, of people in the sample in a particular category will have the probability distribution $\mathrm{Bi}(n_0, \pi)$.

The sample survey yields of course only one data value and the theory is used to make an inference about the population proportion from this single value. The survey does not yield a set of empirically observed data values which collectively would approximate to a binomial frequency distribution. Indeed, it is rare to come across empirical data whose distribution is approximately binomial, other than in experimental type situations where in effect repeated sampling operations are carried out. Even then, unless $n_0$ is small, the normal distribution would be as appropriate.

## 5.14 NEGATIVE BINOMIAL DISTRIBUTION

Technically, the negative binomial is complicated by the fact that its density function can be expressed in different, and usually complex, mathematical forms (which is why it is omitted from Table 5.3) and moreover it has various alternative theoretical origins. At the same time, it is a useful model for frequency variable data and, happily, a familiarity with its mathematical complexities—some of which we touch upon below—is not crucial in the practical applications we shall discuss.

The negative binomial is a two parameter distribution of a frequency variable with a shape similar to that of the Poisson (see Table 5.3). It has the property that the variance is greater than the mean.

One theoretical approach to the negative binomial, and the more helpful for our purpose, is to derive it from the Poisson distribution in the following way. First suppose we have a situation where a random observation is derived from the Poisson distribution $\mathrm{Po}(\lambda_0)$ say. Then repeated random observations of this variable will collectively have a relative frequency distribution closely resembling $\mathrm{Po}(\lambda_0)$. Now suppose that, instead of the Poisson parameter $\lambda$ remaining constant, equal to $\lambda_0$, at each observation, $\lambda$ itself varies randomly in a fashion defined by some other probability distribution, which we shall call the parent distribution. If the parent distribution has a very small variance, that is if $\lambda$ does not vary very much between observations of of the frequency variable, the resulting distribution of the frequency variable will be very close to the Poisson and conversely if the parent distribution of $\lambda$ has a very large variance.

A distribution of this kind is called a *compound Poisson distribution.* Its exact form will depend upon the parent distribution but it can be shown that if the parent distribution is a gamma distribution,[1] then the resulting compound Poisson distribution is in fact the negative binomial distribution.

An example from the statistical analysis of literature will illuminate this theoretical procedure. The first row of Table 5.5 shows a frequency distribution of the number of occurrences of the word 'an' in 262 equal length blocks of words in a literary text.

A theoretical modelling might run as follows. Each block of text constitutes a primary data matrix of a large number of words; each word constitutes a Bernoulli trial with a very small probability of the basic event—the occurrence of 'an'. If this probability—and, hence, the expected number of 'ans'—were constant from one block of text to another, then the frequency distribution would be a Poisson distribution. The second row of the table shows the frequencies which would be given by the Poisson distribution $Po(0.90)$ having the same mean as the data. The correspondence between the two rows is not very good, particularly for the frequencies at variable values 0 and 1.

However, we can improve the modelling by supposing that the occurrence of 'an' is still modelled by a Poisson distribution within each block of text, but the author varies his style from one block of text to another so that the Poisson parameter, $\lambda$, varies according to another distribution, the parent distribution. Then, from the theory summarised above, *if* this parent distribution is a gamma distribution, the occurrences of 'an' over blocks of text would exhibit a negative binomial distribution.

The third row of Table 5.5 shows the frequencies for that negative binomial distribution with the same mean as the data (0.90) and the same variance (1.20) and the correspondence with the data now is much improved.

To sum up, in theoretical terms, the negative binomial often offers a more plausible model than the Poisson. Moreover, in practical terms, its two parameters give more flexibility and allow a closer correspondence with data when the data variance is greater than the mean.

1. The gamma distribution is included in Table 5.1. However, its technical properties are not needed for the present discussion.

**Table 5.5** Frequence Distribution of Occurrences of the Word 'An' in 262 Blocks of Text

| | 0 | 1 | 2 | 3 | 4 | 5 | 6 | *Occur-rences Mean* | *Variance* |
|---|---|---|---|---|---|---|---|---|---|
| *Observed* | 122 | 77 | 40 | 14 | 8 | 0 | 1 | 0.90 | 1.20 |
| Estimated from: | | | | | | | | | |
| Poisson | 106 | 96 | 43 | 13 | 3 | 1 | 0 | 0.90 | 0.90 |
| Negative binomial | 122 | 80 | 37 | 15 | 5 | 2 | 1 | 0.90 | 1.20 |

*Source:* Mosteller and Wallace (1964)

The calculation of the negative binomial frequencies of Table 5.3 requires knowledge of the density function to which we now turn.

Our notation for the distribution is NB($\mu$, $k$) and the density function is

$$f(x) = \frac{\Gamma(x+k)}{\Gamma(k)\,x!}\left(\frac{k}{\mu+k}\right)^k\left(\frac{\mu}{\mu+k}\right)^x$$

where $x = 0, 1, 2, \ldots$ and $k > 0$, $\mu > 0$
and mean $= \mu$, variance $= \mu(\mu + k)/k$ **(5.14.1)**

The $\Gamma$ notation is explained at Section 0.28.4. It can be shown that

$$\Gamma(x+k)/\Gamma(k) = (x+k-1)(x+k-2)\ldots(k+1)k$$

so that the density function can be calculated by ordinary algebra. Alternatively, we note that:

$$f(0) = (k/(\mu+k))^k$$

$$\text{and } f(x+1) = f(x)\frac{\mu}{(\mu+k)}\,\frac{(x+k)}{(x+1)} \qquad \textbf{(5.14.2)}$$

So, for known values of $\mu$ and $k$ one can first calculate $f(0)$, then $f(1) = f(0+1)$, then $f(2) = f(1+1)$ and so on, using (5.14.2).

For an alternative and quite different approach to the negative binomial distribution we return to the notion of a sequence of independent Bernoulli trials with constant probability $\pi$ of observing the basic event. However, instead of counting the frequency of occurrence in a fixed number of trials (as for the binomial distribution), the trials are continued until a fixed number, $k$, of basic events has occurred. The negative binomial variable is then defined as the number of trials in the sequence where the basic event did *not* occur.[2] This approach has been applied to models of disease incidence and product failure.

It can be shown that the resulting density function is:

$$f(x) = \binom{x+k-1}{k-1}\pi^k(1-\pi)^x$$

where $x = 0, 1, 2, \ldots$
and $k = 1, 2, 3, \ldots;\ 0 < \pi < 1$
mean $= k(1-\pi)/\pi$; variance $= k(1-\pi)/\pi^2$ **(5.14.3)**

2. There are other slightly different versions. For example, the negative binomial can also be defined in terms of the total number of trials required for $k$ basic events to occur.

The parameter $k$ is common to both specifications at (5.14.1) and (5.14.3) but it is important to note that, in the latter, $k$ is restricted to positive integral values. There is no such restriction in the compound Poisson approach which makes (5.14.1) a more flexible specification in practice. Apart from this difference, the density function at (5.14.3) equals that at (5.14.1) on putting

$$\pi = k/(\mu + k) \text{ (since } \mu = k(1 - \pi)/\pi)$$

The negative binomial approaches the Poisson distribution for large values of $k$ and small values of $\mu$. The *geometric distribution* (see Table 5.3) is a particular case of the negative binomial with $k = 1$.

## 5.15 COMPARISON OF BINOMIAL, POISSON, AND NEGATIVE BINOMIAL DISTRIBUTIONS

All three distributions are natural candidates for modelling the distribution of a frequency variable. For any particular set of data, which of the three is the more appropriate will depend upon what connection there is between the theoretical and practical data contexts. There is, however, a simple distinguishing feature of these distributions which helps a suitable choice. It depends upon comparing the mean and variance. The ratio mean/variance is always $> 1$ for the binomial, equal to 1 for the Poisson, and $< 1$ for the negative binomial distribution. A calculation of this ratio for the data values will therefore indicate on purely empirical grounds which distribution is likely to give the better fit.

Of the three, the negative binomial tends to be the more useful. Being a two parameter distribution, it is more flexible than the Poisson and, since non-experimental frequency variables more often exhibit a variance larger than the mean, it is in empirical terms more appropriate than the binomial.

## 5.16 DISCRETE UNIFORM DISTRIBUTION

This is the discrete counterpart of the continuous uniform distribution reviewed in Section 5.8. It has $n_0$ feasible values, equally spaced between the smallest, $a$, and the greatest, $b$. Each value has equal relative frequency and the density function equals $1/n_0$ for each feasible value (see Table 5.3). The feasible values are usually integral, but this is not necessary. Our notation for the distribution is $U(a, b, n_0)$. It is a three parameter distribution. The density function and summary statistics can also be defined in terms of the constant interval, $d$, between successive feasible values but $d$ is a function of the others, namely $d = (b - a)/(n_0 - 1)$.

An empirical example is given by the variable 'last digit of a car registration number'. The feasible values are 0, 1, 2, . . . 9; $a = 0$, $b = 9$, and the density function has a constant value 1/10.

## 5.17 CONCLUSION

In this chapter we have reviewed the more important theoretical distributions and considered their properties and appropriateness as models for univariate data sets. However, some essential practical questions remain. The central one is: given the choice of a particular distribution family as a model for a data set, how do we decide which particular member of the family is the best model? In other words we need to know how to estimate, from the data, numerical values for the parameters of the chosen distribution. This forms the central theme of Chapter 6.

## EXAMPLES

**E 5.1** It is proposed to use the lognormal distribution as a model for the following data:

| | | | | | | | | | |
|---|---|---|---|---|---|---|---|---|---|
| 0.14 | 0.16 | 0.21 | 0.26 | 0.34 | 0.41 | 0.50 | 0.68 | 0.76 | 0.77 |
| 0.80 | 0.84 | 0.88 | 0.89 | 0.92 | 0.94 | 0.99 | 1.04 | 1.07 | 1.12 |
| 1.16 | 1.23 | 1.34 | 1.38 | 1.77 | 1.86 | 2.32 | 3.39 | 4.95 | 6.23 |
| 7.17 | 7.39 | | | | | | | | |

(i) Construct a frequency distribution choosing suitable class intervals.

(ii) Draw the histogram and verify that its shape approximates to that of the lognormal.

(iii) Transform the data by taking logarithms of each value, construct a frequency distribution of the transformed data and draw its histogram. Verify that its shape is approximately normal.

(iv) Calculate the mean and variance of the transformed data. What relationship do these values have to the parameters of the proposed lognormal distribution?

**E 5.2** Sketch the frequency curves of the Pareto distributions

Pa(5, 1.5), Pa(5, 2.0), Pa(5, 3)

What is the effect of the value of $\alpha$ on the shape of the distribution?

**E 5.3** Show that, for the Poisson distribution

$$\frac{f(x+1)}{f(x)} = \frac{\lambda}{x+1} \text{ and } f(0) = e^{-\lambda}$$

Use these relationships to calculate the values of the density function for $x = 0, 1, 2, 3, 4, 5, 6$ for the distributions Po(2) and Po(3.5). Sketch the line histograms. What is the connection between the mode and the value of $\lambda$?

**E 5.4** Consider the negative binomial distribution with mean 2 and variance 3. Calculate the value of its parameter $k$, using (5.14.1).
Calculate $f(x)$ for $x = 0, 1, 2, 3, 4$ using (5.14.2). Sketch the line histogram and compare it with that of Po(2)

**E 5.5** Consider the data represented by the following frequency distribution

| $x$ | 0 | 1 | 2 | 3 | 4 |
|---|---|---|---|---|---|
| Frequency | 12 | 20 | 13 | 4 | 1 |

Is the negative binomial distribution a plausible model?

*This chapter depends upon: Sections 0.13—0.17, Chapters 2, 4, and 5*

# 6

# Fitting Theoretical Distributions to Data and Estimation Methods

## 6.1 INTRODUCTION

In Chapter 4 we discussed simple linear models based upon a systematic component and a random component modelled by the normal distribution. We showed how such models can be fitted, that is, how the data can be used to estimate the parameters of the model.

We now turn to consider more general methods of parameter estimation. Discussion will be restricted to the modelling of univariate data and examples will be drawn from the theoretical distributions reviewed in Chapter 5. It will be seen that there are a number of standard estimation methods which are of general applicability to a wide range of models. There are also *ad hoc* procedures which are specific to the fitting of particular distributions.

The present chapter is concerned with model specification and model estimation. A discussion of model checking and validation will follow in Chapter 7.

## 6.2 MODEL SPECIFICATION

In the present context, model specification is concerned with the choice of an appropriate family of theoretical distributions for modelling the univariate data set. Model estimation then amounts to the identification of a particular member of the family—with numerical values put to the parameters—as the chosen model.

The first stage of model specification is to identify the nature of the variable whose values comprise the univariate data set: is it a frequency variable (or otherwise discrete) or is it effectively a continuous variable? If the former, then, of those covered in Chapter 5, the possibilities are binomial, Poisson, negative binomial and discrete uniform distributions. If the latter, possibilities are the normal, lognormal and Pareto distributions and continuous uniform.

At the second stage one may ask if the context or origin of the data matches the context of a particular theoretical distribution as described in Chapter 5. Thus, the Poisson distribution is a model for data comprising frequency counts of the occurrence of a particular value of a categorical variable where the value has a constant but small probability of occurrence. If the origin of the data can plausibly be described in these terms, then the Poisson distribution would be a natural first choice as a model.

However, it may be that there are no *a priori* reasons of this kind which imply the specification of a particular distribution. If so, one must proceed to the third stage of model specification. At this stage, statistical properties of the data are compared with those of the potential candidates amongst the theoretical distributions to identify the distribution which gives the best match. For example, in the case of a frequency variable, the ratio mean/variance easily discriminates between the binomial, Poisson and negative binomial distributions. Similarly, if the variable is continuous, a consideration of the symmetry or asymmetry of the data would help resolve the choice of distribution. The normal distribution would be suggested by roughly symmetrically distributed data. The lognormal or the Pareto distribution would be possible candidates if the data are asymmetrically distributed. An additional guide to the choice of a distribution may be provided by the use of probability plots (see Section 6.8).

These procedures are summarised in the decision tree shown in Figure 6.1. It must however be stressed that the choice of model will not necessarily be a final one. After it has been estimated, the model will be subject to checks and validation by methods which will be discussed in Chapter 7, as a result of which the model may be revised and a different distribution tried. An iterative process of this kind is often the rule rather than the exception.

In Figure 6.1, we have used the notation for each of the distributions which was introduced in Chapter 5. And the decision tree draws attention to the possibility that discrete data might be approximated by some continuous theoretical distribution.

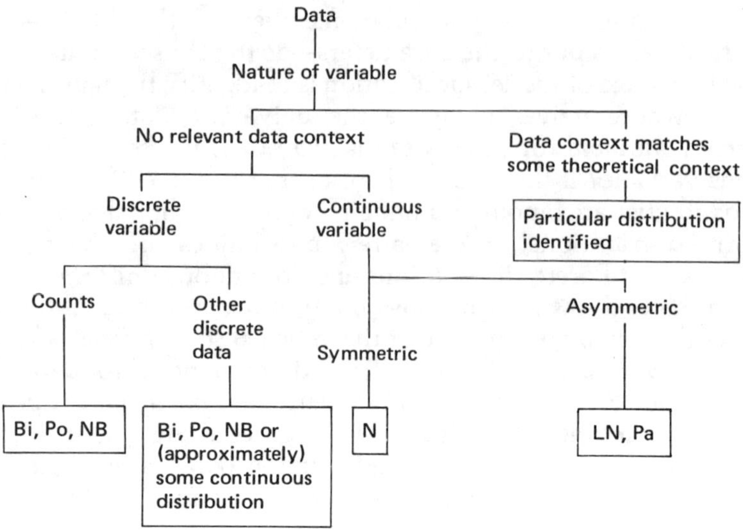


**Figure 6.1** Decision Tree for Choice of Model

## 6.3 MODEL ESTIMATION

We have already encountered, informally, some of the principles of estimation. In Section 4.2 we used the data mean, $a$, as an estimate of the systematic parameter $\alpha$. A single value such as the data mean is known as a *point estimate*; similarly, the regression coefficients are point estimates of the systematic parameters in the regression model.

By contrast, a confidence interval is an example (and the most usual form) of an *interval estimate*. In fact, the two strands of point estimation and interval estimation are brought together in the form of a confidence interval. The central value of a confidence interval is typically a point estimate and the confidence interval range is an interval estimate. The construction of confidence intervals for the systematic parameters of a linear model was considered in Chapter 4 and, for the purposes of this text, is required no more.

In the following sections three estimation procedures are described. Two technical terms first need to be defined. An *estimator* is the formula based upon the data, for calculating a point estimate of a parameter. Thus, the formula $\Sigma x_r/n$ is a possible estimator of the model mean and the least squares regression coefficient formula is an estimator for a systematic parameter of the regression model. The term *estimate* is strictly used for a particular numerical value of an estimator, but the distinction between an estimator and an estimate is a fine one and we shall not bother too much about it.

## 6.4 METHOD OF MOMENTS

The word 'moments' is a generic word for a certain class of summary statistics. The so-called '$k$th moment about the mean' is defined as $(1/n)\Sigma(x_r - \bar{x})^k$; when $k = 2$, we have the variance, when $k = 3$ we have a summary statistic related to a measure of skewness (see Section 2.10) and when $k = 4$ one related to a measure of kurtosis (see Section 2.12). Also, the '$k$th moment about zero' is defined as $(1/n)\Sigma x_r^k$ which gives the mean when $k = 1$; $k$ is known as the *order of the moment.* For present purposes the only moments we shall be concerned with are in fact the mean and the variance.

Moments are similarly defined for theoretical distributions; in particular the mean is, again, the first moment about zero and the variance is the second moment about the mean. Formulae for the moments of theoretical distributions are invariably functions of the parameters. the formulae for the mean and variance of the distributions introduced in Chapter 5 are given in Tables 5.1 and 5.3.

The parameters of a theoretical distribution can usually be written in terms of the moments of that distribution. The *method of moments* makes use of this fact and proceeds as follows. First, the data moments are used to provide estimates of the corresponding theoretical moments; the parameters can then be estimated by using the formulae connecting the theoretical moments and the parameters.

## 6.5 LEAST SQUARES ESTIMATION

Least squares estimation (usually abbreviated to LS) is particularly appropriate for leading to estimators of the systematic parameters in a linear model based upon a systematic plus a random component.

The method has been used already in connection with Model C of Chapter 4 (the bivariate regression model). The model (see Section 4.9) is

$$y_r = \alpha + \beta x_r + u_r \text{ where } u_r \sim \text{N}(0, \sigma^2) \quad (r = 1, \ldots, n)$$

Now suppose the least squares (LS) estimators of $\alpha$ and $\beta$ are indicated by $\hat{\alpha}$ and $\hat{\beta}$. Thus the residuals, $e_r$, are defined by $e_r = y_r - (\hat{\alpha} + \hat{\beta}x_r)$ and the sum of squares of the residuals, $S$, is

$$S = \Sigma e_r^2 = \Sigma(y_r - (\hat{\alpha} + \hat{\beta}x_r))^2 \qquad \textbf{(6.5.1)}$$

The LS principle of estimation says that we choose as estimators of $\alpha$ and $\beta$ those values of $\hat{\alpha}$ and $\hat{\beta}$ which make $S$ a minimum. It is a matter of calculus to work out the formulae for $\hat{\alpha}$ and $\hat{\beta}$, in terms of the data values. These formulae are known as the LS estimators and they are identical to those given in Section 0.9.

Thus, the application of the LS principle leads to an estimated model which best fits the data in the sense of minimising the sum of squared residuals. The LS principle is of wide applicability to models based upon a random component.

There is one important feature of the LS estimation. In the above example, and indeed in general, the model assumption that the random component is normally distributed was not used. The LS estimators depend only upon the form of the systematic component and they will be the same whatever the model assumption about the distribution of the random component—subject only to the condition that it has a mean of zero.

## 6.6 MAXIMUM LIKELIHOOD ESTIMATION

Unlike least squares estimation, maximum likelihood estimation (ML in short) depends completely upon the particular probability distribution. We illustrate the method in the context of the Poisson distribution.

Suppose that a univariate data set of frequency counts is to be modelled by the Poisson distribution. We assume that the data have been put in the form of a distribution showing the frequency of occurrence of the values 0, 1, 2, 3, ... The density function of the Poisson distribution is $\exp(-\lambda)(\lambda^x/x!)$ (see Table 5.3). For any given value of the parameter $\lambda$, this expression equals the probability of occurrence of the value $x$ (where $x$ takes the possible values 0, 1, 2, 3, ...). For each variable value, the data frequency and the density function are simultaneously exhibited in Table 6.1.

From Table 6.1, the probability of obtaining a particular set of data, given algebraically in the table, can be calculated as follows. First, the

**Table 6.1** Modelling of Data by Poisson Distribution

| Variable values | 0 | 1 | 2 | 3 | ... Total |
|---|---|---|---|---|---|
| Data frequencies | $n_0$ | $n_1$ | $n_2$ | $n_3$ | ... $n$ |
| Density function | $\exp(-\lambda)$ | $\lambda\exp(-\lambda)$ | $\frac{\lambda^2}{2}\exp(-\lambda)$ | $\frac{\lambda^3}{3!}\exp(-\lambda)$ | ... 1 |

probability of a value 0 is $\exp(-\lambda)$. So the probability of observing $n_0$ 0's is $[\exp(-\lambda)]^{n_0}$ since the observations are presumed independent and the occurrence of a number of independent events is the product of the separate probabilities of each event (see Section 0.16). Similarly, the probability of observing $n_1$ 1's is $[\lambda \exp(-\lambda)]^{n_1}$, of $n_2$ 2's is $[(\lambda^2/2)\exp(-\lambda)]^{n_2}$ and so on. Hence the probability for the data set as a whole, since again the observations are mutually independent, is given by the product of expressions of this kind:

$$[\exp(-\lambda)]^{n_0}\,[\lambda \exp(-\lambda)]^{n_1}\,[\frac{\lambda^2}{2}\exp(-\lambda)]^{n_2}\ldots \qquad (6.6.1)$$

This expression is known as the *likelihood function.* The likelihood function is denoted by the symbol L (and can usually be algebraically simplified). Now this likelihood function depends upon two things: the data values, $n_0$ (the number of 0's), $n_1$ (the number of 1's), etc., and $\lambda$, the parameter of the model. For a given set of data, L (the probability of observing the data for this model) will vary with the different possible values of $\lambda$. *The maximum likelihood estimator of $\lambda$ is defined as that value of $\lambda$ which maximises* L. In other words, the ML estimator of $\lambda$ is the value of $\lambda$ which maximises the probability of the observed data. For the Poisson model, it can be shown by calculus that L is a maximum when $\lambda$ is equal to the data mean. That is, the ML estimator of $\lambda$ (which, we recall is the mean of the Poisson distribution) is equal to the data mean.

This method is readily generalised to any distribution. The procedure is first to write down the expression for L, the likelihood function. This is done by successively substituting each data value in the density function in place of the general $x$ symbol and multiplying the terms together. L will again be a function of the data values and the parameter (or parameters if there is more than one). Secondly, mathematical methods (usually calculus) are used to determine a formula for the ML estimator of each parameter. The value of the ML estimator can be calculated for a given set of data.

Maximum likelihood estimators can also be calculated for the parameters of continuous distributions though the method is a little different from the one described above for a discrete distribution.

ML estimation has been described mainly to give an idea of its origin and perspective. It is not usually necessary to proceed from first principles in calculating ML estimators. Where appropriate, the ML estimators are specified below in the sections concerning estimation methods for individual distributions.

## 6.7 THE CHOICE OF ESTIMATORS

Three general procedures for arriving at an estimator have now been introduced. These may lead to different estimators and a problem of choice then arises. In practice, the choice may be more apparent than real because different estimation procedures often yield the same estimators.

For example, LS and ML estimators of the systematic parameters of linear models (such as those of Chapter 4) are identical, given the assumption that the random component is normally distributed.

Where choice does arise, there is a variety of principles (sometimes conflicting) discussed in more advanced texts. One such principle is that of *unbiasedness.* An estimator is said to be unbiased if it has the property that, in the long run, it would give estimates whose average would exactly equal the parameter being estimated. If the long-run average is different from the parameter, an estimator is said to be biased. The data mean is always an unbiased estimator of the model distribution mean. The data variance $s^2$ (with a divisor of $(n - 1)$) is similarly an unbiased estimator of the variance of the model distribution. This is one reason why we have chosen to use $s^2$ rather than $S^2$ (with a divisor of $n$) in Chapter 4.

Unbiasedness is one desirable property that an estimator might have. However, it is not the only one and another possible desirable property is illustrated in Figure 6.2. Consider as an example the use of the regression coefficient $b$ as an estimator of the bivariate regression

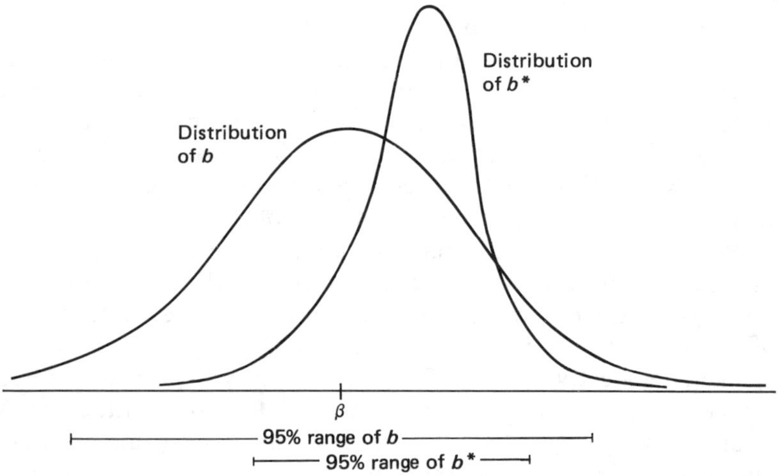


**Figure 6.2** Distribution of an Unbiased and a Biased Estimator

systematic parameter $\beta$. The left-hand curve shows its probability distribution in relation to $\beta$: it is unbiased and centred above the value of $\beta$. The right-hand curve shows the distribution of a biased estimator, call it $b^*$; it is not centred above the value $\beta$. However, *if* (as in the diagram) the standard deviation of $b^*$ is sufficiently small it will be a better estimator than $b$ in the sense that it is more likely to be closer to the parameter $\beta$. So this biased estimator might well be preferred.

For some distributional models, the application of a general procedure such as maximum likelihood leads to mathematical and computational difficulties. It may then be necessary to use simpler *ad hoc* procedures. Some of these will be illustrated below (see, e.g. Section 6.12 dealing with the negative binomial distribution).

**Table 6.2** Estimators of Parameters of Linear Models based upon Normal Distribution.

| *Distribution or Model* | *Parameters* | *Estimators* | *Properties of estimators* |
|---|---|---|---|
| Normal | $\mu$ | $\bar{x}$ | Unbiased |
| | $\sigma^2$ | $s^2$ | Unbiased |
| Bivariate regression | $\alpha$ | $a$ | Unbiased, LS and ML |
| | $\beta$ | $b$ | Unbiased, LS and ML |
| | $\sigma^2$ | $\Sigma e_r^2/(n-2)$ | Unbiased |

*Notation* for bivariate regression model, see Sec. 4.9 et seq.

Table 6.2 sets out the estimators of the parameters of linear models based on the normal distribution, together with their properties. One point to which the reader's attention is drawn concerns the choice of an estimator for the variance of the random component in the bivariate regression model. The chosen estimator, $\Sigma e_r^2/(n - 2)$, is selected because it is unbiased; the ML estimator can be shown to be $\Sigma e_r^2/n$ which, in fact, is a biased estimator.

The discussion which follows will focus upon the problems encountered in choosing and fitting individual distributions to univariate data. This will involve the use of analytical methods of general utility as well as procedures which are specific to the distribution concerned.

## 6.8 NORMAL DISTRIBUTION FITTING

If the data are available as individual values $x_1, \dots x_n$, the estimators of $\mu$ and $\sigma^2$ are, respectively, $\bar{x}$ and $s^2$. These are both unbiased estimators (see Table 6.2).

However, the data may be available only as grouped data in the form of a frequency distribution. Section 0.4 sets out the formulae for calculating the data mean and the data variance (based upon a divisor of $n$) in this case. We denote these by $\bar{x}_g$ and $S_g^2$ respectively; $s_g^2$ stands for the corresponding variance formula with a divisor of $(n - 1)$. For grouped data, $\bar{x}_g$ is the preferred estimator of $\mu$; it is unbiased. It is however a consequence of grouping that $s_g^2$ is biased upwards. That is to say, $s_g^2$ will typically be greater than the variance $s^2$ calculated from individual data values. This can be corrected by applying what is known as *Sheppard's correction* to the calculation of $S_g^2$. The correction requires that the class intervals are of equal width, say, $h$. Sheppard's correction amounts to reducing $S_g^2$ by an amount equal to $h^2/12$. As a consequence, the estimator of $\sigma^2$ becomes $n(S_g^2 - h^2/12)/(n - 1)$.

The *normal probability plot* provides a graphical method of estimating the normal distribution parameters. Also, and perhaps more importantly, it serves as an informal check on the validity of the normal distribution as a model for the data—a matter to be discussed substantively in Chapter 7.

An intuitive understanding of the method is best gained by assuming that the data set is first standardised with mean zero and unit variance (though prior standardisation is not necessary in practice). Now if the standard normal well fits the data, one would expect a close agreement between the data value corresponding to a particular cumulative relative frequency and the normal distribution value determined by the same cumulative relative frequency (i.e. cumulative area). A probability plot is a graph of points with each data value as the '$y$-coordinate' and the corresponding standard normal value (having the same cumulative relative frequency) as the '$x$-coordinate'. For a good fit, the points will lie roughly on a straight line because the $x$ and $y$ coordinates will be approximately equal.

The practical procedure is as follows. The first stage is to arrange the data values in rank order and, as shown in Section 2.7, to calculate the quantile order $q_r$ of each ranked data value, $x_{(r)}$, where $q_r = r/(n + 1)$. (Note that, for technical reasons, the data quantile order is defined slightly differently from the data cumulative relative frequency $r/n$.) The second stage is to calculate the standard normal distribution quantile point, $z_r$, determined by each data quantile order $q_r$. Thus, if a particular $q_r$ value were 0.95, the $z_r$ value would be 1.96 (see Section 3.9). The probability plot is the plot of the values $(z_r, x_{(r)})$.

The closeness of fit of the points of the plot to a straight line gives a visual appreciation of how well a normal distribution fits the data. Furthermore, if the data are *not* standardised, the best fitting straight line can be used to estimate the mean and standard deviation of the normal distribution. The *intercept* of this line estimates the mean and the slope estimates the standard deviation.

Figure 6.3 shows a normal probability plot of the Theban jars data of Table 4.1. The fit appears reasonable except for the smallest data value; we shall see in Chapter 7 that formal analysis confirms the validity of the normal distribution as a model. The fitted regression line (for which the calculations are not exhibited) has an intercept of 15.4 and a slope of 2.07. These compare with the formal estimates in Section 4.5 of 15.4 (for $\mu$) and 1.85 (for $\sigma$).

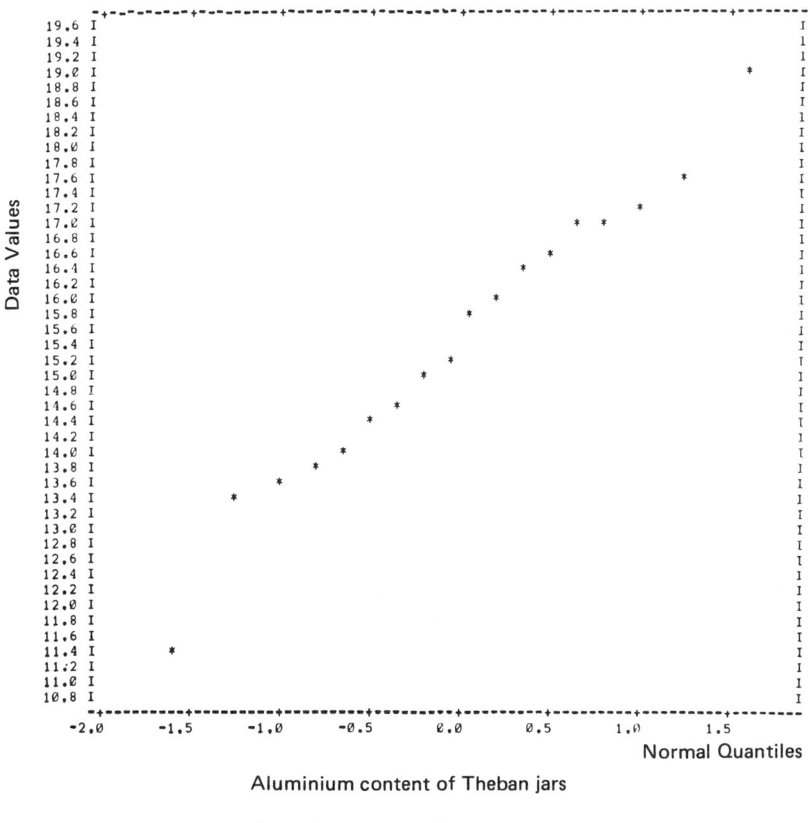


**Figure 6.3** Normal Probability Plot

Probability plotting can be conveniently used for grouped data. The class *boundaries* (excluding the lower boundary of the smallest class) are denominated as the ranked data values $x_{(1)}$, $x_{(2)}$, ... The quantile orders $q_1$, $q_2$, ... are calculated from the corresponding cumulative relative frequencies multiplied by $n/(n + 1)$. It is a bonus that the method still works if the smallest class is open-ended with its lowest boundary value unspecified.

Probability plots can be set up for distributions other than the normal. Further examples will be met below.

## 6.9 LOGNORMAL DISTRIBUTION FITTING

The lognormal distribution was reviewed in Section 5.6. It is a two parameter distribution and our notation for it is LN($\mu$, $\sigma^2$). It will be recalled that if a set of values has a lognormal distribution, its logarithmic transformed values (to the base *e*) have a normal distribution; the parameters $\mu$ and $\sigma^2$ are respectively the mean and variance of the transformed data. This relationship leads to a simple procedure for fitting the lognormal: first carry out a logarithmic transformation on the data and then fit a normal distribution to the transformed values.

Table 6.3 sets out data on the sizes (in terms of membership) of American labor unions. For size data of this kind, there is some *a priori* reason why the lognormal is a potential model, as suggested in Section 5.6. We therefore use these data to illustrate the fitting of the lognormal. The first column of Table 6.3 shows the data values and the second column the transformed values. It also includes a number of summary statistics which will be employed in the fitting process. The mean and variance of the transformed data are respectively 5.7 and 0.57. Hence our first fit is the distribution LN(5.7, 0.57).

Alternatively, we can graph a normal probability plot of the transformed data. This is shown in Figure 6.4. A regression line fitted to this plot gave an intercept of 5.7 and a slope of 0.79. This second method therefore gives the fitted model LN(5.7, $0.79^2$), that is LN(5.7, 0.62).

It is also possible to estimate the lognormal parameters directly from the original data set without any transformation. The formulae are:

$$\text{Estimator of } \mu = 2\ln(\bar{x}) - \tfrac{1}{2}\ln(S^2 + \bar{x}^2)$$
$$\text{Estimator of } \sigma^2 = \ln(S^2 + \bar{x}^2) - 2\ln(\bar{x})$$

The data summary statistics are shown in Table 6.3 and substitution in these formulae gives the fitted distribution LN(5.7, 0.61).

**Table 6.3** Labor Union Membership

| $x_{(r)}$ | $\ln x_{(r)}$ | $\ln(1-q_r)$ | $x_{(r)}$ | $\ln x_{(r)}$ | $\ln(1-q_r)$ |
|---|---|---|---|---|---|
| *a* | *b* | *c* | *a* | *b* | *c* |
| 93 | 4.53 | . . | 260 | 5.56 | −0.65 |
| 100 | 4.61 | . . | 265 | 5.58 | −0.69 |
| 109 | 4.69 | . . | 277 | 5.62 | −0.74 |
| 119 | 4.78 | . . | 300 | 5.70 | −0.79 |
| 135 | 4.91 | −0.02 | 330 | 5.80 | −0.85 |
| 135 | 4.91 | −0.05 | 365 | 5.90 | −0.90 |
| 145 | 4.98 | −0.07 | 420 | 6.04 | −0.97 |
| 150 | 5.01 | −0.10 | 432 | 6.07 | −1.03 |
| 150 | 5.01 | −0.13 | 446 | 6.10 | −1.10 |
| 150 | 5.01 | −0.15 | 483 | 6.18 | −1.17 |
| 153 | 5.03 | −0.18 | 502 | 6.22 | −1.25 |
| 165 | 5.11 | −0.21 | 510 | 6.23 | −1.34 |
| 174 | 5.16 | −0.24 | 575 | 6.35 | −1.44 |
| 177 | 5.18 | −0.27 | 627 | 6.44 | −1.54 |
| 195 | 5.27 | −0.30 | 699 | 6.55 | −1.66 |
| 197 | 5.28 | −0.34 | 750 | 6.62 | −1.79 |
| 200 | 5.30 | −0.37 | 820 | 6.71 | −1.95 |
| 211 | 5.35 | −0.41 | 917 | 6.82 | −2.13 |
| 211 | 5.35 | −0.44 | 924 | 6.83 | −2.35 |
| 227 | 5.42 | −0.48 | 1300 | 7.17 | −2.64 |
| 228 | 5.43 | −0.52 | 1358 | 7.21 | −3.04 |
| 238 | 5.47 | −0.56 | 1889 | 7.54 | −3.74 |
| 252 | 5.53 | −0.60 | | | |

*Summary statistics:* Original data: mean 408, SD 379;
Transformed data: mean 5.7, SD 0.76.

*Notation:* $x_{(r)}$ = ranked value of labor union membership in thousands
$q_r = r/(n+1)$, that is, $r/41$ (see Section 6.8).
*Note:* in fitting Pareto distribution, first four values are omitted.
*Source:* *Statistical Abstract of the United States*, 1979, Table 703, 'Membership in large unions', 1976.

The three estimation methods we have used thus give very similar results and it would not have much mattered which single method might have been chosen. However, the probability plot gives additional information regarding how good a model the lognormal is for data. If the reader looks at Figure 6.4, it will be seen that there is a bend in the probability plot and the points corresponding to the higher transformed values lie above the line which quite well fits the lower values. This implies that the normal distribution fit deteriorates for

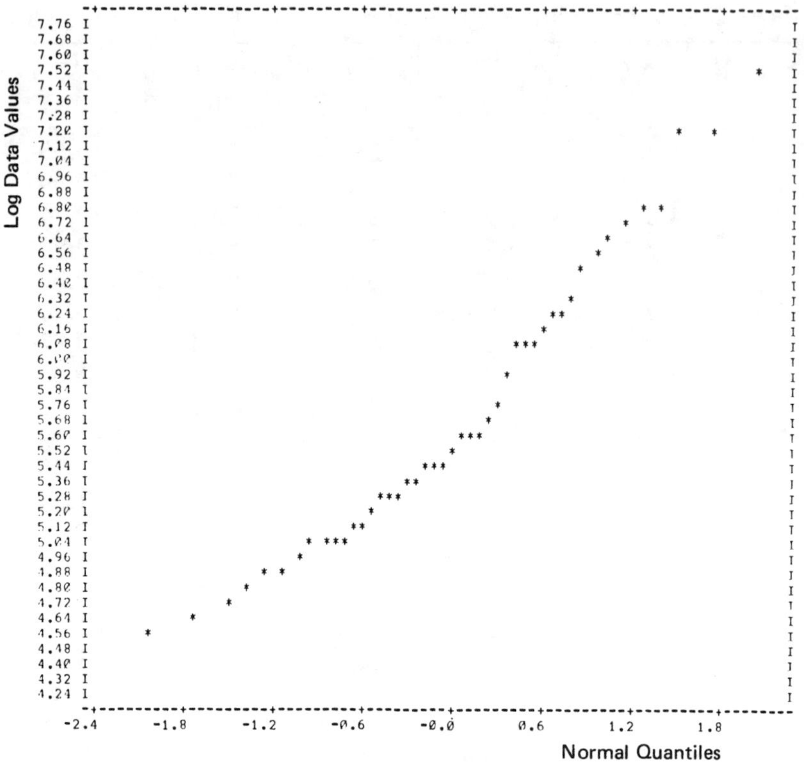


**Figure 6.4** Probability Plot for Lognormal Distribution

higher values and similarly for the lognormal fit to the data values themselves. In other words, the distribution of the data is 'thicker' in the right-hand tail than the lognormal and, following the discussion in Section 5.7, this suggests the Pareto distribution might be a better model for the higher values of the data set. This possibility is explored in the following section.

## 6.10 PARETO DISTRIBUTION FITTING

The Pareto distribution, denoted by Pa($k$, $\alpha$), is a two parameter distribution in which $k$ specifies the smallest (positive) value of the distribution (see Section 5.7).

The distribution function of the Pareto provides one approach to fitting. Result (5.7.2) stated that the relative frequency of values

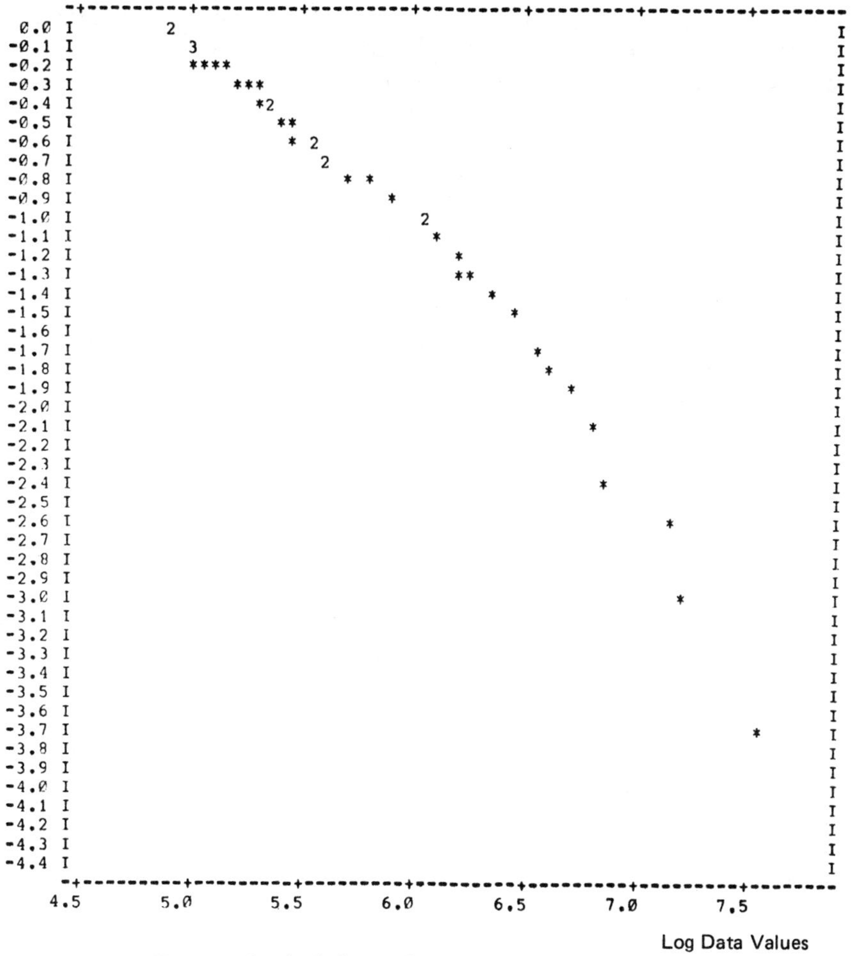


**Figure 6.5** Pareto Probability Plot

*greater than x*, that is $(1 - F(x))$, equals $(k/x)^{\alpha}$. Taking logarithms we have:

$$\ln[1 - F(x)] = \alpha \ln k - \alpha \ln x \tag{6.10.1}$$

Hence, if a variable has the Pareto distribution, a graph of the values of $\ln[1 - F(x)]$ against the corresponding values of $\ln x$ will give a straight line with intercept $\alpha \ln k$ and slope $-\alpha$.

A probability plot for the Pareto distribution makes use of result (6.10.1). On the vertical axis is plotted for each data value the logarithm of the relative frequency greater than that value and on the horizontal axis the logarithm of the data value. These are included in Table 6.3 and a computer probability plot for the labor union data is shown in Figure 6.5. The parameter $k$ was chosen to be 135 as some preliminary analysis suggested that the smallest four values in the data set would not accord with the Pareto fit. (In other applications, it may well be necessary to exclude more of the lower values.)

Apart from the very highest data values, the plot appears fairly linear. A regression line fitted to the data gives a slope of $-1.2$, so the estimate of $\alpha = 1.2$. The intercept is 6.0 ($= \alpha \ln k$) which yields an estimate of $k = 136$. There is of course a slight absurdity in estimating $k$ when it is inevitably known and prespecified as the smallest value of the data set. This could be avoided by calculating that line of best fit which passes through the point on the plot corresponding to the smallest data value.

An alternaive method of estimating the value of the $\alpha$ parameter is by maximum likelihood. The ML estimator of $\alpha$ in the Pareto distribution equals

$$1/[\bar{L} - \ln k]$$

where $\bar{L}$ is the mean of the logarithms (to the base $e$) of the data values. For the labor union data, $\bar{L} = 5.80$ and $\ln(135) = 4.90$, so the estimate of $\alpha = 1/(5.8 - 4.9) = 1.1$. Correct to one decimal place, this is identical to the estimate from the probability plot.

We can now sum up the results of fitting the lognormal and Pareto distributions to the labor union data. The lognormal gave the better fit to roughly the lower half of the data (say the values 93 to 300). The Pareto performed reasonably over the range of values 135 to 820 (which excludes the first four and last four values).

## 6.11 POISSON DISTRIBUTION FITTING

The Poisson distribution Po($\lambda$) is a single parameter distribution and the mean of the distribution is itself equal to $\lambda$ (see Section 5.12). The parameter $\lambda$ is estimated by the data mean $\bar{x}$. There are two justifications for this. First, the data mean can be shown to be the maximum likelihood estimator of $\lambda$ and second, the data mean is also a moment estimator (since it is being used to estimate the distribution mean $\lambda$).

As an example, we shall consider the fitting of the Poisson distribution to data on the number of strikes occurring in successive weekly periods presented in Table 6.4. The variable is a frequency variable and the use of the Poisson for this data has some *a priori* justification in terms of the modelling of rare events occurring over periods of time (see Section 5.12).

**Table 6.4** Number of Strikes in Each Weekly Period of 1974–77, Modelled by Poisson and Negative Binomial Distributions

| $x$ | $f$ | $p$ | $p \mid \text{Po}$ | $f_e \mid \text{Po}$ | $p \mid \text{NB}$ | $f_e \mid \text{NB}$ |
|---|---|---|---|---|---|---|
| *a* | *b* | *c* | *d* | *e* | *f* | *g* |
| 0 | 13 | 0.083 | 0.023 | 3.7 | 0.082 | 12.7 |
| 1 | 21 | 0.135 | 0.088 | 13.7 | 0.143 | 22.4 |
| 2 | 24 | 0.154 | 0.165 | 25.7 | 0.164 | 25.5 |
| 3 | 31 | 0.199 | 0.206 | 32.2 | 0.154 | 24.0 |
| 4 | 20 | 0.128 | 0.194 | 30.2 | 0.129 | 20.1 |
| 5 | 10 | 0.064 | 0.146 | 22.7 | 0.100 | 15.7 |
| 6 | 10 | 0.064 | 0.091 | 14.2 | 0.074 | 11.5 |
| 7 | 11 | 0.070 | 0.049 | 7.6 | 0.052 | 8.2 |
| 8 | 4 | 0.026 | 0.023 | 3.6 | 0.036 | 5.6 |
| 9 | 4 | 0.026 | 0.010 | 1.5 | 0.024 | 3.8 |
| 10– | 8* | 0.051 | 0.005 | 0.8 | 0.042 | 6.4 |
| Total | 156 | 1.000 | 1.000 | 156.0 | 1.000 | 156.0 |

*The frequencies for the values 10, 11, 12, 13, 14 separately are 3, 3, 1, 0, 1 respectively, higher values having zero frequencies.

*Source:* authors' data from UK *Department of Employment and Productivity Gazette* (various issues) of figures of 'prominent stoppages' in British industry.

The data, arranged as a frequency distribution, are given in columns $a$ and $b$ of Table 6.4; column $c$ shows the relative frequencies. The data mean, $\bar{x} = 3.76$ provides the estimate of $\lambda$ and so the data are to be fitted by the distribution Po(3.76).

How well the distribution Po(3.76) fits the data is another matter. This question will be considered substantively in the next chapter. Some preliminary analysis can however be made at this point and is demonstrated in Table 6.4. It depends upon comparing the data frequency distribution with that of the theoretical distribution Po(3.76).

The first stage is to compute the Poisson relative frequencies for each of the variable values 0, 1, 2, . . . by substitution in its density function $\exp(-3.76)\ 3.76^x/x!$. (This is the Poisson density function shown in Table 5.3 with $\lambda$ put equal to 3.76.) For example, the third row figure in column $d = \exp(-3.76)\ 3.76^2/2 = 0.165$. The heading to the column, $p \mid \text{Po}$, simply indicates that the relative frequencies are calculated from the Poisson distribution.

Column $e$, headed $f_e \mid \text{Po}$, is calculated by multiplying each of the relative frequencies in column *d* by the number of data values, $n$ = 156. These $f_e$ values are known as the *expected* frequencies as distinct from the *observed* frequencies of column *b*. The expected frequencies can be looked upon as the average number of times one would observe each of the variable values in the long run with repeated sets of 156 random observations from the distribution Po(3.76). A quick glance down columns *b* and *e* shows that there are substantial discrepancies between model and data. How far these discrepancies are 'significant' is a question we defer to Chapter 7.

An important fact about the data set has so far been withheld from the reader. It is that the data variance $s^2 = 8.04$. This value is much greater than the data mean, 3.76, which suggests, on purely empirical grounds, that the negative binomial distribution might be a better model than the Poisson. (Recall that, for the Poisson distribution, the variance equals the mean but, for the negative binomial, the variance is greater than the mean, (see Section 5.15).) We therefore proceed in the next section to explain how the parameters of the negative binomial distribution are estimated again using the same data set.

## 6.12 NEGATIVE BINOMIAL DISTRIBUTION FITTING

The negative binomial is a two parameter distribution, our notation for it being NB($\mu$, $k$) (see Section 5.14). Its mean is $\mu$ itself and its variance = $\mu(\mu + k)/k$ (see result (5.14.1).

The estimation of the parameter $\mu$ presents no problems. The data mean, $\bar{x}$ is both the ML estimator and the moment estimator, and is also unbiased. The estimation of $k$ is not so straightforward. The method of moments cannot be applied directly. Furthermore, for mathematical reasons it is not possible to derive a formula for the ML estimator (though it is possible to work it out numerically if the appropriate computer program is available.) However, the method of moments can be applied in an indirect fashion as follows.

The equation which expresses the variance in terms of $\mu$ and $k$ can be re-expressed giving $k$ in terms of $\mu$ and $\sigma^2$, namely: $k = \mu^2/(\sigma^2 - \mu)$. An estimate of $k$ is then given by substituting in the right-hand side of this equation $\bar{x}$ instead of $\mu$ and $s^2$ instead of $\sigma^2$.

We have now arrived at estimators for the negative binomial parameters:

$$\text{Estimator of } \mu = \bar{x} \text{ and estimator of } k = \bar{x}^2/(s^2 - \bar{x}) \qquad \textbf{(6.12.1)}$$

For the strike data of Table 6.4, the estimates are respectively 3.76 and 3.30 and so the fitted distribution is NB(3.76, 3.30).

As was done for the Poisson distribution in the last section, the negative binomial expected frequencies can now be calculated. We first need to calculate the density function of NB(3.76, 3.30) for each variable value 0, 1, 2, ... These are shown in column *f* and can be calculated by the method set out in Section 5.14. The expected frequencies are shown in column *g*.

A casual comparison of columns *b*, *e*, and *g* shows that the expected frequencies based upon the negative binomial distribution are closer to the data frequencies than those of the Poisson. The negative binomial appears to be the better model, but a formal analysis must await the next chapter.

## 6.13 OUTLIERS

We are now in a position to take up a question which was left unresolved in the discussion on outliers in Section 2.13. We said there that the classification of any data value as an outlier depends ultimately upon a statistical model for the data. At a descriptive level, outliers can be identified only in terms of pragmatic and somewhat arbitrary rules, such as any value more than four quartile deviations from the mean. On the other hand, given a model for the data, an outlier can be described as a data value which is, in some sense, 'inconsistent' with the model.

It is not easy in general terms to offer a clear-cut procedure which checks for inconsistency. However, in practice some advance can be made. The probability plot offers the most ready means for identifying obvious outliers. Consider for example the Pareto probability plot of the labor union data in Figure 6.5. The four largest data values are the furthest from the straight line plot and the two largest values (1,358 and 1,889) are particularly discordant. These two values therefore would be said to be outliers, not because they are extreme values but because they appear inconsistent with the Pareto model. Indeed, if the sizes of the two largest labor unions had been greater—more extreme—they would have accorded better with the Pareto model.

Having identified outliers, it would normally be prudent to refit the model having excluded them from the data set, since the parameter estimates may be unduly sensitive to the outlying values. In such a case, the model could not then be said to encompass all the data values unless it were possible to extend the model in some way which would account for the discordancy with the initial model. Rather than excluding the outliers from the fitting procedure, it would be less drastic to adopt a trimming or a Winsorisation procedure such as was briefly described in Section 2.14. Such a procedure would give less weight to the outlying values in the estimates of the model parameters.

## EXAMPLES

**E 6.1** Construct normal probability plots for the following two data sets:

| | | | | | | | |
|---|---|---|---|---|---|---|---|
| *A* | −1.97 | −1.08 | −0.33 | −0.15 | −0.06 | 0.07 | 0.25 |
| | 0.29 | 0.74 | 1.84 | | | | |
| *B* | −2.65 | −1.08 | −0.70 | −0.34 | 0.03 | 0.11 | 0.31 |
| | 0.77 | 1.01 | 2.14 | | | | |

Comment on the appropriateness of the normal distribution as a data model.

**E 6.2** Fit a lognormal distribution to the data given in E 5.1 using the following methods:

(i) fit a normal distribution to the transformed data
(ii) estimate the lognormal parameters direct from the formulae in Section 6.9.

## FURTHER WORK

**F 6.1** Choose a suitable data set and examine the fit of the lognormal and/or Pareto distributions. Suggestions for data: city population sizes, sizes of organisations, length of sentences in a particular author's work, distribution (nationally) of personal incomes.

**F 6.2** Fit negative binomial and Poisson distributions to a suitable data set. Suggestions: number of 'disasters' in a fixed time period—number of earthquakes, wars, major airline disasters per year, meteorological data, sentence length.

*This chapter depends upon: Chapters 4, 5 & 6*

# 7

# Testing the Validity of Models

## 7.1 INTRODUCTION

No statistical model will ever perfectly fit the data and there will be some discrepancies, large or small, between the data and their model. How to assess these discrepancies, and how to come to a conclusion about the adequacy of a proposed model, will form the central themes of this chapter.

We shall consider sequentially the following cases:

(i) where the univariate data relate to a frequency variable, or some other discrete variable;

(ii) where the variable is continuous and the proposed model is (a) a normal distribution, and (b) some other distribution; and

(iii) where the data is bivariate and the proposed model is the regression model (Model C of Chapter 4).

In the course of the discussion, we shall introduce a theoretical distribution which plays an important part in model validation: the $\chi^2$-distribution. Statistical hypothesis test procedures are basic to the material of this chapter and will also be discussed.

## 7.2 MODELS OF DISCRETE VARIABLES: PRELIMINARY ANALYSIS

We shall base the discussion of this and some later sections upon the data on numbers of strikes in weekly periods introduced in Chapter 6, (Section 6.11). In that chapter, two alternative models were proposed for the data: the Poisson and the negative binomial distributions. Both these distributions were fitted to the data and Table 6.4 exhibited (a) the data arranged as a frequency distribution and (b) the expected frequencies of the two fitted models. These are reproduced in Table 7.1, columns $a-c$ and, for the negative binomial expected frequencies, column $f$. The significance of the other columns will be considered below.

We can now address ourselves to the main question: how adequate are these distributions as models of the data?

The adequacy of the fitted Poisson distribution, Po(3.76) will be considered initially. A comparison of columns $b$ and $c$ gives a preliminary notion. The distribution Po(3.76), for $x = 0$, has an expected frequency of 3.7, while the data frequency is 13, and so on. Some of the differences are quite large and one may start to wonder whether the Poisson model is a valid one.

At this point, however, we need to recall a crucial fact. Even if the

**Table 7.1** Number of Strikes in Each Weekly Period of 1974–77 Modelled by Poisson and Negative Binomial Distributions

| $x$ | $f$ | $f_e \mid \text{Po}$ | $(f-f_e) \mid \text{Po}$ | $(f-f_e)^2/f_e \mid \text{Po}$ | $f_e \mid \text{NB}$ | $(f-f_e)^2/f_e \mid \text{NB}$ |
|---|---|---|---|---|---|---|
| *a* | *b* | *c* | *d* | *e* | *f* | *g* |
| 0 | 13 | 3.7 | 9.3 | 23.4 | 12.7 | 0.0 |
| 1 | 21 | 13.7 | 7.3 | 3.9 | 22.4 | 0.1 |
| 2 | 24 | 25.7 | −1.7 | 0.1 | 25.5 | 0.1 |
| 3 | 31 | 32.2 | −1.2 | 0.0 | 24.0 | 2.0 |
| 4 | 20 | 30.2 | −10.2 | 3.4 | 20.1 | 0.0 |
| 5 | 10 | 22.7 | −12.7 | 7.1 | 15.7 | 2.0 |
| 6 | 10 | 14.2 | −4.2 | 1.2 | 11.5 | 0.2 |
| 7 | 11 | 7.6 | 3.4 | 1.5 | 8.2 | 1.0 |
| 8− | 16* | 5.9 | 10.1 | 17.3 | 15.8 | 0.0 |
| Total | 156 | 156.0 | 0.0 | 57.9 | 156.0 | 5.4 |

*Values equal to 8 and more are grouped together for reason given in Section 7.4.

*Source:* Table 6.4

model were valid, that is, *even if the data were really random observations from the distribution Po(3.76)*, there would still be chance differences between the Poisson frequencies and the data frequencies. So discrepancies between columns *b* and *c* are not in themselves sufficient to invalidate the model. The question is rather: are the discrepancies small enough to be attributable to chance—in which case the Poisson model would still be judged a valid one—or are they so great that the Poisson distribution as a model for the data must be rejected? The subsequent calculations of Table 7.1 are concerned with producing a single summary figure which will resolve this question.

The first stage is to calculate column *d* showing the magnitude of the difference between observed and expected frequency, algebraically, $(f - f_e)$. Thus $9.3 = 13.0 - 3.7$. A glance down column *d* shows for which variable values there is the greatest discrepancy between data and model.

A final stage, demonstrated in column *e*, leads to a single value which measures the discrepancy between model and data. In each row of column *e*, we calculate $(f - f_e)^2/f_e$ (e.g. $24.0 = 9.3^2/3.6$). The total of these values, $\Sigma(f - f_e)^2/f_e$, is our single measure of discrepancy. It is based upon the difference between observed and expected frequencies, which in turn reflect discrepancies between data and model. Hence large values of this total measure will suggest an invalid model, and small will suggest an acceptable model. But, how large is 'large' and how small is 'small'? To answer this we need some more theory concerning another theoretical distribution, the $\chi^2$-distribution.

## 7.3 $\chi^2$-DISTRIBUTION

The $\chi^2$- (or chi-squared) distribution is introduced in the present chapter because it will be used mainly in testing the validity of certain models. It is a theoretical distribution of a continuous variable which can take only positive values.

The chi-squared distribution is a one parameter distribution like the *t*-distribution, the parameter being given the Greek letter $\nu$ and called the degrees of freedom of the distribution. Our notation for the distribution is $\chi^2(\nu)$. The parameter $\nu$ takes only positive integral values 1, 2, .... The shape of the distribution is shown in Figure 7.1. For all values of the degrees of freedom, except $\nu = 1$, the distribution is unimodal with an indefinitely long right-hand tail. When $\nu = 1$, there is no mode and the density function takes higher and higher values as the variable value gets smaller and approaches zero.

For the purpose of describing the relation between variable values

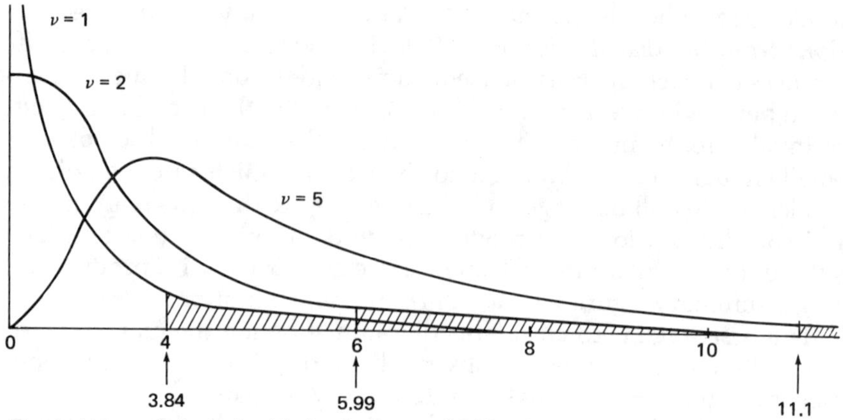


The shaded areas indicate the right-hand tail areas of 0.05 and the associated critical values are also shown

**Figure 7.1** Chi-Squared Distribution

and areas under the $\chi^2$-distribution curve, we shall use a similar notation to that introduced for the *t*-distribution in Section 4.4. Specifically $\chi^2_p(\nu)$ stands for the variable value (i.e. quantile point) having a quantile order of $p$ (i.e. a cumulative area of $p$).

A selection of values is given in Table 7.2. It shows for example that $\chi^2_{0.95}(1) = 3.84$. That is to say, for the $\chi^2$-distribution with the degrees of freedom parameter equal to 1, the value of 3.84 has a cumulative area of 0.95, and consequentially a right-hand tail area of 0.05; this is shown pictorially in Figure 7.1. The $\chi^2$-distribution is generally used as a probability distribution. So one can say that, for the distribution $\chi^2(1)$, there is a probability of 0.95 that an observed value will be less than 3.84 and only 0.05 that it will be greater than 3.84.

Table 7.2 does not give information about central areas; in applications interest is almost always focused upon right-hand tail areas.

## 7.4 THE $\chi^2$ GOODNESS OF FIT TEST FOR A DISCRETE VARIABLE

It is now possible to make a formal assessment of the validity of the fitted model Po(3.76) on the basis of the summary value $\Sigma(f - f_e)^2/f_e = 57.9$ calculated in Section 7.2.

This summary value is known as the $\chi^2$ goodness of fit statistic (or

**Table 7.2** Quantile Orders and Variable Values for Particular $\chi^2$-Distributions

| | | | |
|---|---|---|---|
| Quantile order | $p$ | 0.95 | 0.99 |
| RH tail area | $(1-p)$ | 0.05 | 0.01 |
| $(\nu = 1)$ | $\chi_p^2(1)$ | 3.84 | 6.63 |
| $(\nu = 2)$ | $\chi_p^2(2)$ | 5.99 | 9.21 |
| $(\nu = 5)$ | $\chi_p^2(5)$ | 11.1 | 15.1 |
| $(\nu = 7)$ | $\chi_p^2(7)$ | 14.1 | 18.5 |
| $(\nu = 10)$ | $\chi_p^2(10)$ | 18.3 | 23.2 |

$\chi^2$ statistic in short). It can be shown that, *if the Poisson model is valid for the data,* the calculated $\chi^2$ statistic will have a $\chi^2$ probability distribution. Furthermore, in the present case, the appropriate value of the degrees of freedom parameter is given by the number of rows in the data frequency distribution less two, that is, $(9 - 2) = 7$. So, if the Poisson model is valid, the probability distribution of the $\chi^2$ statistic is $\chi^2(7)$. (Note that this does not equal the number of observed variable values less two, since those values greater than or equal to 8 were grouped together. We return to this matter of grouping below.) Consequentially, as shown in Table 7.2, there is a probability of 0.05 that the $\chi^2$ statistic has a value greater than $\chi^2_{0.95}(7) = 14.1$.

It is now possible to state a 'decision rule'—a procedure for deciding whether the model is valid or not. It is as follows:

> If the calculated value of the $\chi^2$ statistic is greater than 14.1, the model Po(3.76) is taken to be invalid; if it is less than 14.1 it is taken to be valid. **(7.4.1)**

In fact, the calculated value was 57.9 and so the decision rule says 'reject validity of model Po(3.76)'.

The decision rule has been stated dogmatically, but we can see how it makes sense. First, we note that it deals with the question posed at the end of Section 7.2: 'large' is defined to be greater than 14.1, 'small' less than 14.1. Secondly, although large values of the $\chi^2$ statistic suggest an invalid model, it is possible to obtain a large value by chance even if the model is valid. However, this eventuality has a small probability. Indeed the probability is only 0.05 that the $\chi^2$ statistic will be greater than 14.1 if the model is valid, so the decision rule is unlikely to lead us astray.

The $\chi^2$ goodness of fit test has been demonstrated for the particular case of a Poisson model, but it can readily be generalised for testing any other distribution as a data model.

First, one carries out the calculations as in Table 7.1. That is, the proposed distribution is first fitted by estimating its parameter or parameters. The theoretical relative frequencies are then calculated and entered in column *d*. The rest of the calculations proceed as before and lead to a calculation of the $\chi^2$ goodness of fit statistic $\Sigma(f - f_e)^2/f_e$. This $\chi^2$ statistic as before has a $\chi^2$ probability distribution *if the model is valid*. The appropriate value of the degrees of freedom parameter in fact depends upon the number of rows in the data frequency table and also upon the number of parameters of the fitted distribution. Specifically:

> For the $\chi^2$ goodness of fit test, the degrees of freedom equal $(k - r - 1)$, where $k$ is the number of rows in the data frequency table and $r$ is the number of parameters estimated in fitting the distribution. **(7.4.2)**

In our example, $k = 9$, $r = 1$ (because the Poisson is a single parameter distribution), and $(k - r - 1) = 7$.

Secondly, as at (7.4.1), the decision rule is set up by looking up in the $\chi^2$-distribution table, the value $\chi^2_{0.95}(k - r - 1)$. This is known as the *critical value* of the goodness of fit test. The model is judged valid or invalid according to whether the calculated $\chi^2$ statistic is less than or greater than the critical value.

The goodness of fit test is a 'large sample' test in that it requires that the *expected* frequencies for each variable value should not be too small. 'Too small' has been traditionally taken as less than 5, but recent work suggests that this can be reduced to 3 without much risk.[1] This constraint can be overcome by grouping together variable values as was done in Table 7.1.

Sometimes, one may wish to test the adequacy of a model when the parameters have been prespecified rather than estimated from the data. For example, a previous study for another country might have found that, say, Po(4.5) was a good model. The question might then he asked: is the model Po(4.5) valid for the present data? The analysis would proceed as before with just two exceptions. First, the column *c* expected frequencies would be calculated from the Po(4.5) distribution. Secondly, application of the degrees of freedom specification at (7.4.2) would give $\nu = (9 - 0 - 1) = 8$, where $r = 0$ because no parameters have been estimated from the data.

The validity of the negative binomial model can similarly be tested.

1. See Lawal, H.B., 'Tables of percentage points of Pearson's goodness-of-fit statistic for use with small expectations, *Applied Science, 29*, No. 3, 1980.

The analytical stages are: (i) calculate the expected values $f_e$, reproduced in Table 7.1, column *f*, (ii) calculate $\Sigma(f - f_e)^2/f_e = 5.4$ (Table 7.1 column *g*), (iii) compare the calculated $\chi^2$ statistic, 5.4, with the critical value $\chi^2_{0.95}(9 - 2 - 1) = \chi^2_{0.95}(6)$. This critical value equals 12.6 and so the negative binomial is judged a valid model.

For many reasons therefore the negative binomial is a better model than the Poisson for the strike data which have been studied. The negative binomial variance is greater than the mean and so it is for the data. Comparison of the observed data frequencies and the two sets of expected frequencies in Table 7.1 (columns *b*, *c* and *f*) suggests that the overall fit is better for the negative binomial model. Finally, the negative binomial passes the $\chi^2$ goodness of fit test whilst the Poisson does not.

## 7.5 THE $\chi^2$ GOODNESS OF FIT TEST FOR A CONTINUOUS VARIABLE

The procedure outlined above may also be used for testing the goodness of fit of a continuous distribution. The data frequency distribution remains the basis of the analysis and, unless it has already been done, the data will need to be grouped in class intervals.

As an example, we use the data on the aluminium content of Theban jars set out in Table 4.1 and test the goodness of fit of the normal distribution N(15.4, $1.8^2$) (where the mean and variance have been estimated from the data). Table 7.3 sets out in columns *a* and *b*

**Table 7.3** Theban Jars Data: Goodness of Fit of Normal Distribution

| *Class interval* | $f$ | $p \mid N$ | $f_e \mid N$ | $\frac{(f - f_e)^2}{f_e} \mid N$ |
|---|---|---|---|---|
| *a* | *b* | *d* | *e* | *f* |
| 14.2 | 5 | 0.25 | 4.5 | 0.055 |
| 14.2–15.4 | 4 | 0.25 | 4.5 | 0.055 |
| 15.4–16.6 | 4 | 0.25 | 4.5 | 0.055 |
| 16.6 | 5 | 0.25 | 4.5 | 0.055 |
| Total | 18 | 1.00 | 18 | 0.22 |

*Source:* Table 4.1

the chosen class intervals and frequencies for this data. In columns *d*, *e* and *f* are set out the same kinds of calculation as were exhibited in Table 7.1.

The class boundaries have been chosen so that the values within each class interval have equal probabilities for the presumed model $N(15.4, 1.8^2)$. (This is a conventional recommendation but it is not crucial.) Having settled the class intervals, the calculation proceeds exactly as for the discrete variable case and the calculated value of the $\chi^2$ statistic equals 0.22. Result (7.4.2) tells us that the appropriate degrees of freedom value is $(4 - 2 - 1) = 1$, since the number of rows in the data frequency table is 4, and two parameters of the normal distribution were estimated. Reference to Table 7.2 shows that the critical value $\chi^2_{0.95}(1) = 3.84$ and so the normal model is judged a valid one.

It will be clear that this procedure can equally be applied to any other distribution for which the probabilities associated with predefined class intervals can be calculated or looked up in tables.

## 7.6 HYPOTHESIS TESTING: PRELIMINARY DISCUSSION AND TERMINOLOGY

The procedures presented in Sections 7.4 and 7.5 are examples of what are known as hypothesis testing procedures. We had, to begin with, a hypothesis to be examined, namely that a particular data set is correctly modelled by a specified distribution, for example Po(3.76) or $N(15.4, 1.8^2)$. The hypothesis under examination is known as the *null hypothesis* and is sometimes indicated symbolically by $H_0$. We calculated a *test statistic* (in these examples, the $\chi^2$ goodness of fit statistic). In general, a test statistic has two properties; first, its distribution is known under the assumption that the null hypothesis is true and secondly, extreme values indicate that the model is invalid. In the examples so far, 'extreme' means large positive values lying in one tail of the distribution; such tests are known as *one-tailed tests*. We will encounter cases where 'extreme' additionally includes large negative values (or very small values close to zero) lying in the other tail of the test statistic distribution; such tests are *two-tailed tests*.

We then specify a *decision rule*. This involves two stages:

(i) With reference to the known distribution of the test statistic a particular range of extreme values is identified whose collective probability is some specified small number: conventionally 0.05. This range is known as the *critical region*. (Other values not in the critical region constitute the *acceptance region*). The small probability is known as the *size of the test* and one speaks of the test being carried out at the *5% level of significance.*

(ii) We note whether the calculated value of the test statistic falls within the critical region or not. If it does, the model is judged invalid and the null hypothesis is said to be rejected at the 5% level of significance. Otherwise the null hypothesis is *accepted* at the 5% level of significance.

The 1% level of significance is an alternative conventional choice. In the context of the example of Section 7.4 we would have made a test at this level if we had used $\chi^2_{0.99}(7)$ as the critical value.

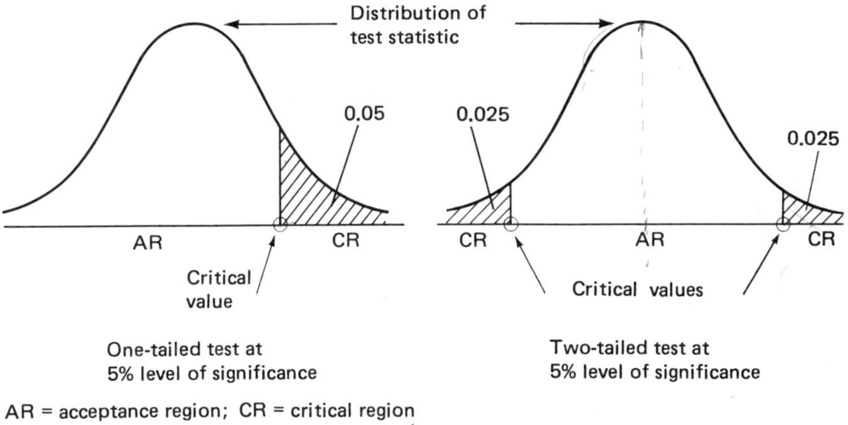


**Figure 7.2** Hypothesis Test Procedure

The level of significance determines the probability of drawing a mistaken conclusion that the null hypothesis is invalid if in fact it is valid. For suppose the null hypothesis is valid and the test is made at the 1% level of significance. Then there is a probability of 0.05 that the test statistic will have a value within the critical region which, if this happens, will cause the hypothesis to be rejected. A mistaken conclusion of this nature is called a *Type I error.*

## 7.7 SOME TESTS FOR NORMALITY

Where the null hypothesis is that a normal distribution is a valid model there are some particular tests available, apart from the $\chi^2$

goodness of fit test of Section 7.5, which can be used for any proposed distribution.

One such test is known as Geary's test and depends upon the ratio of the mean deviation to the standard deviation, $S$. This ratio gives as the test statistic

$$G = \frac{\Sigma(\,|\,x_r - \bar{x}\,|\,)/n}{\sqrt{[\Sigma(x_r - \bar{x})^2/n]}}$$

The distribution of $G$ is tabulated under the assumption that the null hypothesis is valid. Like the $\chi^2$ test statistic, $G$ is always positive, but the nature of the critical region is a little different in that both extreme 'small' values (near to zero) and extreme 'large' values of $G$ point to the maintained hypothesis of normality being invalid.

If the null hypothesis of normality is valid, $G$ may be expected to be around 0.80. Skewed and leptokurtic distributions tend to have values of $G$ less than 0.80 while platykurtic distributions tend to have values greater than 0.80. So extreme small values and extreme large values of $G$ suggest that the data would be better modelled by some distribution other than the normal.

The probability distribution of $G$, for data correctly modelled by a normal distribution, can be evaluated. Table A.5 shows the critical values for a test at the 5% level of significance. Its use can be demonstrated in terms of the Theban jars data. We have: mean deviation $= 1.51$, standard deviation $= 1.80$, and $G = 0.84$. For a data set with $n = 16$, Table A.5 gives a lower $2\frac{1}{2}$% critical value of 0.70 (that is, if the model is correct the probability is only 0.025 that a data value of $G$ will be in the critical region less than 0.70) and similarly an upper $2\frac{1}{2}$% critical value of 0.90. The calculated value of $G$, 0.84, is within the acceptance region and the normal distribution again is shown to be an acceptable model.

This is an example of a two-tailed test because the critical region comprises values lying in both tails of the distribution of the test statistic: very small values in the LH tail and very large values in the RH tail. The level of significance is still 5% because the probability is $0.025 + 0.025 = 0.05$ that the test statistic will lie in the critical region if the normal model is valid.

Another test of normality may be made by using as a test statistic the computed measure of skewness, $g_1$ (see formula (2.10.1)). $g_1$ is zero for the normal distribution and both large negative and large positive values constitute the critical region for rejecting the hypothesis of normality. It is therefore a two-tailed test. Critical values may be found in Pearson and Hartley (1970).

The normal probability plot (see Section 6.8) offers a third method of checking the assumption of normality. Although it is an informal graphical procedure, it has the advantage that it makes it possible to identify those parts of the data set, or individual outlying data values, which appear inconsistent with the normal distribution model.

## 7.8 TESTING THE FIT OF CONTINUOUS DISTRIBUTIONS GENERALLY

A test of general applicability has already been met in Section 7.5: the $\chi^2$ goodness of fit test for continuous distributions. It requires that the data values should first be arranged in the form of a frequency distribution, but otherwise its use is relatively straightforward.

Another procedure is the *Kolmogorov—Smirnov* test. Its approach is similar to that of probability plotting in that it depends upon relating the cumulative relative frequencies (crf's) of the data set to the crf's of the theoretical distribution model. However, there is a difference between the two. The probability plot is based upon a comparison of data values and theoretical distribution values having common crf's, whereas the Kolmogorov-Smirnov test is based upon comparing data crf's and theoretical crf's having common variable values.

The test requires that the model distribution should be completely specified with numerical values given to all parameters. It is best described in terms of a simple example. Column *a* of Table 7.4 headed $x_{(r)}$, shows 5 (hypothetical) data values and the model to be tested is N(2, 1). Column *b*, headed $S_r$, shows the data crf's, and $S_r = r/n$ (which equals $r/5$ in this example). Column *c*, headed $F(x)$, shows for

**Table 7.4** Kolmogorov—Smirnov Test of the Model N(2.1)

| $r$ | $x_{(r)}$ | $S_r$ | $F(x)$ | $S_r - F(x)$ | $S_{r-1}$ | $F(x) - S_{r-1}$ |
|---|---|---|---|---|---|---|
| | *a* | *b* | *c* | *d* | *e* | *f* |
| 1 | 1.3 | 0.2 | 0.24 | . . | 0.0 | 0.24 |
| 2 | 2.1 | 0.4 | 0.55 | . . | 0.2 | *0.35* |
| 3 | 2.5 | 0.6 | 0.70 | . . | 0.4 | 0.30 |
| 4 | 2.6 | 0.8 | 0.75 | 0.05 | 0.6 | 0.15 |
| 5 | 3.2 | 1.0 | 0.98 | 0.02 | 0.8 | 0.18 |

*Source:* hypothetical data

each value in column *a* the crf (that is the cumulative area or the distribution function value) of N(2, 1); thus 1.3 has a standardised z-value of −0.7 which from the normal tables has a cumulative area of 0.24.

Figure 7.3 shows superimposed graphs of the empirical cumulative distribution based upon columns *a* and *b* and of the theoretical distribution function based upon columns *a* and *c*. The former has a stepwise appearance similar to that of the distribution function of a discrete distribution (see Figure 5.3). For a good fit, the two graphs would be close to each other. The test depends upon identifying the largest distance (in the vertical direction) between the two graphs over *all feasible values* of the variable. Inspection shows that this occurs at the value $x_{(3)} = 2.5$, or more exactly, the feasible value infinitesimally smaller than 2.5 where the data crf equals 0.2 and the theoretical crf equals 0.55. The largest distance is therefore 0.35.

A convenient computational method is set out in the columns *d, e* and *f* of Table 7.4. Column *d* calculates the distance for those data values where $S_r > F(x)$, that is where the theoretical curve lies below the empirical cumulative distribution. When the theoretical curve is above the empirical cumulative distribution, our example shows that the greater distance is given by calculating, not $(F(x) - S_r)$, but $(F(x) - S_{r-1})$. These latter values are shown in column *f*. The Kolmogorov test statistic is defined as the largest of the numbers in columns *d* and *f* which is, as before, 0.35.

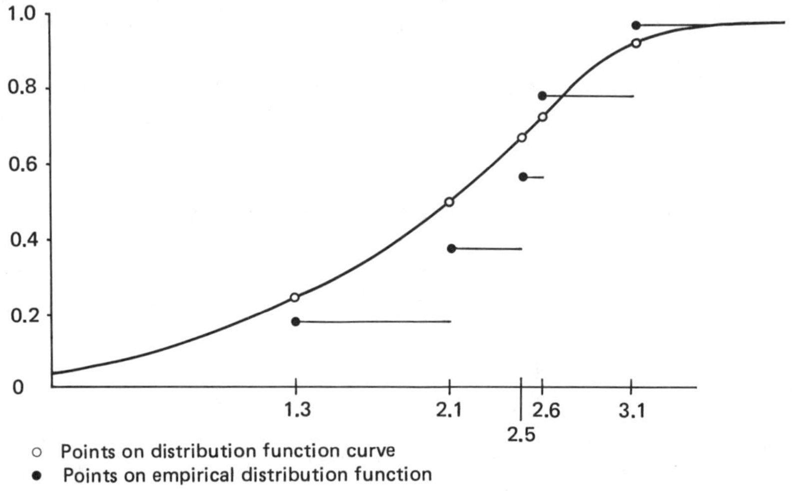


**Figure 7.3** N(2,1) Distribution Function and Empirical Cumulative Distribution

The test is a two-tailed test, and Table A.6 shows that when, as in this example, $n = 5$, the 5% critical value equals 0.56. The data is therefore consistent with the proposed model.

## 7.9 TESTING THE SPECIFICATION OF LINEAR MODELS

The various approaches developed so far can also be brought to bear upon linear models of the kind discussed in Chapter 4 and we illustrate them for the more important Model C, the bivariate regression model. In the present situation, we consider the validation of the model specification as a whole.

The regression model may be invalid in two ways. First, the random component may not be normally distributed. Secondly, the underlying relation between $Y$ and $X$ may be curvilinear rather than linear which would imply that the specification of the systematic component as $(\alpha + \beta x_r)$ is incorrect, and should instead be specified as, for example, $(\alpha + \beta x_r^2)$. It will be seen that a study of the residuals themselves throws light on both these possibilities.

An informative procedure is based upon what is known as the *residual plot.* The residual plot is a scatterplot, for each data point, of the residual, $e_r$, against the corresponding fitted value of the dependent variable, $\hat{y}_r$. Figure 7.4 demonstrates this pictorially. The first diagram shows a hypothetical scatterplot where the regression model appears to be a valid one and the points are randomly distributed around the regression line. The second diagram shows the corresponding residual plot. The plot shows no pattern at all because there is no connection between the fitted Y-value of a data point and its residual value.

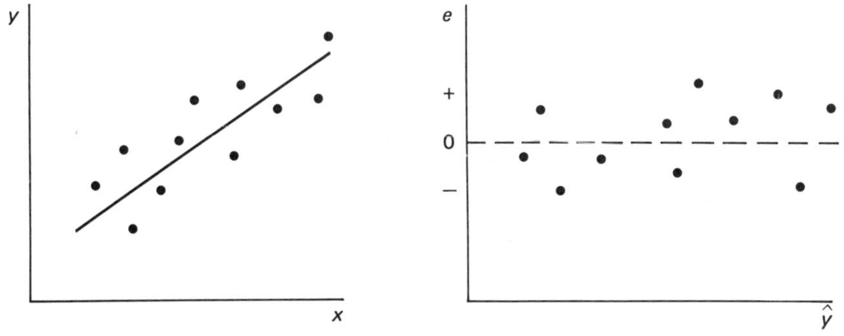


**Figure 7.4** Residual Plot where Systematic Component is Linear

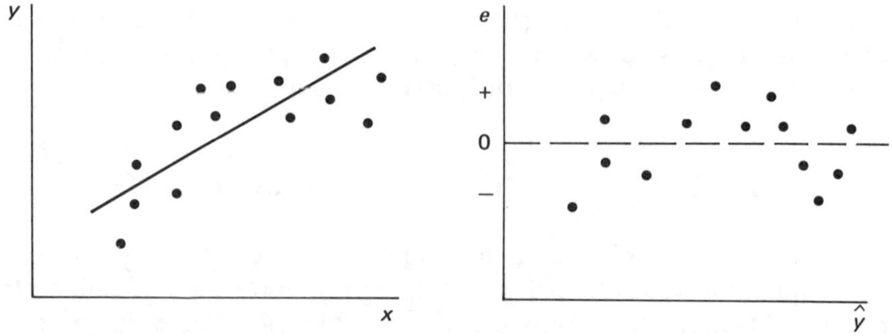


**Figure 7.5** Residual Plot: Systematic Component Curvilinear

In contrast, Figure 7.5 shows a scatterplot where the relation is curvilinear. If a regression line is fitted, the resulting residual plot will show a systematic pattern as in the second diagram. The residual values tend to be positive for dependent variable values in the centre of its range and negative at the two ends of its range.

In general, the existence of a pattern in the residual plot will suggest that the systematic component is incorrectly specified. The residual plot also highlights those data points which are outliers in terms of the model, that is, those which have extreme residual values.

Finally, the residual plot also throws light on the validity of the assumption that the residual component is normally distributed. It depends upon the following theoretical result:

> If the linear model is valid, the residuals have a normal probability distribution. **(7.9.1)**

In other words, the residuals should have a roughly symmetrical distribution about their mean of zero and the residual plot should exhibit a symmetrical pattern above and below the horizontal line $e = 0$. Alternatively, one can apply the normality test procedures to the residuals as described in Section 7.7.

The residuals have therefore an important role in the analysis of linear models. Computer regression programs typically calculate them and can be made to print out a residual plot.

## 7.10 TESTS ON THE SYSTEMATIC PARAMETERS OF LINEAR MODELS

At this stage, it is presumed that the chosen linear model specification is a valid one and, given this assumption, the question now to be considered is: are the data consistent with the parameter values taking any prespecified numerical values? There are two closely related procedures: one based upon confidence interval estimation and the other based upon hypothesis testing procedures. We illustrate both these approaches in terms of the bivariate regression model.

Result (4.9.2) sets out the 95% CI for $\beta$ as:

$$b \pm t_{0.975}(n-2) \times (\text{SE of } b) \qquad \textbf{(7.10.1)} = \textbf{(4.9.2)}$$

which, for the practical example of Section 4.8, was $0.37 \pm (2.09)(0.08)$. Now consider the question: are the data consistent with the model where $\beta = \beta_0$, $\beta_0$ being a prespecified numerical value? The CI approach is to answer 'yes' if $\beta_0$ is within the 95% CI and 'no' if it is not. (For this example, $\beta_0 = 0.20$ would be acceptable and $\beta_0 = 0$ would not.)

A more formal hypothesis testing procedure is as follows. The null hypothesis is $\beta = \beta_0$ and, if this is true, it can be shown that the test statistic is

$$\frac{(b-\beta_0)}{\text{SE}(b)} \sim t(n-2) \qquad \textbf{(7.10.2)}$$

This test statistic has the usual properties. Its distribution, $t(n-2)$, is known assuming the hypothesis is valid. Furthermore, extreme numerically large negative and positive values of the test statistic would indicate an invalid model. So the critical region comprises the values in the LH and RH tails of the distribution. For a two-tailed test at the 5% level of significance, the two critical values are $\pm\, t_{0.975}(n-2)$: Hence the hypothesis $\beta = \beta_0$ would be accepted if the numerical value of $(b - \beta_0)/\text{SE}(b) > t_{0.975}(n-2)$. If this inequality is true, it is an algebraic consequence that $\beta_0$ must lie within the 95% CI at (7.10.1). So the two-tailed test at a 5% level of significance and the 95% confidence interval method are equivalent.

If the hypothesis is that $\beta = 0$, the test statistic at (7.10.2) takes the simple form $b/\text{SE}(b)$. This ratio is called the *t-ratio* or *t-value* and is usually shown in computer printout. In the printout in Section 4.11, the $t$-values are reproduced in the '$T$' column and $b/\text{SE}(b) = 0.37/0.08 = 4.65$. Since $t_{0.975}(19) = 2.08$ and $4.65 > 2.08$, the hypothesis $\beta = 0$, is rejected at the 5% level of significance.

Sometimes it is more appropriate to carry out a one-tailed test on a systematic parameter value. Suppose for example that by reason of external information it is known that $\beta$ cannot possibly be negative. Then the only possible alternative to the null hypothesis being valid is that $\beta > 0$. This alternative is known as the *alternative hypothesis*,

**Table 7.5** Two-Tailed and One-Tailed Tests for Null Hypothesis $\beta = 0$ (at 5% level of significance)

| *Alternative hypothesis* | *Critical value(s)* | *Critical region* |
|---|---|---|
| $\beta \neq 0$ (i.e. $\beta < 0$ or $\beta > 0$) | $\pm\, t_{0.975}(n-2)$ | LH and RH tail |
| $\beta < 0$ only | $-\, t_{0.95}(n-2)$ | LH tail |
| $\beta > 0$ only | $t_{0.95}(n-2)$ | RH tail |

which, in general, is simply a statement of what would happen if the null hypothesis were not valid. The last row of Table 7.5 shows that the critical region is restricted to one tail of the $t$-distribution since only large positive values of the test statistic suggest the null hypothesis is invalid.

A similar procedure can be applied to testing whether the difference between the means of two data sets is statistically significant. Assuming that Model B(I) of Section 4.7 is valid, the 95% confidence interval estimate for $(\alpha_1 - \alpha_2)$ is given at (4.7.1). For a formal test of the null hypothesis $(\alpha_1 - \alpha_2) = 0$, it can be shown that the appropriate test statistic is:

$$T = \frac{(a_1 - a_2)}{s\sqrt{(1/n_1 + 1/n_2)}}, \text{ where } T \sim t(n_1 + n_2 - 2) \quad \textbf{(7.10.3)}$$

## 7.11 HYPOTHESIS TESTING REVIEWED

We now consider a number of practical and conceptual problems in the light of the various applications of hypothesis tests earlier in this chapter.

One problem concerns the choice of level of significance. It was pointed out at the end of Section 7.6 that the level of significance determines the probability of making a Type I error, that is, the probability of wrongly rejecting the null hypothesis when it is valid. This would suggest that it is safer and better to make the level of significance very small indeed. Unfortunately, the smaller the level of significance, the greater the probability of making another kind of mistake: of concluding that the null hypothesis is valid when in fact it is not. This is known as a *Type II error*. (The reason is that the smaller the level of significance, the smaller is the range of values of the test statistic in the critical region and therefore the greater is the range of values included in the acceptance region. So if the null hypothesis is invalid, the greater is the probability of wrongly accepting it as valid.)

In some areas of application, it is possible to decide on the 'best' level of significance to use by calculating the costs (in monetary or other terms) of making Type I and Type II errors and deciding on an optimal balance between the two. However this is rarely possible in the kinds of applications of concern in the present text. This is why conventional levels of significance, such as 5% or 1%, are so often used.

Another problem arises from the fact that whether a hypothesis is rejected or not depends upon the choice of level of significance. Thus, if the calculated value of the test statistic happened to lie between the critical value for a test at the 5% level and the critical value for the 1% level, the hypothesis would be rejected at the 5% level but accepted at the 1% level of significance.

This difficulty can be overcome to some extent by calculating and reporting, not the result of a test at an arbitrary (or conventional) level of significance but, instead, what is known as the *prob-value* of the test. The prob-value is defined as that level of significance which would have a critical value exactly equal to the calculated test statistic. However, in practice it is often not possible to calculate a prob-value because this requires tables of the distribution of the test statistic more detailed than are usually available, though prob-values are sometimes given in computer output.

Whatever the outcome of a formal hypothesis test, it is important to keep in mind what are the alternative possibilities, particularly when testing hypotheses relating to the numerical values of model parameters. On the one hand, if it turns out that, at some chosen level of significance, the hypothesis is accepted, it would be rash to assume that the preferred model is the one incorporating the hypothesised value. Other values of the parameter around the hypothesised value may well give a better fit to the data. For this and other reasons, it is better to conclude that the hypothesis is 'not rejected' rather than 'accepted'. For simplicity, we shall use the term 'accepted', but the above reservations need to be kept in mind.

On the other hand, if a hypothesis is formally rejected, care is still required in interpretation. Suppose for example that, in the bivariate regression model, the hypothesis $\beta = 0$ is rejected and the calculated regression coefficient is deemed to be 'statistically significant', that is, significantly different from zero. Although such a test result would have ruled out the possibility that $\beta = 0$, this does not necessarily mean that other values of $\beta$ close to zero would also be ruled out. In terms of the subject context of the analysis, these other values might not be 'significantly' different from zero *from the context point of view*. For example, the hypothesis $\beta = 0$ might be rejected while, say, the hypothesis $\beta = 0.02$ is accepted although, in the subject matter context, 0.02 may be insignificantly different from zero. In short, it is important to distinguish between 'statistical significance' and 'subject significance'. On the whole, a clearer message results from calculating a confidence interval estimate of a parameter than from a single hypothesis test of the proposed numerical value.

The problem that a hypothesis test might make a very fine distinction between the validity of two values of a parameter very close to each other is particularly likely to arise for large data sets. The reason is as follows, again taking as an example a test of the value of the parameter in the bivariate regression model. We first recall from Section 7.10 that the hypothesis $\beta = \beta_0$ will be rejected if $\beta_0$ falls outside the range $b \pm t_{0.975} (n - 2 \times (SE(b))$. Now as $n$ gets larger and larger, any standard error calculation takes on smaller and smaller numerical values because its formula has $\sqrt{n}$ as a denominator. So the larger $n$, the smaller the range of values around $b$ for 'acceptable' values of $\beta_0$.

To sum up, a hypothesis test is best treated as giving some indication of the best direction for further investigation rather than as a procedure which yields a final simple truth: accept or reject. From this point of view, the level of significance is not of crucial importance. The justification for using conventional levels of significance such as 5% or 1% is simply that they are conventional and they make for better communication of analytical results to others.

## 7.12 TESTING THE INDEPENDENCE OF BIVARIATE DATA AND THE CORRELATION COEFFICIENT

We start by considering the bivariate regression model in which $y_r = \alpha + \beta x_r + u_r$ (where $u_r \sim N(0, \sigma^2)$). The probability distribution of each $y_r$ depends upon $x_r$, the corresponding value of $X$. In fact, each $y_r \sim N(\alpha + \beta x_r, \sigma^2)$. In such a situation, where the probability distribution of the variable depends upon the value taken by another, the two variables are said to be dependent. Otherwise, they are said to be independent. For the bivariate regression model $Y$ and $X$ are not independent *unless* $\beta = 0$. If $\beta = 0$, the bivariate model is reduced to Model A (see Section 4.2).

This suggests the first approach to testing the independence of two variables in a data set, namely, to postulate a bivariate regression model for the data and test the null hypothesis that $\beta = 0$. The procedure has been described in Section 7.9 and the test statistic, $T$, equals $b/\mathrm{SE}(b)$.

A second approach is via the correlation coefficient, $r$. There is an algebraic identity connecting the regression coeeficient, $b$, and the correlation coefficient:

$$r = b \times s_x/s_y \text{ (where } s_x \text{ and } s_y \text{ are respectively the standard deviations of } X \text{ and } Y) \quad \textbf{(7.12.1)}$$

(Note that if the data have been standardised, that is, if $s_x = 1 = s_y$, then $r = b$). It can be shown that, as a consequence of this identity, the test statistic $b/\mathrm{SE}(b)$ can be written as:

$$T = \frac{r\sqrt{(n-2)}}{\sqrt{(1-r^2)}} \quad \textbf{(7.12.2)}$$

So the hypothesis $\beta = 0$ can also be tested using the value of $r$. If $r$ is sufficiently close to zero to make the numerical value of $T$ less than the critical value, the hypothesis is accepted. Otherwise the hypothesis is rejected and the value of $r$ is said to be significantly different from zero. As a rule of thumb when $n$ is large, the correlation coefficient has a standard error of approximately $1/\sqrt{n}$ and has the distribution N(1, $1/n$); a correlation coefficient value is significant if it is numerically greater than $2/\sqrt{n}$.

At this point, however, a most important qualification needs to be made. Whilst it is the case that if two variables in a bivariate model are independent, their correlation coefficient is zero, the converse is not necessarily true. It is possible for the correlation coefficient to be zero but the variables are not independent. An idealised example is shown in Figure 7.6. It shows a curvilinear relation between $X$ and $Y$. The line

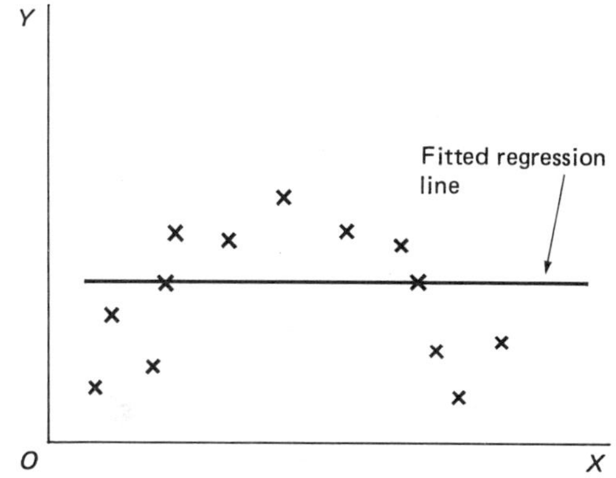


**Figure 7.6** Curvilinear Relation where r = 0

of best fit, the regression line, is horizontal and so both $b$ and $r$ equal zero. But the two variables are certainly not independent.

What practical conclusions can then be drawn from a test of the null hypothesis $\beta = 0$? First, suppose that the hypothesis is rejected. In this case we can conclude that $X$ and $Y$ are dependent in terms of the model and the correlation coefficient is statistically significantly different from zero. However, the distinction between statistical significance and contextual significance made in Section 7.11 needs to be kept in mind. A correlation coefficient may be large enough to be statistically significant but still be too small to matter in its context.

Secondly, suppose that the null hypothesis is accepted. Two alternative conclusions are now possible. Either the data are consistent with a model in which the variables are independent, or the variables are dependent but the nature of the dependence is something other than linear. Visual inspection, by means of, for example, a residual plot (see Section 7.9) may suggest whether some alternative specification is required.

## EXAMPLES

**E 7.1** Carry out Geary's test of the normal distribution fit for the two data sets in E 6.1.

**E 7.2** Make a formal test of the goodness of fit of the lognormal distribution to your chosen data set in F 6.1. (Note that the test will, in effect, examine the fit of the log transformed data to the normal distribution).

**E 7.3** A computer printout of a regression analysis gives the following information:

| | *Coefficient* | *Standard error* | *t-value* |
|---|---|---|---|
| constant | 2.5 | 1.0 | 2.50 |
| $X$ | 0.8 | 0.3 | 2.67 |
| $n = 12$ | | | |

Calculate a 95% confidence interval for the model parameters $\alpha$ and $\beta$. Test the hypothesis that $\beta = 0$ at some chosen level of significance. Would any different conclusions be drawn if it were known that $\beta$ could not be less than zero?

## FURTHER WORK

**F 7.1** Fit the bivariate regression model to a data set of your choice. Identify, on the computer printout, the values of the coefficients and their standard errors and hence calculate 95% confidence intervals for the systematic model parameters.

**F 7.2** Obtain a scatterplot of the data in F 7.1 and consider whether there appear to be any obvious departures from the assumptions of the linear regression model. Compare the scatterplot with plot of the random against the fitted values.

Make a formal test of the assumption that the random component is normally distributed.

*This chapter depends upon: Chapter 4, Sections 6.5, 7.9–7.12*

# 8

# Multiple Regression Analysis

## 8.1 INTRODUCTION

This chapter and the next will return to the theme of linear modelling; in particular, we will be concerned with the formulation, estimation and checking of linear models whose systematic component depends upon several explanatory variables rather than just one. We assume that there is a data matrix containing information on $n$ individuals and that one column of this contains values of a dependent variable. Some or all of the other columns contain values of explanatory variables whose variation we know from past experience or informed common sense is likely to be useful in explaining the variation in the dependent variable. As always, we will not capture all this variation, so the model will contain a random component also.

Let us turn first to a very basic point, namely what purposes might be served by fitting a regression model. Table 8.1 sets out some plausible instances in which multiple regression analysis would be helpful. In case A, regression analysis will provide a convenient summary of the way in which a child's weight depends upon his or her age and various body features. Analysis of case B might attempt to go further and establish not only a description, as in case A, but also a basis for making a prediction of university performance on the basis of school performance in the final year. Case C might similarly lead to

**Table 8.1** Instances of Multiple Regression Analysis

| | *Dependent variable* | *Explanatory variables* | *Individuals* |
|---|---|---|---|
| A. | Body weight | Various body measurements (height, waist, etc.)<br>Age | Schoolchildren |
| B. | Overall grade in university degree examination | Performance in individual subjects in final year at school | Students |
| C. | Tobacco consumption | Consumers' income<br>Price of tobacco<br>Expenditure on tobacco advertising | National figures for a sequence of individual years |
| D. | Heart disease mortality rate | Consumption of various foodstuffs | Countries |

a method for predicting tobacco consumption in a future year and might also be used by, say, a tobacco company to estimate the effect of an increase in advertising expenditure upon tobacco sales. Regression analysis of case D may be directed towards a first check on a hypothesis that dietary factors are an important determinant of heart disease. Data bearing on this will be used to illustrate the discussion in the present chapter.

The possible uses of regression may therefore be summed up as: description, prediction, policy formulation and causal modelling. If the description is found to apply more widely than just for the data set analysed then we may say that we have the beginnings of an empirical law.

## 8.2 THE MULTIPLE REGRESSION MODEL AND THE REGRESSION EQUATION

The tasks which lie ahead may be summarised under the general headings presented at the end of Chapter 3, namely *specification, estimation* (or *fitting*) and *diagnostic checking*. At the most general level, specification is a matter of setting out the following model for some dependent variable $Y$:

$$y_r = \underbrace{\beta_0 + \beta_1 x_{1r} + \beta_2 x_{2r} + \ldots + \beta_k x_{kr}}_{\text{systematic component}} \quad \underbrace{+\, u_r}_{\substack{\text{random}\\ \text{component}}}$$

$$r = 1, \ldots, n \qquad \textbf{(8.2.1a)}$$

where $y_r$ denotes the value of $Y$ for the $r$th individual, $x_{1r}, x_{2r}, \ldots, x_{kr}$ denote the corresponding values of the $k$ explanatory variables and $u_r$ the value of the random component. By its nature, the latter cannot be directly observed and is unpredictable from one individual to another. We shall make the conventional assumption, however, that

$$u_r \sim \mathrm{N}(0, \sigma^2) \qquad \textbf{(8.2.1b)}$$

The model has $k + 2$ parameters, $\beta_0, \beta_1, \ldots \beta_k$ which define the systematic component and $\sigma^2$, the variance of the random component.

Fitting involves using the data to provide the estimates $b_0, b_1, \ldots, b_k$ of the parameters $\beta_0, \beta_1, \ldots, \beta_k$ respectively. Given the assumption that the random component has a normal distribution, maximum likelihood provides a method of estimation, although we do not go into detail. As for bivariate regression (see Table 6.2), the ML procedure yields estimates which are identical to those which the method of least squares gives, and they are again unbiased. In fact, both methods determine parameter estimates as solutions of a set of equations known as the *normal equations.* Although these, and the method for solving them, are not needed in what follows in this chapter, we state them here for the case where $k = 3$:

$$\begin{aligned}
nb_0 &+ b_1\Sigma x_1 &+ b_2\Sigma x_2 &+ b_3\Sigma x_3 &= \Sigma y\\
b_0\Sigma x_1 &+ b_1\Sigma x_1^2 &+ b_2\Sigma x_1x_2 &+ b_3\Sigma x_1x_3 &= \Sigma x_1y\\
b_0\Sigma x_2 &+ b_1\Sigma x_1x_2 &+ b_2\Sigma x_2^2 &+ b_3\Sigma x_2x_3 &= \Sigma x_2y\\
b_0\Sigma x_3 &+ b_1\Sigma x_1x_3 &+ b_2\Sigma x_2x_3 &+ b_3\Sigma x_3^2 &= \Sigma x_3y
\end{aligned} \qquad \textbf{(8.2.2)}$$

The three explanatory variables are here indicated by $X_1, X_2, X_3$; notation such as $\Sigma x_1x_2$ is short-hand for $\sum_1^n x_{1r}x_{2r}$, the sum over all data points of the product of the corresponding $X_1$ and $X_2$ variable values. The extension to any number of explanatory variables will be fairly obvious.

The *regression coefficients,* $b_0, b_1, \ldots, b_k$ determine the regression equation, and, with an obvious extension to the result (4.11.1), we write this as:

$$\hat{y} = b_0 + b_1x_1 + \ldots + b_kx_k \tag{8.2.3}$$

where $\hat{y}$ denotes the fitted value of $Y$ corresponding to a set of values of the explanatory variables. Specifically, for the $r$th data point:

$$\hat{y}_r = b_0 + b_1x_{1r} + \ldots + b_kx_{kr} \tag{8.2.4}$$

Again, as in the discussion of the bivariate case, the value $y_r$ of the data set will, in general, differ from the fitted value $\hat{y}_r$. The difference, $e_r$, is the $r$th residual. Algebraically:

$$y_r - \hat{y}_r = e_r$$
$$\text{and } y_r = b_0 + b_1x_{1r} + \ldots + b_kx_{kr} + e_r \tag{8.2.5}$$

The residuals play an important part in diagnostic checking which we will discuss further in Section 8.9 and beyond.

To sum up, we have set out three fundamental relations. First, there is the model specification at (8.2.1) involving unknown parameters and the unobservable random component. Secondly, as a result of the estimation procedure, we can write down (8.2.5) where each value of $Y$ in the data set is expressed in terms of estimated (and therefore known) coefficients and the residual, the latter being an estimate of the value of the random component. Finally, (8.2.4) defines the fitted values of $Y$.

A preliminary insight into the interpretation of the coefficients of the multiple regression equation is given by considering the hypothetical data set in Table 8.2. The dependent variable $Y$ may be thought of as the overall marks of mathematics students in their degree examination, and $X_1$ and $X_2$ the corresponding school marks in mathematics and English respectively. The computed multiple regression equation for these data is $y = 30 + 0.75x_1 - 0.25x_2$. (The residuals are in fact zero, but this does not affect the argument which follows.)

**Table 8.2** Hypothetical Data Set

| $x_1$ | $x_2$ | $y$ |
|---|---|---|
| 36 | 40 | 47 |
| 52 | 44 | 48 |
| 68 | 56 | 57 |
| 80 | 60 | 75 |
| 72 | 52 | 71 |

The main feature of the regression equation (we shall no longer append the appellation 'multiple') is that it estimates the *marginal* effect of any one explanatory variable on the dependent variable, that is to say it shows how much the dependent variable changes as a result of a particular change in an explanatory variable *assuming that the values of the other explanatory variables remain fixed*. Thus, in the example, the dependent variable *increases* by $\frac{3}{4}$ for a unit increase in the explanatory variable $x_1$ if the other explanatory variable $x_2$ is held fixed and it *decreases* by $\frac{1}{4}$ for a unit increase in $x_2$ with $x_1$ being held fixed. It is indeed generally the case that, as in this example, the magnitude and the sign of the regression coefficient are of interest. The sign of the regression coefficient determines whether the dependent variable increases or decreases for marginal increases in the explanatory variable.

We have seen that the marginal relation between $X_2$ and $Y$ is negative. However, a glance at the data matrix in Table 8.2 shows that there is a *positive* correlation between $X_2$ and $Y$. The reason for this apparent conflict will be explored below but meanwhile we simply note that bivariate analysis may give a different picture from multiple regression analysis.

More generally, it needs to be stressed that the estimation of marginal effects depends upon whatever other explanatory variables are included in the systematic component. Furthermore, outside the experimental sciences, it is rarely possible *actually* to vary only one explanatory variable and hold the others fixed.

## 8.3 FITTING A REGRESSION EQUATION: A PRACTICAL EXAMPLE

To set the scene, consider the data shown in Table 8.3. This is an extension of the data set used in Chapter 4 and it will be recalled that the variable *IHDMR* is a measure of the mortality from heart disease. Throughout this chapter, this will be taken to be the dependent variable. The other variables are potential explanatory variables. They relate to the consumption of various foods with one exception, *AGEMP*, which is an indicator of the importance of agricultural employment in a country. We will refer to this as the 'country mortality' data set. The regression calculations in this chapter have been mainly done using the program IDA (see Appendix B); the output which this provides is, however, typical of other programs.

We first consider a model with just three explanatory variables, namely $X_1$: Total dairy products comsumption (*TDP*); $X_2$: Sugar consumption (*SUG*), and $X_3$: percentage employment in agriculture,

**Table 8.3** Country Data on Mortality Rate from Heart Disease and Consumption of Food Products etc.

| | *TDP* | *MLK* | *SUG* | *BUT* | *FAT* | *IHDMR* | *AGEMP* |
|---|---|---|---|---|---|---|---|
| Australia | 422.7 | 385.0 | 134.6 | 26.6 | 99.6 | 263.2 | 6.6 |
| Austria | 447.6 | 393.9 | 99.0 | 15.2 | 115.0 | 237.5 | 11.8 |
| Bel/Lux | 292.5 | 229.5 | 99.5 | 25.4 | 114.3 | 181.4 | 3.6 |
| Canada | 472.1 | 409.6 | 127.8 | 21.2 | 99.8 | 226.8 | 5.7 |
| Denmark | 526.7 | 343.2 | 131.0 | 26.3 | 130.6 | 297.1 | 9.1 |
| Finland | 860.6 | 789.0 | 114.6 | 46.8 | 103.9 | 240.9 | 12.9 |
| France | 332.7 | 250.1 | 91.7 | 24.9 | 106.2 | 80.9 | 9.6 |
| W. Germany | 317.0 | 209.6 | 88.1 | 23.4 | 116.6 | 171.5 | 6.8 |
| Greece | 240.1 | 171.4 | 44.1 | 3.2 | 41.5 | 66.5 | 28.4 |
| Ireland | 706.9 | 589.0 | 137.6 | 38.0 | 114.4 | 263.9 | 23.1 |
| Italy | 211.0 | 183.1 | 71.3 | 5.2 | 48.6 | 134.0 | 15.9 |
| Japan | 67.1 | 61.8 | 56.7 | 0.8 | 25.1 | 36.4 | 11.9 |
| Netherlands | 477.8 | 315.9 | 124.7 | 9.3 | 95.5 | 182.1 | 6.3 |
| New Zealand | 529.4 | 497.4 | 111.2 | 51.2 | 135.8 | 243.0 | 11.6 |
| Norway | 559.1 | 477.4 | 117.4 | 13.1 | 103.4 | 262.9 | 9.0 |
| Portugal | 121.4 | 114.5 | 61.4 | 1.7 | 43.7 | 77.2 | 32.5 |
| Spain | 220.3 | 206.6 | 65.4 | 0.5 | 40.5 | 46.6 | 20.7 |
| Sweden | 501.8 | 330.4 | 111.4 | 21.5 | 91.7 | 333.9 | 6.1 |
| Switzerland | 445.9 | 398.5 | 116.9 | 18.3 | 91.6 | 105.3 | 8.5 |
| UK | 454.5 | 400.4 | 124.9 | 24.8 | 110.3 | 293.0 | 2.7 |
| USA | 490.4 | 356.7 | 114.0 | 7.1 | 98.5 | 326.1 | 3.6 |

*Notation:* *TDP* = Total Dairy Products, *MLK* = Whole Milk, *SUG* = Sugar, *BUT* = Butter, *FAT* = Animal Fat, (all foregoing in grams per head per day, 1965–1969), *IHDMR* = Ischaemic Heart Disease Mortality Rate 1973–4, *AGEMP* = Percentage employment in Agriculture, Fishing and Forestry.

*Sources:* *Food Consumption Statistics 1955/1973*, OECD 1975; *United Nations Demographic Yearbooks*, United Nations; *OECD Observer*, OECD 1978

fishing and forestry (*AGEMP*). In the formal terms of (8.2.1), the model is:

$$IHDMR_r = \beta_0 + \beta_1 TDP_r + \beta_2 SUG_r + \beta_3 AGEMP_r + u_r$$
$$u_r \sim N(0, \sigma^2), \quad r = 1, \ldots, 21 \qquad \textbf{(8.3.1)}$$

In line with the notation of (8.2.3), the multiple regression equation may be written as $\widehat{IHDMR} = b_0 + b_1 TDP + b_2 SUG + b_3 AGEMP$.

An extract from the computer output for the three explanatory

**Table 8.4** Regression Coefficients and Standard Errors: Computer Printout

| *VARIABLE* | *B(STD.V)* | *B* | *STD.ERROR(B)* | *T* |
|---|---|---|---|---|
| *TDP* | 0.3499 | 1.7752E−01 | 1.2144E−01 | 1.462 |
| *SUG* | 0.4112 | 1.4070E+00 | 9.7516E−01 | 1.443 |
| *AGEMP* | −0.1820 | −2.1241E+00 | 2.1368E+00 | −0.994 |
| *CONSTANT* | 0 | 1.6200E+00 | 8.6156E+01 | 0.019 |

*Source:* IDA printout. As in computer output generally, the number after the symbol E indicates the number of places the decimal point is to be moved to the right (if E+) or to the left (if E−).

variable model is shown in Table 8.4. From this, values of $b_0$, $b_1$, $b_2$, and $b_3$ are easily identified in the column headed *B*. The regression equation is therefore

$$\widehat{IHDMR} = 1.62 + 0.18TDP + 1.41SUG - 2.12AGEMP$$

For this model, therefore, *TDP* and *SUG* have marginal positive effects and *AGEMP* a marginal negative effect on the dependent variable. Given these coefficient values we may calculate fitted values of the dependent variable for each country and the corresponding residuals, $e_r$, from which an estimate of the variance, $\sigma^2$, of the random component can be made. We will return to a study of the fitted values and the residuals in Section 8.4.

As in the case of the bivariate regression model (Section 4.9) we may also make interval estimates of the structural parameters in terms of the point estimates and their associated standard errors. We remind the reader that a confidence interval for a systematic parameter in a linear model can be expressed in terms of the point estimate, its standard error and an appropriate *t*-distribution value, as at result (4.6.1). Interpreting this for a multiple regression equation with *k* explanatory variables, and therefore $k + 1$ systematic parameters, we have:

> The 95% CI of a systematic parameter, say $\beta_j$, in a multiple regression model is equal to
>
> $$b_j \pm t_{0.975}(\nu) \times \mathrm{SE}(b_j)$$
> $$\text{where } \nu = n - (k + 1) \qquad \textbf{(8.3.2)}$$

For the current model, $\nu$ therefore equals 17, $t_{0.975}(17)$ is 2.11 and Table 8.4 gives the standard errors for each of the regression coefficients. For example, the 95% CI for the parameter $\beta_2$ (the coefficient on *SUG*) is $1.41 \pm 2.11 \times 0.98$, that is $1.41 \pm 2.07$, the range ($-0.66$, 3.48).

We may also formally test the hypothesis that a systematic parameter, $\beta_j$, in the given model is zero. By analogy with (7.12.2) for the bivariate model, the test statistic is $b_j/\text{SE}(b_j)$ which is generally known as the *t-ratio*. Under the null hypothesis, this has the distribution $t(n-k-1)$. Using $\beta_2$ as an example once more, the test statistic is 1.41/0.98 which equals 1.44. It will be seen that this figure appears in the final column, headed *T*, of Table 8.4. Since 1.44 is less than the critical value of 2.11 the null hypothesis would be accepted. However, the cautionary remarks about the interpretation of hypothesis testing made in Section 7.11 should be borne in mind in interpreting conclusions of this kind.

It is normal practice, when presenting a regression equation, to give either the *t*-ratios or the standard errors in brackets. Ours will be to report *t*-ratios, and so the results of fitting the three explanatory variable model are:

$$\widehat{IHDMR} = \underset{(0.02)}{1.62} + \underset{(1.46)}{0.18TDP} + \underset{(1.44)}{1.41SUG} - \underset{(0.99)}{2.12AGEMP} \tag{8.3.3}$$

It is important to note that the magnitude of an estimated parameter depends partly upon the units in which the variable is measured. If the measurement scale of any explanatory variable is changed by *dividing* each value by a constant *c*, for example, the only effect is that the corresponding regression coefficient will be *multiplied* by *c*. The standard error is similarly multiplied by *c* and so the *t*-ratio is unchanged. Of particular interest is the regression based upon variables standardised so that each has zero mean and unit variance. The effect on the regression coefficients is as follows. If $b_j$ is the regression coefficient before standardisation, then $b_j^*$(the coefficient after standardising both the dependent and the explanatory variable) is given by

$$b_j^* = \frac{\text{SD}(X_j) \times b_j}{\text{SD}(Y)} \tag{8.3.4}$$

The $b_j^*$'s are known as the *beta coefficients.* These are given in the first column of the computer printout in Table 8.4. The beta coefficients have the advantage of not being affected by a change in a possibly arbitrary scale of measurement, but we shall in fact make no use of them.

## 8.4 EXPLAINED, RESIDUAL AND TOTAL SUM OF SQUARES

In the discussion of the bivariate regression model, we explained in Section 4.11 how the total sum of squares (TSS) can be expressed as the sum of the explained (or regression) sum of squares (ESS) and the residual sum of squares (RSS). Algebraically:

$$\underset{\text{TSS}}{\Sigma(y_r - \bar{y})^2} = \underset{\text{ESS}}{\Sigma(\hat{y}_r - \bar{y})^2} + \underset{\text{RSS}}{\Sigma e_r^2} \quad \textbf{(8.4.1)}$$

Verbally, TSS represents the sum of squared deviations of the dependent variable $Y$ values from their mean, ESS the sum of squared deviations of the fitted values of $Y$ around their mean, and RSS the sum of the squared residuals.

We recall that when the bivariate regression relation is a good fit, ESS will be large relative to RSS and conversely for a poor fit. The nearer the ratio ESS/TSS to one, the better the fit. This ratio is identically equal to $r^2$, the square of the correlation coefficient between $Y$ and $X$ (see result (4.11.8)).

The calculation of the three sums of squares depends simply upon the values of the $Y$-variable, the $y_r$'s, the values of the fitted values of $Y$, the $\hat{y}_r$'s calculated from the regression equation, and the residuals, the $e_r$'s. Exactly the same calculations can be made in multiple regression analysis *and the identity at (8.4.1) is still true.*

In the multiple regression situation, the ratio ESS/TSS is written as $R^2$ and $R$ is called the multiple correlation coefficient. Like $r^2$, $R^2$ is a measure of goodness of fit and it takes values in the range (0, 1).

The identity at (8.4.1) is illustrated numerically in Table 8.5. It shows the results of fitting four separate models to the data of Table 8.3. As before the dependent variable is *IHDMR* each time. The sums of squares are given in the columns headed *SS*; the value of TSS is opposite TOTAL, of ESS opposite REGRESSN, and of RSS opposite RESIDUAL.

The reader can easily check that, for each of the four regressions, TSS = ESS + RSS. Furthermore, a comparison of the four sets of results gives additional insight into this kind of analysis. First we see that for all four sets, TSS = 180972; this is because TSS is calculated only from the values of the dependent variable. Secondly, we note that as one passes through the four sets of tables I to IV, where the numbers of explanatory variables brought into the analysis successively increase from just *SUG* in I to all variables in IV, the ESS increases at each stage and the RSS decreases at each stage. For example, as a result of including *TDP* in the regression analysis of II, ESS

**Table 8.5** ANOVA Tables for Four Sets of Explanatory Variables

| | *I : SUG* | | | | *II : SUG, TDP* | | | |
|---|---|---|---|---|---|---|---|---|
| | *SS* | *DF* | *MS* | *F* | *SS* | *DF* | *MS* | *F* |
| REGRESSN | 114848 | 1 | 114848 | 33.0 | 119654 | 2 | 59827 | 17.6 |
| RESIDUAL | 66124 | 19 | 3480 | | 61318 | 18 | 3407 | |
| TOTAL | 180972 | 20 | | | 180972 | 20 | | |

| | *III : SUG, TDP, AGEMP* | | | | *IV : SUG, TDP, AGEMP, MLK, BUT, FAT* | | | |
|---|---|---|---|---|---|---|---|---|
| | *SS* | *DF* | *MS* | *F* | *SS* | *DF* | *MS* | *F* |
| REGRESSN | 123022 | 3 | 41007 | 12.0 | 130629 | 6 | 21771 | 6.1 |
| RESIDUAL | 57950 | 17 | 3409 | | 50343 | 14 | 3596 | |
| TOTAL | 180972 | 20 | | | 180972 | 20 | | |

*Source:* IDA output of regression analysis of *IHDMR* on stated explanatory variables. Data at Table 8.3

increases from 114848 to 119654 and RSS decreases from 66124 to 61318.

At the same time, the $R^2$ values also increase from case I to case IV. (They are 0.63, 0.66, 0.68 and 0.72.) This is reasonable in the sense that the more explanatory variables there are, the better the fit.

These features are typical of any regression analysis and may be generalised as follows:

> If a further explanatory variable is added to a model, the explained (or regression) SS increases, the residual SS decreases and $R^2$ increases.[1] **(8.4.2)**

It is another matter whether or not the increase in the goodness of fit is 'statistically significant'. We consider this aspect in the following sections.

1. It is just possible that ESS, RSS and $R^2$ will remain unchanged, but this is so unusual in practice that it can be ignored.

## 8.5 ANOVA AND THE F-DISTRIBUTION

ANOVA is shorthand for *analysis of variance*. Each of the four tables in Table 8.5 is an ANOVA table and we have already noted one feature: the first column shows a decomposition of TSS into its two components ESS and RSS.

When testing hypotheses, it is sometimes the case that, if a particular null hypothesis is valid, two alternative estimates, say $s_1^2$, and $s_2^2$, of a model variance parameter $\sigma^2$ can be calculated from the data. If $s_1^2$ and $s_2^2$ are approximately equal, that is to say if $s_1^2/s_2^2$ is close to 1, the null hypothesis is then supported.

This idea can be applied in regression analysis, and we first illustrate it in the context of Table 8.5, case I. This relates to estimating the model $IHDMR_r = \beta_0 + \beta_1 SUG_r + u_r$, where $u_r \sim N(0, \sigma^2)$. Consider the null hypothesis $\beta_1 = 0$, that is, the hypothesis that *SUG* has no statistically significant role in the explanation of the *IHDMR* variable. Then the theory tells us that, if indeed $\beta_1 = 0$, we can estimate $\sigma^2$ independently in two ways. The first, $s_1^2$, is given in the first row of figures by dividing the ESS value 114848 by the number in the DF column, 1, to give 114848, shown in the MS column. The second, $s_2^2$, is given by the second row: RSS (66124) is divided by DF (19) to give MS (3480). The ratio $s_1^2/s_2^2 = 114848/3480 = 33.0$ is given under the heading *F*. This *F*-value is the test statistic for the null hypothesis. One can see intuitively that large values of *F* will suggest rejection of the null hypothesis since the larger ESS is, relative to RSS, the bigger is the *F*-value. Furthermore, only relatively large values of ESS as compared to RSS indicate rejection of the null hypothesis, so that any test is going to be a one-tailed test.

The two numbers in the column *MS* (mean square) are thus the two estimates of $\sigma^2$ (the variance of the random component). The numbers in the *DF* column (degrees of freedom) give the appropriate numbers to divide the sums of squares in the *SS* column by to yield the two variance estimates. The degrees of freedom values can be rationalised as follows. The value 20 opposite TOTAL equals $(n - 1)$, there being $n = 21$ countries in the data matrix. The residual mean square estimate of $\sigma^2$ is given by dividing RSS ($= \Sigma e_r^2$) by 19 which equals $(n - 2)$; two degrees of freedom have been used up since the regression analysis I estimates two parameters $\beta_0$ and $\beta_1$. The DF value in the first row, 1, equals the number of explanatory variables in the regression analysis. Note that the DF values add up in the same way as the SS values.

The same procedure can be used to derive a test statistic for the multiple regression case. In case III, the regression model is

$IHDMR_r = \beta_0 + \beta_1 SUG_r + \beta_2 TDP_r + \beta_3 AGEMP_r + u_r$. The ANOVA procedure tests the hypothesis that $\beta_1 = \beta_2 = \beta_3 = 0$, that is, that the regression relation *as a whole* is insignificant. The degrees of freedom entries change because there are now three explanatory variables but the arithmetic process is the same. The test statistic, the *F*-ratio, is 12.0.

Finally, we need to know how to assess the value of our test statistic, the *F*-ratio. How big does it have to be to lead to rejection of the null hypothesis? The answer lies in an important theoretical distribution, the *F*-distribution. It is a two parameter distribution and our notation for it is $F(\nu_1, \nu_2)$. Like the *t*-distribution, the parameters are called the degrees of freedom parameters.

The *F*-distribution takes only positive variable values because it is essentially a ratio of sums of squares. Table 8.6 shows selected $F_{0.95}(\nu_1, \nu_2)$ values for various values of the parameters. Thus for the distribution $F(1, 19)$, it shows that $F_{0.95}(1, 19) = 4.4$. So there is a probability of 0.95 that a variable with this distribution will be less than 4.4 and a probability of 0.05 that it will be greater than 4.4.

In the regression situation, the *F*-ratio calculated in the ANOVA table can be shown to have an *F*-distribution, with parameters defined by the numbers in the *DF* column, *if the null hypothesis is valid.* The *F*-distribution thus gives us the necessary critical value for a hypothesis test.

Specifically, in case I, if $\beta_1 = 0$, the *F*-ratio has the distribution $F(1, 19)$. For a test at the 5% level of significance, the critical value from Table 8.6 is $F_{0.95}(1, 19) = 4.4$; 33.0 is greater than 4.4 and so the hypothesis is rejected. Similarly in case III, the distribution of the *F*-ratio is $F(3, 17)$. $F_{0.95}(3, 17) = 3.2$ which is less than the observed 12.0 and again the null hypothesis $\beta_1 = \beta_2 = \beta_3 = 0$ is rejected at the 5% level of significance.

**Table 8.6** Abstract from *F*-Distribution Table
$F_{0.95}(\nu_1, \nu_2)$ values

| | $\nu_1$ | | | | | |
|---|---|---|---|---|---|---|
| $\nu_2$ | 1 | 2 | 3 | 4 | 5 | 6 |
| 15 | 4.5 | 3.7 | 3.3 | 3.1 | 2.9 | 2.8 |
| 17 | 4.4 | 3.6 | 3.2 | 3.0 | 2.8 | 2.7 |
| 19 | 4.4 | 3.5 | 3.1 | 2.9 | 2.7 | 2.6 |

## 8.6 THE SIGNIFICANCE OF ADDED EXPLANATORY VARIABLES

The ANOVA test of the significance of the whole regression discussed in the last section is not often useful in practical terms. For one thing, a regression relation as a whole is usually significant. Its importance is that it can be developed to answer the question—is one or more particular explanatory variable worth adding to an existing model?

Suppose we are at the stage of case II in Table 8.4. The variables *SUG* and *TDP* have already entered the analysis and we are wondering whether *AGEMP* adds any further explanatory power. We know that the addition of *AGEMP* is bound to increase $R^2$. (The increase was from 0.66 to 0.68.) But has this increase any significance? The answer lies in setting up an ANOVA table derived from those in Table 8.5.

The first ANOVA table A in Table 8.7 is based upon case III of Table 8.5 (the regression on *SUG, TDP* plus *AGEMP*) supplemented by case II (the regression *SUG, TDP* without *AGEMP*). The key to the analysis is the calculation (in the second row) of the increase in the explained sum of squares due to the addition of *AGEMP*, 3,368. This is simply the difference between two regression sums of squares: one with *AGEMP* included (123,022 from case III) and one with *AGEMP* excluded (119,654 from case II). The degrees of freedom entry is similarly calculated by subtraction. The test statistic (the *F*-ratio) is the ratio of the second row mean square to the residual mean square and equals 0.99.

The test statistic has the distribution $F(1, 17)$ defined by the corresponding degrees of freedom and so the 5% level of significance critical value is $F_{0.95}(1, 17) = 4.4$ (see Table 8.5). The null hypothesis that *AGEMP* provides no significant added explanatory power is therefore supported.

This procedure is easily extended to testing the collective significance of the addition of a number of explanatory variables to an existing regression. Part B of Table 8.7, for example, tests the significance of adding *MLK, BUT* and *FAT* to an existing regression with *SUG, TDP* and *AGEMP*. The reader will see that the *F*-ratio 0.71 (which is clearly not significant) is based upon the ANOVA tables III and IV of Table 8.5. However, these calculations can more easily be done in terms of values of $R^2$. Let $k_0$ be the number of explanatory variables in the existing regression and $k_1$ the total number as a result of adding $(k_1 - k_0)$ further explanatory variable. Further, suppose $R_0^2$ is the value corresponding to the existing regression and $R_1^2$ that corresponding to the regression with the additional explanatory variables. Then the test statistic can be written:

**Table 8.7** ANOVA Tables for Added Explanatory Variables

| *Source of variation* | *Sum of squares* | *Degrees of freedom* | *Mean square* | *F-ratio* |
|---|---|---|---|---|
| A. *Addition of one explanatory variable* | | | | |
| Regression on: | | | | |
| *SUG, TDP* | 119,654 | 2 | 59,872 | |
| Addition of *AGEMP* | 3,368 | 1 | 3,368 | 0.99 |
| *SUG, TDP, AGEMP* | 123,022 | 3 | | |
| Residual | 57,950 | 17 | 3,409 | |
| Total | 180,972 | 20 | | |
| B. *Addition of a number of explanatory variables* | | | | |
| Regression on: | | | | |
| *SUG, TDP, AGEMP* | 123,022 | 3 | 41,007 | |
| Addition of *MLK, BUT, FAT* | 7,607 | 3 | 2,536 | 0.71 |
| All variables | 130,629 | 6 | | |
| Residual | 50,343 | 14 | 3,596 | |
| Total | 180,972 | 20 | | |

*Source:* Table 8.5

$$\left[\frac{n - k_1 - 1}{k_1 - k_0}\right]\left[\frac{R_1^2 - R_0^2}{1 - R_1^2}\right] \tag{8.6.1}$$

This test statistic, under the null hypothesis, has the distribution $F(k_1-k_0, n-k_1-1)$.

These ANOVA tests are framed in terms of testing the significance of the increase in the explained sum of squares, but they are also equivalent to testing the significance of regression coefficients. Part A of Table 8.7 in effect tests the significance of the regression coefficient of *AGEMP* in the regression on *SUG, TDP* and *AGEMP* and part B tests the hypothesis that all three coefficients of *MLK, BUT, FAT* are zero in the regression on all variables. In the former case, we have already described in Section 8.3 the $t$-ratio test for the significance of a regression coefficient. For a test on the significance of an individual regression coefficient, the *F*-test of the ANOVA table and the *t*-test of a regression coefficient always lead to the same conclusion (for tests at the same level of significance).

$R^2$, the squared coefficient of multiple correlation, has been defined as ESS/TSS (Section 8.4). This can be re-expressed in terms of RSS and TSS:

$$R^2 = \frac{(\text{TSS} - \text{RSS})}{\text{TSS}} = 1 - \frac{\text{RSS}}{\text{TSS}} \qquad (8.6.2)$$

An alternative measure of goodness of fit is called $\bar{R}^2$. Specifically

$$\bar{R}^2 = 1 - \frac{(n/(n-k-1))\ \text{RSS}}{(n/(n-1))\ \text{TSS}}$$

which can be simplified to:

$$\bar{R}^2 = \frac{(n-1)}{(n-k-1)} R^2 - \frac{k}{(n-k-1)}$$

$$= 1 - \frac{(n-1)[1-R^2]}{(n-k-1)} \qquad (8.6.3)$$

$\bar{R}^2$ can be looked upon as a 'better' estimate of the model goodness of fit. Unlike $R^2$, it does not inevitably increase as the number of included explanatory variables increases. It is possible for $\bar{R}^2$ to have a negative value, even though it is an estimate of something essentially positive; however, this only happens if $R^2 < k/(n-1)$.

## 8.7 MULTICOLLINEARITY

We have so far evaded a difficult conceptual and practical problem in multiple regression analysis: that of 'multicollinearity'. Multicollinearity is concerned with the statistical behaviour of the values of the explanatory variables. When two or more explanatory variables are correlated with each other they are said to be 'multicollinear' and the data are said to exhibit 'multicollinearity'. Multicollinearity is therefore almost an inevitable feature of non-experimental data sets. As we shall see, multicollinearity—depending upon its degree—can cause several practical problems. In particular, the standard errors of regression coefficients may be very large and so the precision of estimates of model coefficients very low, that is, the confidence interval range is relatively large. As a result it may be difficult to disentangle the marginal effects of individual explanatory variables upon the dependent variable.

**Table 8.8** Country-Mortality Data: Correlation Matrix (extract)

| | | | | |
|---|---|---|---|---|
| *TDP* | 1.00 | | | |
| *SUG* | 0.79 | 1.00 | | |
| *AGEMP* | −0.30 | −0.60 | 1.00 | |
| *IHDMR* | 0.73 | 0.80 | −0.53 | 1.00 |
| | *TDP* | *SUG* | *AGEMP* | *IHDMR* |

The source of multicollinearity, its effect on the estimation process and its implications for interpretation are well exemplified in the country-mortality data set. As a basis for discussion we shall assume there are just the three explanatory variables, *SUG*, *TDP* and *AGEMP*. The correlation coefficients between all pairs of variables (including the dependent variable) are shown in Table 8.8. Such a table is known as a correlation matrix. It has the same structure as a table of mileages between towns; like the mileages, there is no point in showing the top right-hand part of the table since it would merely repeat the bottom left-hand part. Some of the correlations in the matrix are quite high and the correlation coefficient between two explanatory variables, *TDP* and *SUG*, is 0.79, so problems associated with multicollinearity may be expected. Unfortunately, the absence of high correlations between pairs of variables does not in itself rule out the possibility of multicollinearity problems, since the latter can arise as a result of relatively weak correlations within a set of explanatory variables.

The effects of multicollinearity are best seen by looking at how the regression relation changes as different explanatory variables are included or excluded. Table 8.9 shows the regression coefficients (with the *t*-ratios in brackets) for all possible sets of explanatory variables, row by row. The $R^2$ and $\bar{R}^2$ values are also included. In cases 1 to 3, there is only one explanatory variable included, in cases 4 to 6 two out of three are included and finally case 7 shows the results with all the chosen variables.

The essence of the multicollinearity problem is brought out by comparing cases 1 to 3 and case 7. Case 7 shows the highest $R^2$ value (which incidentally is inevitable since $R^2$ never decreases as more explanatory variables are added to a regression). Yet in case 7, none of the regression coefficients is significant at any conventional level, while in cases 1 to 3 the coefficients are significant and are much more precisely estimated. If only the case 7 regression had been carried out, the naive and mistaken conclusion might have been that none of the three explanatory variables was significant.

What has happened is this. The *TDP* and *SUG* variables are highly

**Table 8.9** Country-Mortality Data: Regression Results for Subsets of Variables

| | Regression coefficients (with t-ratios) | | | | |
|---|---|---|---|---|---|
| | *TDP* | *SUG* | *AGEMP* | $R^2$ | $\bar{R}^2$ |
| Case 1 | 0.37 (4.6) | . . | . . | 0.53 | 0.51 |
| Case 2 | . . | 2.7 (5.7) | . . | 0.63 | 0.62 |
| Case 3 | . . | . . | −6.2 (2.7) | 0.28 | 0.25 |
| Case 4 | 0.13 (1.2) | 2.0 (2.6) | . . | 0.66 | 0.62 |
| Case 5 | 0.32 (4.2) | . . | −4.0 (2.3) | 0.64 | 0.60 |
| Case 6 | . . | 2.5 (4.2) | −1.0 (0.5) | 0.64 | 0.60 |
| Case 7 | 0.18 (1.5) | 1.4 (1.4) | −2.1 (1.0) | 0.68 | 0.62 |

correlated with each other and the regression analysis finds it difficult to disentangle with any precision their separate effects on the dependent variable though, as cases 1 and 2 show, the dependence is undoubtedly there. The technical explanation is that the standard error of a regression coefficient depends not only upon the random component variance but also upon a function of the explanatory variable values; it can be shown that the value of this function increases in the presence of multicollinearity.

Multicollinearity has been discussed in the context of a particular set of data but the points made are of general applicability. The severity of its presence may usually be identified by inspection of the correlation matrix and the effect on the regression coefficients and their standard errors as a result of changing the mix of explanatory variables. Another method is to identify variables which may be a source of multicollinearity by regressing in turn each explanatory variable on all the rest; a high $R^2$ indicates a warning sign.

A change in the coefficient of an individual explanatory variable as others are added to or subtracted from the model is inevitable if there is any degree of multicollinearity. Now *if the model is correct*, the regression coefficient is an unbiased estimator of the structural parameter (Section 8.2). Conversely, if we are estimating an incorrect model due to the omission of some relevant explanatory variable, the regression coefficients will generally no longer be unbiased estimators for the variables which *are* included.

It remains to be considered what can be done if a significant degree of multicollinearity has been identified. First, it must be said that there is no easy solution. The difficulty stems from intercorrelations among the explanatory variables, that is to say from the available data. One obvious possibility is to seek more data, but this would be useless if the additional data merely repeated the intercorrelation pattern of the original data. A second possibility is to drop from the regression analysis variables which are a source of the multicollinearity. In our example it might imply reporting the case 5 regression in which *SUG* is excluded. This suggestion, however, is of no comfort if one wishes to assess the marginal dependence on each explanatory variable. A third possibility is to produce one or more derived aggregative variables based upon the given explanatory variables. For example, to deal with the intercorrelation between *TDP* and *SUG* (respectively consumption of total dairy products and sugar), they might be replaced by a weighted average of the two; the weights could be proportional to their calorific content if there were a theory that this is an underlying cause of heart disease.

An extension of this last possibility is the multivariate procedure of applying principal components analysis to the explanatory variables. This leads to a sequence of principal component variables each of which is a weighted combination of the explanatory variables. This approach is discussed in Section 11.10.

To sum up, there are two consequences of multicollinearity which should be emphasised. First, its presence allows several models which include different sets of explanatory variables to be broadly consonant with the data. Secondly, there can be no clear cut interpretation of the regression coefficients as measures of marginal effects. Moreover, the coefficients may well be unstable in the sense that small data revisions could have a disproportionate effect on the calculated coefficients.

## 8.8 SOURCES OF BREAKDOWN OF REGRESSION MODEL

We now consider some of the ways in which a data set may fail to match the formal regression model at (8.2.1). First, however, we need to make explicit the full implications of the statement that each value of the random component is modelled as a random observation from the distribution $N(0, \sigma^2)$. This in turn implies the following model assumptions:

(i) each value of the random component is statistically independent of every other value;
(ii) the variance of the random component is constant (equal to $\sigma^2$);
(iii) the probability distribution of the random component is independent of the values taken by the explanatory variables.

**(8.8.1)**

These assumptions (8.8.1) together with the specification at (8.2.1) constitute what we shall call the *basic regression model.* If, for a particular data set, one or more of the assumptions do not hold, the model is said to be misspecified.

In subsequent sections of the present chapter, attention will be focused upon the checking of the specification of the systematic component. We shall consider some informal graphical checks of the random component specification by examination of the residuals, but a more formal discussion is deferred until Chapter 9.

## 8.9 POST FITTING CHECKS

Having completed the first steps of specifying and fitting a proposed model, we need to stand back and appraise the results. The statistical assumptions which have been made are catalogued above and we will be seeking any evidence that some (or all) of these are violated by the fitted model which is currently being entertained. The outcome of this post fitting check will perhaps also suggest ways of correcting for these failures and generally improving the model by appropriate respecification.

It will be convenient to set out under the following headings what we may hope to detect (the section references indicating the source of further consideration):

(i) Evidence of incorrect specification of the systematic component, by the omission of useful explanatory variables (Section 8.10).

(ii) Evidence that individual explanatory variables do not enter linearly into the systematic component (Section 8.11).

(iii) Suggestions that the individuals whose values have been used in the fitting process do not constitute a homogeneous group; this may mean either that there are a few individuals who are outliers and should be excluded from the data used to make the fit, or that there are two or more distinct groups whose existence needs to be taken into account in specifying the systematic component (this section and Chapter 9).

(iv) Breakdowns of the assumptions concerning the random component, in particular the assumptions of *normality*, of *independence* and of *constant variance* (Chapter 9).

To detect unambiguously even one of these failures may prove difficult and indications gathered under one heading may have been caused by trouble elsewhere. Furthermore, if evidence of a problem can be found, it may not always prove easy to put it right. Perfection

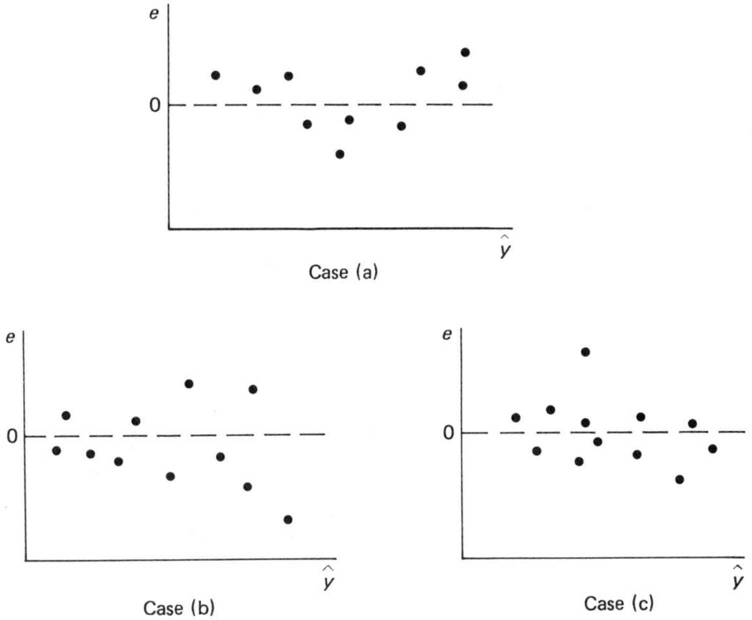


**Figure 8.1** Patterns of Residuals

cannot be achieved with real data, but we can often carry out some patching.

Our first line of enquiry focuses upon the residuals and the pattern of their relationship to the fitted values of $Y$. It will be recalled from (4.11.5c) that it is an inevitable consequence of regression analysis (bivariate or multiple) that the residuals are uncorrelated with the fitted values. However, this does not imply that a plot of one against the other carries no useful information. Patterns in these plots such as those shown in Figure 8.1 indicate respectively that:

(i) a linear systematic component has been used to model an essentially curvilinear relation. The failure is illustrated by the curved shape in Case (a) which reflects the residual values being persistently negative in the middle of the range;

(ii) the random component does not have a constant variance (Case (b));

(iii) there is one or more outliers (Case (c)).

In our experience these plots are, in practice, rarely as clear-cut as textbook diagrams suggest, but they are usually straightforward to produce as part of the output from the computer program and are worth having.

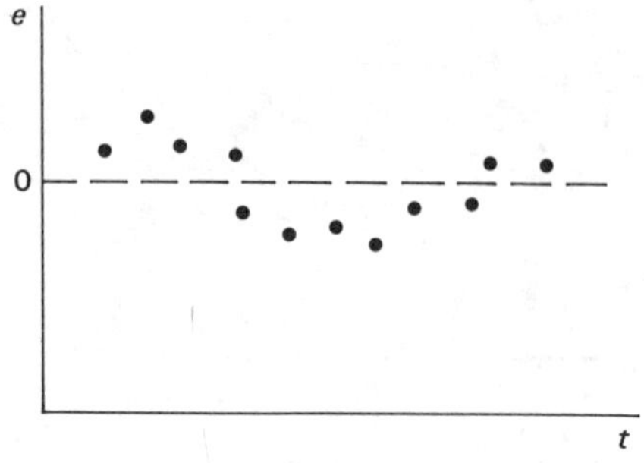


**Figure 8.2** Serial Correlation of Residuals

If the data set possesses a natural ordering, as happens when the data are observed over time, then a plot of the residuals against time may indicate a failure of the assumption that the random component is satisfactorily modelled as a sequence of *independent* drawings from a normal distribution. This failure produces *serial correlation* in the residuals and Figure 8.2 shows, again in idealised form, what to look for here. The ordered residuals typically show runs of positive or negative values.

Finally, a normal probability plot of the residuals (see Section 6.8) may draw attention to any failure of the normality assumption; other model assumptions being valid, the failure will usually show up in terms of a few outliers at one end or the other. The individuals corresponding to these points might then be flagged for closer study and the model refitted after excluding them.

Some work by Hoaglin and Welsch (1978) suggests another useful diagnostic line of enquiry. This is particularly directed towards identifying individuals whose values have a crucial effect in determining the coefficients of the regression equation. Removal of such individuals from the analysis is likely to change substantially the coefficients or their standard errors. Individuals whose residual values are outliers are one possible instance. Less obviously, individuals which have relatively extreme explanatory variable values may also have a substantial effect.

The diagnostic check amounts to calculating, for each individual, an 'influence statistic' which, if it is unduly large, serves as a warning signal. As a rule of thumb, unduly large can be taken to mean greater than $2k/n$ where, as before, $k$ is the number of explanatory variables. The effect, if any, may be assessed by refitting the model after excluding the individual.

To illustrate this effect, we computed the 'influence statistic' for each country for the regression on *TDP*, *SUG* and *AGEMP* (for which the regression coefficients are given in Table 8.9). The highest was for Finland and exclusion of this data point from the regression analysis radically changed some of the coefficients: for *TDP* from 0.18 to 0.37, and for *SUG* from 1.41 to 0.33. Judging in terms of the residuals, Finland would not be considered an outlier. By contrast, the exclusion of Austria, whose residual was much the same size as Finland but whose influence statistic was very small, had little effect on the coefficients.

Few programs provide these statistics as part of the output. One which does is Minitab (see Appendix B), which automatically flags individuals according to the rule above.

## 8.10 CHOICE OF EXPLANATORY VARIABLES

We have intentionally left this topic until now. The purpose of the discussion so far has been directed towards fitting, interpreting and checking some specified model. There are a number of issues concerning specification which now need to be discussed however, and it is to these which we now turn.

The starting point for any model building, as we have stressed at several places already, has to be some more-or-less precise theory. There will also be some objectives and the formulation of the systematic component must arise from an interplay between theory, objectives and data. This interplay will not generally be sufficient to give precise answers to such questions as:

(i) which, and how many, variables should be used in specifying the systematic part?—often there will be several which fit the theory and the objectives and the choice between them leads to several plausible specifications of the systematic part;

(ii) given a possible set of variables, in what form should they be used—'raw' or mathematically transformed?—theory may give a clue, but some preliminary and some post fitting checks may help;

(iii) whether a specified model can be regarded as applying to the data for all individuals or whether some split into two or more 'regimes' or groups is necessary.

These questions will serve to provide three headings for the discussion of specification, namely *choice of variables*, *transformations* and *dummy variables*. The first two of these are the subject of this and Section 8.11 respectively; the third is postponed until the next chapter.

The interplay referred to above, between theory and objectives will, we may assume, narrow down the list of potential explanatory variables. A choice from these to formulate the systematic part will often be partly pre-empted; in an analysis which has a potential policy application, then it will be important to include variables which can be (and, preferably, have been) controlled, for example tax rates or subsidy levels fixed by economic policy makers. If the objective is to make predictions of the dependent variable, only those explanatory variables whose values can be determined in some way will be useful. Unless the theory well identifies the explanatory variables there may still be a wide choice, however. One line of attack is to make a selection by means of purely data based criteria, some of which we shall now consider.

In practice there may be an embarrassingly large number of models to choose between and examples of this were met in Section 8.7. In fact, with $k$ explanatory variables there are $2^k - 1$ different possible choice sets of variables. Some programs allow the automatic calculation of all possible regression equations but, without any clear guidance for sieving and selection, this only makes the analyst's task more difficult.

A procedure which automatically produces a smaller number of models to choose from is known as *forward selection.* In forward selection, the single best explanatory variable is chosen and then a succession of models is fitted each with one more explanatory variable than the last. At each stage, the variable selected for addition is the one which contributes most to increasing the explained sum of squares. For the country-mortality data, the forward selection procedure adds variables in the following order: *SUG, TDP, MLK, AGEMP, FAT, BUT.*

Forward selection has the advantage that it reduces the models to be scrutinised from $2^k - 1$ (63 in our example) to $k$ (6). A possible objection to the forward selection procedure is that it rules out possible combinations of explanatory variables which on statistical grounds are equally convincing. For example, the three variable forward selection regression equation is:

$$\widehat{IHDMR} = -61.10 + \underset{(2.31)}{1.77}\ SUG + \underset{(1.75)}{0.50}\ TDP - \underset{(1.39)}{0.40}\ MLK$$
$$R^2 = 0.70$$

Compare this with the following:

$$\widehat{IHDMR} = -44.87 + \underset{(2.66)}{0.29}\ TDP - \underset{(1.53)}{2.48}\ BUT + \underset{(2.57)}{1.80}\ FAT$$
$$R^2 = 0.66$$

The latter $R^2$ is slightly lower, but the coefficients are on the whole no less well determined.

It may be felt that these difficulties of judging between competing models are avoided by including all explanatory variables. For the country-mortality data, this is:

$$\begin{aligned}\widehat{IHDMR} = &- 28.95 + \underset{(0.98)}{1.07}\ SUG + \underset{(1.31)}{0.43}\ TDP - \underset{(0.73)}{0.26}\ MLK \\ &- \underset{(0.55)}{1.31}\ AGEMP + \underset{(0.61)}{0.62}\ FAT - \underset{(0.47)}{0.94}\ BUT \\ &\qquad\qquad R^2 = 0.72\end{aligned}$$

As a result of multicollinearity, none of the variables now attains conventional levels of significance.

The fact is that there are no purely statistical criteria which, by themselves, will lead to an unequivocal choice of model. Some comfort may be derived from observing which explanatory variable coefficients have a consistency of sign irrespective of the other explanatory variables in the model. From an examination of all 63 variants for the country-data model specification, it was found that the *TDP*, *SUG*, and *FAT* coefficients were consistently positive and the *AGEMP* coefficient consistently negative. So although we cannot be confident about the magnitude of their marginal contributions to *IHDMR*, the direction of their effect for all models is fairly clear. Bearing in mind that it is people and not countries who consume dairy products and sugar and who die of heart disease, the results of this aggregate analysis may serve at least to justify relating it to further research at a micro level.

The data set used to illustrate this chapter was suggested by the work of Knox (1977) and Segall (1977), which explored, though in a rather different fashion, possible connections between mortality and diet.

## 8.11 TRANSFORMATIONS AND FUNCTIONAL FORM

'Multiple *linear* regression model' is the full title of the model under discussion in the present chapter. As we remarked in Section 4.1, 'linear' refers to the role of the parameters in the model; a linear model does not contain terms such as $\beta_1^2$ or $\beta_2\ \beta_3$ or $\log \beta_4$. The model does not exclude the possibility of the variables, dependent or explanatory, being transformed in a non-linear fashion. Thus, the methods discussed so far can equally be applied to models such as

$$y_r = \beta_0 + \beta_1 x_{1r}^2 + \beta_2 x_{2r}^3 + u_r$$
$$\text{or } \log(y_r) = \beta_0 + \beta_1 \log(x_{1r}) + \beta_2 \log(x_{2r}) + u_r$$

These models are estimated simply by prior transformation of the data values and by calculating the regression equation using the transformed values. Computer programs normally allow for such transformations of the data to be made before proceeding with the regression analysis.

Different transformations of the same explanatory variable can enter the same model. The model $y_r = \beta_0 + \beta_1 x_r + \beta_2 x_r^2 + u_r$ is based on a two variable data matrix, but an additional explanatory variable has been generated by a square transformation. The right-hand side is a quadratic functional form of the variable $X$ and the model would be appropriate if there were thought to be a quadratic relation between the two variables. Models where basic variables enter as non-linear transformations are known as *curvilinear regression models.*

Sometimes, the systematic component of a model may be formulated in such a way that the parameters are non-linear, but a prior transformation of the model itself will yield an expression linear in the parameters which can in turn be estimated by the standard regression method. An example is the widely used constant elasticity function in economics. With two explanatory variables, this function takes the form $y = \alpha x_1^{\beta_1} x_2^{\beta_2}$, implying that changes in $Y$ are in constant proportion to changes in $X_1$ or $X_2$. In this theoretical model, $\alpha$, $\beta_1$ and $\beta_2$ are parameters to be estimated and $Y$, $X_1$, and $X_2$ are economic variables. (As an example, $Y$ might be an individual firm's output, $X_1$ a measure of the labour input and $X_2$ of the capital input). Now if we take logarithms of both sides, the relation becomes:

$$\log y = \log \alpha + \beta_1 \log x_1 + \beta_2 \log x_2$$

This is now linear in the parameters (with the trivial exception that the constant term is expressed as a logarithm) and the addition of a random component leads to a regression model.

Sometimes, the background theory, as in the above example from economics, may suggest the kind of transformations to be applied to some or all of the variables. More often, while the theory or background knowledge may suggest the existence of a statistical relationship, it provides no guidance as to the functional form: whether it is linear, logarithmic, quadratic and so on. One way out of the difficulty may be to fit a small number of models of different functional form and choose between them on the basis of either formal significance tests or descriptive procedures such as residual plots. For example, a quadratic form may be compared to a linear form by testing the significance of adding the term in $X^2$.

The competition is often between the normal linear model and the log-linear model in which some or all the variables have been transformed logarithmically. The log-linear model may be a plausible candidate, partly because of its proportionate effect interpretation and partly because it tends to remove skewness.

The *logit transformation* is sometimes advocated when the dependent variable is expressed as a proportion, $p$, say. The logit transformation replaces the values $p$ of the dependent variable by their transformed values $\ln(p/(1 - p))$. This has the effect of allowing the dependent variable to take any value in the range $(-\infty, \infty)$ rather than be constrained to the range (0, 1). Thus the regression model, with a normally distributed random component, may be better justified.

## 8.12 CONCLUSION

This chapter has dealt with a number of aspects of regression analysis, both theoretical and practical, and further issues remain to be discussed in Chapter 9. The reader may well be left with the impression that a large amount of research effort is necessary simply to determine a satisfactory specification of a single regression equation.

To put the size of the problem into perspective, the following points may be made. First, the amount of effort to be expended must depend upon the purposes of the analysis and the resources and time available. Secondly, the analyst is not generally taking on a series of new unfamiliar investigations each day or each week. Therefore, on encountering some new data he will be guided by previous experience in deciding choice of variables and how far model assumptions are worth checking. Finally, it may be that the data cannot bear any detailed analysis because, for example, it is known to be prone to errors of measurement or to be a less than satisfactory measure of the theoretical concepts. In such a case, some simple analysis may still be useful in suggesting further lines of enquiry.

## EXAMPLES

**E 8.1** Using the information in Table 8.4, calculate the 95% confidence intervals for the parameters of *TDP*, *SUG* and *AGEMP*.

**E 8.2** Given that the standard deviations of *TDP*, *SUG*, *AGEMP* and *IHDMR* are respectively 187, 27.8, 8.2, 95.1, verify (using result (8.3.4)) the entries in the column headed *B(STD. V)* of Table 8.4.

**E 8.3** Using the information in Table 8.5, write down the estimates of the random component variance for the models specified by the four sets of explanatory variables.

**E 8.4** (i) What are the values of $\nu_1$ and $\nu_2$ for testing the significance of the case II and case IV regressions for which ANOVA tables are given at Table 8.5?
(ii) What are the corresponding critical values of the *F*-ratio for tests at the 5% level of significance?
(iii) For each case, test the hypothesis that the structural parameters are all zero.

**E 8.5** Section 8.6 shows, in particular, how the significance of adding *AGEMP* to *TDP* and *SUG* can be established.
Show, using the information in Table 8.7, that the same conclusion may be obtained by using changes in $R^2$ (see result (8.6.1)). (Note that, in practice, since the differences between $R_1^2$ and $R_0^2$ may be small, problems can arise if the $R^2$ values are over-rounded.)

## FURTHER WORK

**F 8.1** As noted in Section 8.10, there are a number of possible models to explain the variation in *IHDMR*. Fit some of these and compare the results with those reported in the text.

**F 8.2** For your chosen model or models, examine the effect of deleting individuals which, on the basis of residual plots, appear to be outliers.

*This chapter depends upon: Chapter 8*

# 9

# Time Series and Further Regression Analysis

## 9.1 THE NATURE OF TIME SERIES

This chapter will take the discussion of the multiple regression model further and, in particular, will pay some attention to the modelling of data which are in the form of a time series.

By contrast with the country mortality data used in the last chapter, time series data possess a natural ordering which adequate description (and modelling) may require us to acknowledge. The fact that successive data values relate to successive time periods may mean that the value taken at any one period is dependent, to some extent, upon the values taken at one or more of the preceding time periods. Often, although by no means always, this dependence results in a roughly systematic increase or decline over several periods if not over the whole time period represented by the data. Examples of time series variables are the population of a country year by year, a sequence of monthly rainfall figures, and the daily closing prices on the New York stock exchange.

A useful preliminary step in time series analysis will often be to make a plot of the values of an individual time series variable on a graph whose horizontal axis represents time. Figure 9.1 shows some typical plots of this kind. In Case A, the values show a general tendency

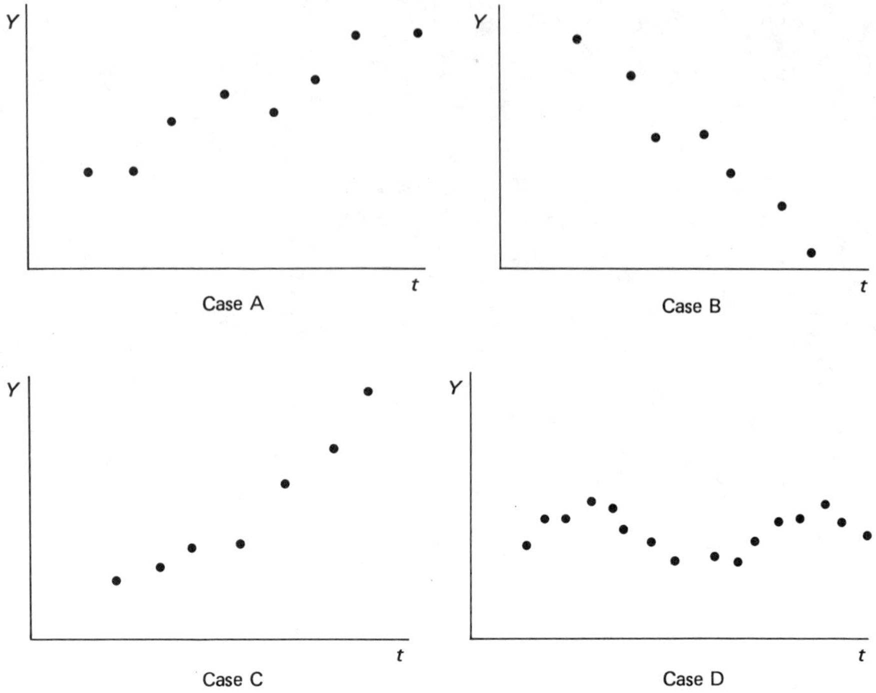


**Figure 9.1** Examples of Idealised Time Series Plots

to rise over time and are well fitted by a linear regression line with time as the explanatory variable. Such a time series is said to exhibit a positive linear trend. Case B is similar, except that the trend is negative. In Case C, the trend is curvilinear and it could be that the values broadly exhibit a constant proportional growth from one period to the next. Case D is more complex in that the time series exhibits an oscillatory pattern; an example would be average monthly temperatures plotted over a period of years.

The word *trend* has been deliberately left somewhat imprecise. Any precise definition must depend upon fitting a particular model to the data, for example, a bivariate regression model.

Before proceeding to regression and other methods of time series analysis, some notation is required. We shall, as before, use the symbols $Y$, $X_1$, . . . , $X_k$ to denote the dependent and explanatory variables. In keeping with common practice, we will use $t$ to denote a particular point of time and therefore $y_t$, $x_{1t}$, . . . , $x_{kt}$ to denote values of the variables at time $t$. We shall also follow the usual practice in time series analysis of indicating the total number of observations by

$T$ (rather than $n$). The basic regression model for time series data corresponding to the specification at (8.2.1) is therefore rewritten as:

$$y_t = \beta_0 + \beta_1 x_{1t} + \ldots + \beta_k x_{kt} + u_t$$
$$u_t \sim \text{NID}(0, \sigma^2) \qquad t = 1, \ldots, T \qquad \textbf{(9.1.1)}$$

We will continue to regard this, together with the assumptions at (8.8.1), as constituting the basic regression model. The notation NID(0, $\sigma^2$) is introduced to emphasise the assumption of the basic model that each value of the random component is statistically independent of every other value (Section (8.8.1)); NID stands for normally and independently distributed.

Table 9.1 shows one of the data sets which will be used for illustrative purposes in the present chapter. A plot of any of the variables in Table 9.1 against time would exhibit the kind of features shown in Figure 9.1. We now make more explicit the notion of dependence within a time series in terms of the variable $RC$.

First consider Table 9.2 which sets out in the row $RC_t$, the values of $RC$ from Table 9.1 and in the row below, labelled $RC_{t-1}$, the corresponding value for the preceding point of time. These latter values are known as the one period *lagged values*. The subscript $t-1$ is standard notation to denote one period lagged values. Note that the first entry in the second row is not known. One way of exploring the dependence between successive values would be to regard these numbers as pairs of values of two different variables, $RC_t$ and $RC_{t-1}$, and calculate the correlation coefficient between them. This procedure yields what is called the *autocorrelation coefficient of lag 1* or, alternatively, the *first order autocorrelation coefficient*. Its value (based on $T - 1 = 23$ observations) is 0.75. In a similar fashion, one could calculate the *second order autocorrelation coefficient* by relating each value of a time series to the corresponding value two periods previously; notationally, this would be relating the variable $RC_t$ with $RC_{t-2}$ (the *two period* lagged variable); for the $RC$ variable its value turns out to be 0.59. Where dependences of this kind exist, the time series is said to be *autocorrelated*.

After presenting in the next section the preliminary analysis of the data matrix at Table 9.1, we shall proceed in Section 9.3 to investigate the implications of time dependence for regression model estimation. Later sections will review other aspects of regression analysis (which may be relevant to both cross section and time series data). The final Section 9.11 returns to time series analysis to consider how a single univariate time series may be modelled solely in terms of its own values.

**Table 9.1** Money, Income and Interest Rates

| | | *M1* | *Y* | *RC* | *RLA* | *T* | *DREG* | *QII* | *QIII* | *QIV* |
|---|---|---|---|---|---|---|---|---|---|---|
| 1969 | I | 9,637 | 14,829 | 8.42 | 7.98 | 1 | 0 | 0 | 0 | 0 |
| | II | 9,360 | 14,900 | 9.11 | 9.14 | 2 | 0 | 1 | 0 | 0 |
| | III | 9,253 | 14,980 | 9.15 | 9.57 | 3 | 0 | 0 | 1 | 0 |
| | IV | 9,419 | 15,085 | 8.87 | 9.22 | 4 | 0 | 0 | 0 | 1 |
| 1970 | I | 9,271 | 14,973 | 8.60 | 9.10 | 5 | 0 | 0 | 0 | 0 |
| | II | 9,400 | 15,359 | 9.23 | 8.13 | 6 | 0 | 1 | 0 | 0 |
| | III | 9,372 | 15,362 | 9.28 | 7.54 | 7 | 0 | 0 | 1 | 0 |
| | IV | 9,491 | 15,540 | 9.59 | 7.18 | 8 | 0 | 0 | 0 | 1 |
| 1971 | I | 9,769 | 15,404 | 9.38 | 7.53 | 9 | 0 | 0 | 0 | 0 |
| | II | 9,607 | 15,649 | 9.24 | 6.66 | 10 | 0 | 1 | 0 | 0 |
| | III | 9.807 | 15,950 | 9.05 | 5.82 | 11 | 0 | 0 | 1 | 0 |
| | IV | 9,832 | 15,957 | 8.59 | 4.82 | 12 | 0 | 0 | 0 | 1 |
| 1972 | I | 10,107 | 16,031 | 8.45 | 4.80 | 13 | 0 | 0 | 0 | 0 |
| | II | 10,372 | 16,295 | 8.98 | 5.13 | 14 | 0 | 1 | 0 | 0 |
| | III | 10,262 | 16,151 | 9.46 | 7.62 | 15 | 0 | 0 | 1 | 0 |
| | IV | 10,416 | 16,803 | 9.63 | 7.91 | 16 | 0 | 0 | 0 | 1 |
| 1973 | I | 10,135 | 17,528 | 9.95 | 10.09 | 17 | 1 | 0 | 0 | 0 |
| | II | 9,501 | 17,301 | 13.20 | 9.19 | 18 | 1 | 1 | 0 | 0 |
| | III | 9,953 | 17,503 | 11.27 | 11.73 | 19 | 1 | 0 | 1 | 0 |
| | IV | 9,847 | 17,455 | 11.86 | 14.11 | 20 | 1 | 0 | 0 | 1 |
| 1974 | I | 9,278 | 16,620 | 13.37 | 15.66 | 21 | 1 | 0 | 0 | 0 |
| | II | 8.870 | 17,779 | 14.72 | 13.33 | 22 | 1 | 1 | 0 | 0 |
| | III | 8,885 | 18,040 | 15.37 | 13.13 | 23 | 1 | 0 | 1 | 0 |
| | IV | 9,133 | 17,827 | 16.77 | 12.63 | 24 | 1 | 0 | 0 | 1 |

*Key*: *M1* = Money circulation (definition *M1* of Bank of England)
*Y* = income (total final expenditure at 1970 prices)
*RC* = yield on $2\frac{1}{2}$% British Government Consols
*RLA* = 3-month municipal authority rate

*Source*: Abstracted from data in T.C. Mills, 'The functional form of the demand for money', *Applied Statistics*, No. 1, 1978, except for 1973II *M1* value (see F9.1 at end of chapter).

**Table 9.2** Lagged Values of Variable *RC*

| $t$ | *1* | *2* | *3* | ... | *22* | *23* | *24* |
|---|---|---|---|---|---|---|---|
| $RC_t$ | 8.42 | 9.11 | 9.15 | ... | 14.72 | 15.37 | 16.77 |
| $RC_{t-1}$ | .. | 8.42 | 9.11 | ... | 13.37 | 14.72 | 15.37 |
| $RC_{t-2}$ | .. | .. | 8.42 | ... | 11.86 | 13.37 | 14.72 |

*Source*: Table 9.1

## 9.2 THE DATA SET AND PRELIMINARY ANALYSIS

The data of Table 9.1 arise from an investigation of some economic hypotheses concerning the relationship between the demand for money (the dependent variable) and national income and the level of interest rates. This data matrix will form the illustrative material for much of the discussion in this chapter; we shall refer to it as the 'money data'.

Economic theory suggests that, *ceteris paribus*, the demand for money will be positively related to national income and negatively related to interest rates. In the table, the demand for money is measured by the variable *M1* (a particular Bank of England definition of money in circulation), *Y* is national income and there are two interest rates: *RC*, the yield on $2\frac{1}{2}\%$ Government Stock and *RLA*, the three-month rate for municipal authority borrowing.

Table 9.3 exhibits the results of fitting a number of models to the data. (The column headed *d* will be explained in Section 9.3.) Case A, with explanatory variables *Y*, *RC* and *RLA*, will be the main focus of attention in the ensuing discussion. It will be noted that, in all cases, the signs of the coefficients of the explanatory variables are consistent with the economic hypotheses.

In Case B, the time variable *T* has been added with the time periods being given successively the integral values 1, 2, ..., 24. Since the random component in regression analysis will reflect the influence of any relevant variables omitted from the systematic component, the inclusion of time as an additional explanatory variable has the effect of accounting for a common linear trend component of any omitted variables. This is often a device of last resort to introduce at least the trend in variables which cannot be measured: for example, technological change in economic situations. However, insofar as there is a linear trend in the existing variables, further multicollinearity is likely to be introduced. The reader can confirm from Table 9.3 that this appears to have happened with our data.

**Table 9.3** Summary of Regression Results

| | *Dependent variable* | *Explanatory variables* | | | | | | | | |
|---|---|---|---|---|---|---|---|---|---|---|
| | | $M1_{t-1}$ | $Y$ | $RC$ | $RLA$ | $T$ | *Cons* | $R^2$ | $\bar{R}^2$ | $d$ |
| Case A | *M1* | . . | 0.46 | −230 | −25.5 | . . | 4880 | 0.75 | 0.72 | 1.53 |
| | | | (6.0) | (5.3) | (1.0) | | (5.0) | | | |
| Case B | *M1* | . . | 0.15 | −2.52 | − 0.18 | 52.7 | 9360 | 0.81 | 0.77 | 1.24 |
| | | | (1.0) | (6.3) | (0.8) | (2.4) | (4.5) | | | |
| Case C | *M1* | 0.41 | 0.30 | −140 | −32.6 | . . | 2590 | 0.84 | 0.80 | not applic. |
| | | (3.0) | (3.4) | (3.0) | (1.5) | | (2.3) | | | |
| Case D | ln *M1* | | ln *Y* | ln *RC* | ln *RLA* | | Cons | | | |
| | | . . | 0.80 | −0.30 | − 0.01 | . . | 3.57 | 0.75 | 0.71 | 1.54 |
| | | | (5.8) | (5.3) | (0.4) | | (3.3) | | | |

*Source*: IDA output based upon Table 9.1

In Case C, the one period lagged values of *M1* are included as an explanatory variable. A model of this type with lagged values of the dependent variable in the systematic component can be justified by an economic theory of partial adjustment of the money supply from one period to the next. Case D is the log-linear version of the model of Case A and is based upon a hypothesis of proportionate effect; see Section 8.11.

## 9.3 AUTOCORRELATION IN THE RANDOM COMPONENT

We have so far proceeded on the assumption that the random component of the multiple regression model may be modelled as a set of *independent* observations from a normal distribution, notationally $u_t \sim \text{NID}(0, \sigma^2)$ as at (9.1.1). If this assumption is incorrect the model is said to be misspecified. Let us consider the implications of this kind of misspecification, in particular:

(i) in what circumstances might independence be implausible?
(ii) what evidence may be examined in order to decide that it *is* untenable?
(iii) what are the consequences of ignoring evidence of dependence?
(iv) what remedies are available if the consequences are felt to be unacceptable and how effective (and costly) are they likely to be?

Violation of the independence assumption is extremely likely with time series data. We have already noted that the values of a time series variable are often dependent upon earlier values. Furthermore, since the random component in the regression model reflects the effect of time series variables omitted from the specification of the systematic component, it may be unrealistic to model the random component as if its values were mutually independent. In other words, the assumption of an autocorrelated random component may be more realistic.

In practice, the first step is to calculate the residuals from the chosen regression equation. Taking Case A of Table 9.3 as our example, these are set out in Table 9.4, and examination of them is the key to detecting failure of the assumption of independence of the random component. Informally, if this assumption were valid, one would expect the signs of the residuals to change 'randomly'. Examination of Table 9.4 shows some tendency for signs to persist from one time period to the next; in other words one positive value tends to be followed by another and likewise negative follows negative. This kind of pattern is fairly common in time series regression analysis and may lead one to question the independence assumption.

**Table 9.4** Table of Residuals and Fitted Values

| *T* | *Residuals* | *Fitted Values* | *T* | *Residuals* | *Fitted Values* | *T* | *Residuals* | *Fitted Values* |
|---|---|---|---|---|---|---|---|---|
| 1 | 145 | 9492 | 9 | 224 | 9545 | 17 | −180 | 10315 |
| 2 | 24 | 9336 | 10 | −103 | 9710 | 18 | 13 | 9488 |
| 3 | − 99 | 9352 | 11 | −106 | 9913 | 19 | − 5 | 9958 |
| 4 | − 54 | 9473 | 12 | −215 | 10047 | 20 | 107 | 9740 |
| 5 | −217 | 9488 | 13 | − 6 | 10113 | 21 | 305 | 8973 |
| 6 | −143 | 9543 | 14 | 269 | 10103 | 22 | −380 | 9250 |
| 7 | −176 | 9548 | 15 | 398 | 9864 | 23 | −340 | 9225 |
| 8 | − 76 | 9567 | 16 | 302 | 10114 | 24 | 314 | 8819 |

*Source*: Residuals and fitted values of *M1* for regression of *M1* on *Y*, *RC*, *RLA*.

We now present a more formal approach to testing the assumption of independence. In the language of Chapter 7, the null hypothesis is that each $u_t$ is independently and normally distributed. For a formal test, an alternative model needs to be specified. The simplest possibility is that there is a one period dependence between successive values of the random component. Under this alternative hypothesis, the random component $u_t$ is modelled in terms of its one period lagged value $u_{t-1}$ (multiplied by an additional parameter $\rho$) and an additional random component $\varepsilon_t$ whose values are assumed to be independently and normally distributed. Specifically:

$$u_t = \rho u_{t-1} + \varepsilon_t$$

$$\text{where } \rho \text{ is a parameter and } \varepsilon_t \sim \text{NID}(0, \sigma^2) \qquad \textbf{(9.3.1)}$$

To support the alternative hypothesis the estimate of $\rho$ needs to be significantly different from 0. If $\rho$ is zero, the model (9.3.1) is equivalent to the modelling of the random component at (9.1.1).

The *Durbin–Watson statistic, d*, is a test statistic for this situation. Its formula is:

$$d = \frac{\sum_{t=2}^{T} (e_t - e_{t-1})^2}{\sum_{t=1}^{T} e_t^2}$$

and most computer packages will calculate it. A table of critical values of the Durbin–Watson statistic is included in Appendix A. Values close to 2 tend to support the null hypothesis of independence. The critical value depends upon $T$, the number of observations and $k$, the number of explanatory variables. In the present example, $T = 24$ and $k = 3$.

Reference to the Appendix table shows a novel feature, namely that two columns of critical values are given, one headed $d_L$ (here 1.10) and the other $d_U$ (here 1.66). If a calculated value is greater than $d_U$, the null hypothesis is supported; if it is less than $d_L$, the null hypothesis is rejected at the 5% level of significance. If, however, the calculated value falls between $d_L$ and $d_U$, the so-called inconclusive range, the test provides no definitive indication. For our example, the computed value is 1.53, which is in fact in the inconclusive range.

As a general point, it should be noted that the Durbin–Watson statistic is designed to test the independence hypothesis against the alternative hypothesis of one period autocorrelation in the random component. It is possible that a non-significant value could be obtained because the random component dependence *is* present, but in a more complex form or in terms of some longer lag. With data recorded at, for example, quarterly intervals, there might be a fourth order autocorrelation due to seasonal effects. A further limitation is that the Durbin–Watson test has deficiencies when applied to regression models where the systematic component includes lagged values of the explanatory variable. This is the reason why no Durbin–Watson statistic was shown for the Case C model in Table 9.3.

The consequences and cures for autocorrelation of the random component is one which has generated considerable research in recent years, much of it being theoretically and mathematically very complex. In simple terms, however, we may note that, if autocorrelation in the random component is ignored, then the normal estimation procedure will still yield regression coefficients which are unbiased (so long as lagged values of the dependent variable are not included as explanatory variables). Theoretically, however, it can be shown that the calculated standard errors of the regression coefficients will be unreliable.

*If* evidence of autocorrelation is found, it may be worth attempting a respecification which models the autocorrelation. Unfortunately, this is often difficult. The assumption of independence is clear cut; dependence, on the other hand, can be modelled in a large number of ways. We attempt here no more than an indication of what may be done.

The simplest model of autocorrelation is that set out in (9.3.1), which is known as a *first order autoregressive process*. This enables the basic model (9.1.1) to be rewritten. We have:

$$y_t = \beta_0 + \beta_1 x_{1t} + \ldots + \beta_k x_{kt} + u_t$$

and therefore, for the previous time period:

$$y_{t-1} = \beta_0 + \beta_1 x_{1t-1} + \ldots + \beta_k x_{kt-1} + u_{t-1}$$

Multiplying the second equation by $\rho$ and subtracting from the first gives:

$$y_t - \rho y_{t-1} = \beta_0(1 - \rho) + \beta_1(x_{1t} - \rho x_{1t-1}) + \ldots + \beta_k(x_{kt} - \rho x_{kt-1}) + u_t - \rho u_{t-1} \quad \textbf{(9.3.2)}$$

The random component, $u_t - \rho u_{t-1}$, from (9.3.1) is now equal to $\varepsilon_t$ which is, by definition, not autocorrelated.

If we now calculate new variables: $y_t^* = (y_t - \rho y_{t-1})$, $x_{1t}^* = (x_{1t} - \rho x_{1t-1})$ . . . etc., there would appear to be no problem in estimating the systematic parameters. The snag is that we do not know the value of $\rho$. As we shall see below, an estimate of $\rho$ can be made, but model (9.3.2) based on an *estimated* value of $\rho$ will not necessarily be any improvement.

One way of estimating $\rho$ is to carry out a regression analysis using as data the residuals as first computed (see Table 9.4). The residuals themselves ($e_t$) form the dependent variable and their one period lagged values ($e_{t-1}$) the single explanatory variable. The coefficient of the explanatory variable gives the estimated value of $\rho$. Application of this procedure to Case A (Table 9.3) gives an estimate of $\rho$ equal to 0.19. A regression in terms of the transformed variables, as at (9.3.2) shows in fact little change as compared with the original fitted regression. This could have been anticipated since the Durbin–Watson statistic was in the inconclusive range. The analysis might be taken further by more complex modelling of the random component but this would take us beyond the scope of this text.

To sum up, if autocorrelation is the only problem, then ignoring it will have the single consequence that standard errors of regression coefficients will be unreliable and so the coefficients themselves will not be as precisely determined as they might be.

## 9.4 HETEROSCEDASTICITY

In this section we shall be concerned with the consequences of the failure of another assumption, namely that the variance of the random component is a constant value, $\sigma^2$. If this model assumption is not valid, the random component is said to be *heteroscedastic*. In the absence of heteroscedasticity, the random component is said to be homoscedastic.

Our discussion of heteroscedasticity follows the framework set out at the beginning of the last section and we first consider in what circumstances it might arise.

Heteroscedasticity is often exhibited by cross section data, particularly if some of the variables in the data matrix reflect aspects of the 'size' of the individuals. In these circumstances, the random component might also be expected to reflect some aspect of size since it will reflect similar variables omitted from the systematic component of the model. As a consequence, the random component might be expected to have a greater variance for the 'larger' individuals.

To illustrate heteroscedasticity, we shall use the country-mortality data of Chapter 8. Although this scaling feature is not obviously inherent in that cross section data, the plot of residuals against fitted values (for the model with explanatory variables *TDP*, *SUG* and *AGEMP*) shown in Figure 9.2 exhibits heteroscedastic features. The points of the plot tend to be more widely dispersed the higher the fitted values. The plot is similar to the idealised example already presented in Figure 8.1, Case (b) (page 199).

So long as there are no other sources of invalidity in the basic model it can be theoretically shown that ordinary least squares fitting will still produce unbiased regression coefficients but their standard errors will be greater than need be: the standard errors will be inflated by the larger residual variances over part of the data set and the fitted values of the dependent variable may be subject to relatively large uncertainties. In other words, a better job can in principle be done if an appropriate model of the heteroscedasticity can be formulated.

It must first be said that it is not easy to establish formally that the constant variance assumption is invalid, partly because there is no unique way of specifying a heteroscedastic model for the random component and therefore there is a multitude of other possible models to compare with the basic model. The Goldfeld and Quandt test makes the assumption that, insofar as there is heteroscedasticity, the variance of the random component for the $r$th data point is proportional to the corresponding squared value of one of the explanatory variables, say, $X_j$. In other words, the alternative hypothesis to homoscedasticity is that $u_r \sim \mathrm{N}(0, \sigma^2 x_{jr}^2)$.

Under the alternative hypothesis, the residuals corresponding to the data points for the larger values of $X_j$ will tend to be numerically greater than the residuals for the smaller values of $X_j$. With this in mind, the first stage in the Goldfeld and Quandt test procedure is to put the individuals in rank order according to their values of $X_j$ and to delete from

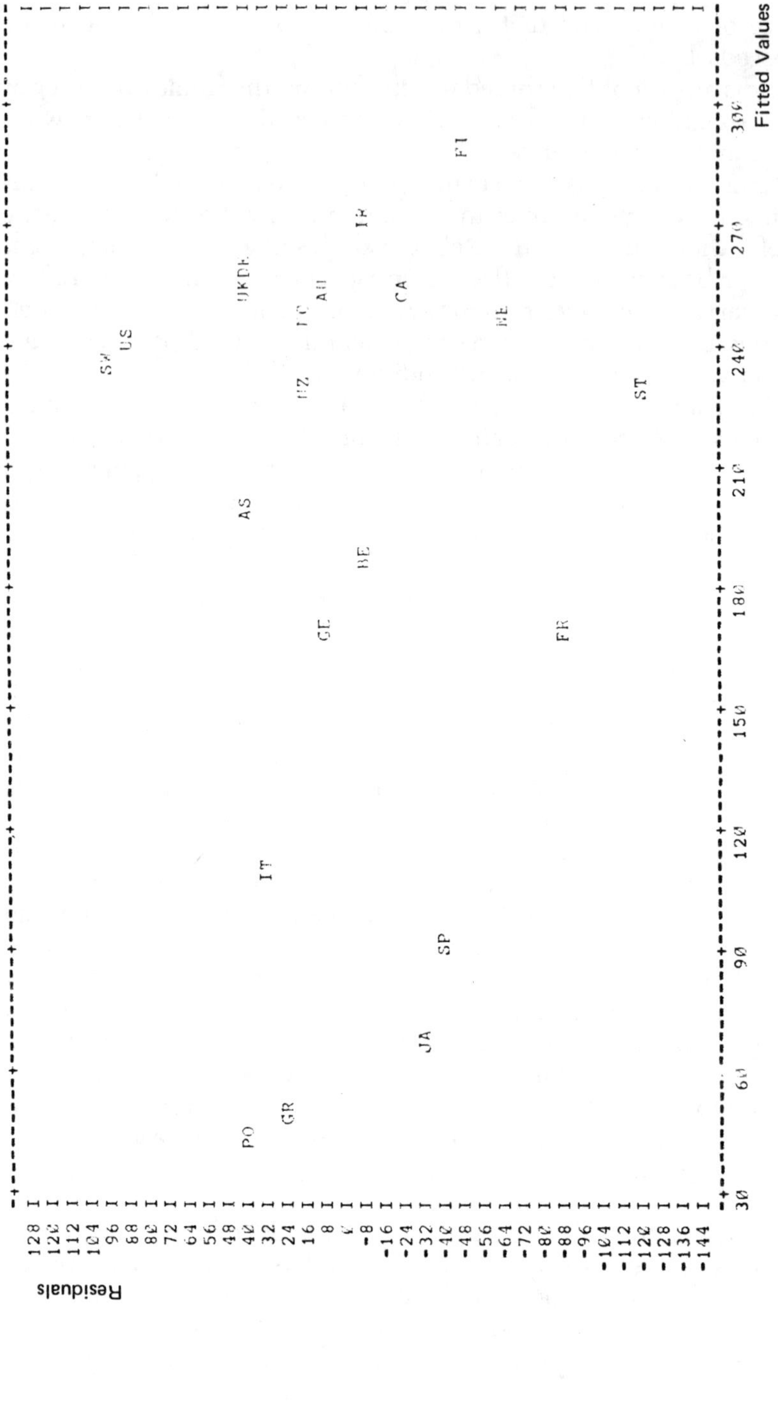


*Source*: Regression analysis of model at (8.3.1)

**Figure 9.2** Plot of Residuals against Fitted Values

subsequent analysis the middle quarter of the observations. (This is a rough empirical rule based upon that in Johnston (1972).) Thus, for the country-mortality data matrix with $n = 21$, one would delete the middle five observations. Two separate regression analyses are then carried out. The first is based upon the $n_0$ individuals having the lower set of values of $X_j$ ($n_0$ in our case equalling 8) and the sum of squared residuals, $s_1^2$ say, is calculated. Similarly $s_2^2$ is calculated for the $n_0$ individuals having the higher set of values of $X_j$. The test statistic is defined as $G = s_2^2/s_1^2$. Under the null hypothesis of homoscedasticity, $G$ has the distribution $F((n_0-k-1), (n_0-k-1))$. Under the alternative hypothesis, $G$ will tend to be large and so the test is a one-tailed test.

An obvious problem with this test is how to choose $X_j$. Insofar as any heteroscedasticity may be related to an aspect of size, then this would suggest the choice of a size-related explanatory variable for $X_j$. This of course does not help with a data set such as the country-mortality data set. In such a case it might be sensible to try as an '$X_j$' variable one which has relatively high explanatory power or is correlated with a subset of variables collectively having high explanatory power; examples are *TDP* or *SUG* (see Section 8.10).

For illustrative purposes only (since the number of data points in each regression is uncomfortably small), we apply the Goldfeld–Quandt test to the country-mortality data, with explanatory variables *TDP*, *SUG* and *AGEMP*. Using *TDP* as the $X_j$ variable gives: $s_1^2 = 9{,}713$ and $s_2^2 = 16{,}031$ and so $G$, the test statistic $= 1.65$. The 5% critical value is $F_{0.95}(4,4) = 6.4$ and the null hypothesis of homoscedasticity is not rejected.

We have remarked that there are many possible formulations of an improved model if heteroscedasticity is identified and we limit ourselves to mentioning just two. The first is simply to apply a log transformation to all the variables. If increasing variance of the random component tends to be positively associated with increasing variable values, this transformation can sometimes be expected to reduce heteroscedasticity.

The second approach is to deflate all the variables by some variable which is an indication of 'size'. This 'deflator variable' will often be an existing explanatory variable. The implications for the fitting procedure can be seen in terms of a bivariate model:

$$y_r = \alpha + \beta x_r + u_r, \text{ where } u_r \sim \mathrm{N}(0, \sigma^2 x_r^2) \qquad \textbf{(9.4.1)}$$

In this model, the residual variance is presumed to be proportional to the square of the value of the explanatory variable. Now if both sides of the model equation are divided by $x_r$, (9.4.1) becomes:

$$\frac{y_r}{x_r} = \beta + \alpha\left(\frac{1}{x_r}\right) + v_r \tag{9.4.2}$$

where the random component $v_r = u_r/x_r$ and so $v_r \sim N(0, \sigma^2)$. The revised model (9.4.2) now has the form of the basic regression model, with the dependent variable defined as $y_r/x_r$ and the explanatory variable as $1/x_r$. A regression of $(y_r/x_r)$ on $(1/x_r)$ yields estimates where the constant term estimates $\beta$ and the slope coefficient estimates $\alpha$.

This procedure is readily extended to multiple regression. All the variables are first deflated by the chosen explanatory deflator variable, say $X_j$. A regression is then carried out including the reciprocal of $X_j$ as an explanatory variable in place of $X_j$ itself. The constant term of this regression provides the estimate of $\beta_j$ in the original model and the coefficient of the reciprocal of $X_j$ the estimate of the constant term in the original model.

## 9.5 DUMMY VARIABLES

So far, regression analysis has been presented as a procedure for examining the relation between numerical variables. In this section we shall see how categorical variables can be used as explanatory variables in the regression model. (The case where the *dependent* variable is categorical is considered in Chapter 15.)

Turning back to the money data at Table 9.1 it is known that a change took place in Central Bank interest rate policy in 1972 and this is, in fact, reflected in the higher level of interest rates from this point onwards. In the light of this knowledge, one might wonder whether, under the different 'regime' (to use a jargon phrase), there might have been some change in the underlying model relation between the variables.

To take a very simple possibility first, let us postulate that, post-1972, the model relation changed only as regards the constant parameter $\beta_0$. Then, in effect, two models are being proposed:

*Period 1 (1969–72)*

$$M1_t = \beta_0 + \beta_1 Y_t + \beta_2 RC_t + \beta_3 RLA_t + u_t$$

*Period 2 (1973–74)*

$$M1_t = (\beta_0 + \beta_4) + \beta_1 Y_t + \beta_2 RC_t + \beta_3 RLA_t + u_t \tag{9.5.1}$$

Note that the period 2 model differs from the period 1 model only in that the constant term is changed by a value represented by the parameter $\beta_4$. Now these two models can be put together into one model specification as follows:

$$M1_t = \beta_0 + \beta_1 Y_t + \beta_2 RC_t + \beta_3 RLA_t + \beta_4 DREG_t + u_t$$

where *DREG* is such that it takes the value zero for observations in period 1 and the value 1 for observations in period 2. **(9.5.2)**

For, when $DREG = 0$, we have the appropriate period 1 model, and when $DREG = 1$, the period 2 model.

Model (9.5.2) still takes the form of a basic regression model with an additional explanatory variable *DREG*. These values of *DREG* have been included in the money data Table 9.1. Hence the corresponding regression coefficient $b_4$ will provide an estimate of any change in the constant term. The regression equation turns out to be:

$$\underset{}{\widehat{M1}_t} = \underset{(1.7)}{3200} + \underset{(4.6)}{0.55Y_t} - \underset{(5.3)}{230RC_t} + \underset{(0.0)}{1.52RLA_t} - \underset{(1.0)}{355DREG_t}$$

In the event, the coefficient of *DREG* is not significant, with a *t*-ratio of only 1.0, and does not suggest a change in the model so far as the constant term goes.

The variable *DREG* is an example of a *dummy variable*. A dummy variable conventionally is assigned only the values 0 or 1. The dummy variable here stands for a dichotomous categorical variable having one of two values 'period 1' and 'period 2' for each observation. The role of the dummy variable in the model equation is to exhibit any differential constant effect on the dependent variable. Other examples are easy to imagine. Thus, some of the observations in an analysis may have occurred with one political party *A* in power and others with political party *B* in power. The dummy variable would take the value 0 for those periods when *A* is in power and the value 1 when *B* is in power.

The dummy variable method can also be used where the categorical variable has more than two categories. Consider for example a model for examining whether there is any constant seasonal fluctuation from one quarter (or one month) to another. Such a model would involve a constant parameter $\beta_0$ for *QI* (first quarter) periods, $(\beta_0 + \beta_{II})$ say for *QII* periods, and $(\beta_0 + \beta_{III})$ and $(\beta_0 + \beta_{IV})$ for *QIII* and *QIV* periods. The additional parameters, $\beta_{II}$, $\beta_{III}$ and $\beta_{IV}$ measure the change in the constant term in these quarters compared to *QI*. As before, a single model will suffice:

$$M1_t = \beta_0 + \beta_1 Y_t + \beta_2 RC_t + \beta_3 RLA_t + \beta_{II} QII_t + \beta_{III} QIII_t + \beta_{IV} QIV_t + u_t \quad \textbf{(9.5.3)}$$

There are three dummy variables here and, for example, *QII* takes the value 1 for all *QII* periods but the value 0 otherwise. The full set of values for these dummy variables is shown in Table 9.1. Such a set is often known as a scoring system, in this case for the categorical variable 'Quarter of the year'. The categorical variable has 4 categories and there are 3 dummy variables. It is worth noting that it would be incorrect to introduce 4 dummy variables, one for each quarter; the calculation would fail because it is only possible to estimate the effect of one quarter compared to another, in this case the first quarter. More generally, if the categorical variable has $c$ values, then $(c-1)$ dummy variables need to be introduced.

The discussion so far has assumed that the only effect of various categorical variable values is on the constant term in the regression model. The dummy variable method can readily be developed to allow for effects on the systematic parameters of one or more explanatory variables. Suppose, for example, that we postulate that $\beta_2$, the coefficient of the interest rate variable *RC*, changes over the period 1973–74, from $\beta_2$ to, say, $\beta_2 + \beta_2^*$. In other words the two proposed models are:

*Period 1*

$$M1_t = \beta_0 + \beta_1 Y_t + \beta_2 RC_t + \beta_3 RLA_t + u_t$$

*Period 2*

$$M1_t = \beta_0 + \beta_1 Y_t + (\beta_2 + \beta_2^*) RC_t + \beta_3 RLA_t + u_t$$

Again these two models can be collapsed into one model using the dummy variable *DREG*:

$$M1_t = \beta_0 + \beta_1 Y_t + \beta_2 RC_t + \beta_2^* (DREG_t) \times (RC_t) + \beta_3 RLA_t + u_t \qquad \mathbf{(9.5.4)}$$

For all observations in period 1, $DREG = 0$ and the combined model becomes the period 1 model; for observations in period 2, $DREG = 1$, and the combined model becomes the period 2 model. In effect, this procedure amounts to adding an explanatory variable, with coefficient $\beta_2^*$, which equals 0 in period 1 and equals *RC* in period 2. The added variable is a simple function of two variables in the data matrix, *RC* and *DREG*, and computer programs typically have facilities for its calculation.

In practice, it would probably be unrealistic to assume that the constant term is not affected by the change of regime. If so, the models (9.5.2) and (9.5.4) can be combined to allow for changes in both the constant term and the parameter of the variable *RC*.

The reader may have been wondering what is the point of introducing dummy variables to examine the possibility that a regression model might be different from one set of observations to another. Why not just fit separate regression models for the sets of observations defined by each of the values of the categorical variable?

The justification for the dummy variable technique — and its limitations — will emerge by looking again at the first dummy variable model introduced at (9.5.2). If we had fitted periods 1 and 2 separately, each of the model equations at (9.5.1) would have been fitted and quite separate estimates, for each period, would have been made of *all* the model parameters. This would be an appropriate procedure if we had no specific view about any changes in the model between period 1 and period 2. The dummy variable model (9.5.2) on the other hand makes the assumption that the only change is in the constant term and that all other model parameters (including the random component variance $\sigma^2$) is constant over both periods. If this assumption is correct, the dummy variable model will yield better estimates of the constant term effect.

## 9.6 PREDICTION

The discussion of regression analysis so far, both in Chapter 8 and the present chapter, has been directed mainly towards its use for description and causal modelling. Regression analysis can also be used for prediction: predicting a value of the dependent variable taking as given a set of values of the explanatory variable. Such a prediction is known as a *conditional* prediction — it is conditional upon the given explanatory variable values. For example, one may ask in the light of the regression analysis of the money data, what would be the money circulation (the value of *M1*) at some future period of time for certain specified values of the explanatory variables? (These specified values may themselves have arisen from some separate forecasting exercise.)

In its simplest terms and, as we shall see, in rather naive terms, a predicted value of the dependent variable can be calculated by substituting in the regression equation the given explanatory variable values. For example, consider the fitted model $\widehat{M1} = 4880 + 0.46Y - 230RC - 25.5RLA$ (Case A of Table 9.3) which was based on observations over the period 1969–74. Suppose that at the end of this period, the following forecasts (indicated by the suffix 0) had been made of the explanatory variables, for the first quarter of 1975:

$$Y_0 = 17{,}500, \; RC_0 = 17.0, \; RLA_0 = 12.5$$

**Table 9.5** Predicted Value and Standard Errors of Prediction

| *Y* | *SE (FITTED)* | *SE PRED. (Y)* |
|---|---|---|
| 8,620 | 178 | 292 |

*Source*: Extracted from IDA output

Then a conditional prediction of *M1* may be obtained by substituting these values into the regression equation, giving $\widehat{M1}_0 = 8{,}700$. (The value given in the computer printout in Table 9.5 is actually 8,620, the apparent discrepancy is due to our having rounded the regression coefficients.)

More generally, for a given set of explanatory variable values ($x_{10}$, $x_{20}$, . . . $x_{k0}$), the conditional prediction of $Y$ is:

$$\hat{y}_0 = b_0 + b_1x_{10} + b_2x_{20} + \ldots b_kx_{k0} \tag{9.6.1}$$

Assuming that the model for the predicted observation is correct, the actual value, $y_0^*$, for 1975 (I) will be:

$$y_0^* = \beta_0 + \beta_1x_{10} + \beta_2x_{20} + \ldots + \beta_kx_{k0} + u_0 \tag{9.6.2}$$

We may note two sources of statistical error: first, $\hat{y}_0$ is based upon the *estimated* model (9.6.1) rather than upon (9.6.2) and, secondly, the future value of the random component, $u_0$, is not known. We need to make an important distinction concerning what is being predicted. There are two possibilities. On the one hand, we may wish to predict the actual observation $y_0^*$ which will arise. In this case the standard error will reflect both sources of error. On the other hand we may wish simply to predict the expected or fitted value of the dependent variable, that is, its value with the random component set equal to zero; the standard error in this second case will be smaller. In both cases the conditional prediction at (9.6.1) is the central estimate.

Table 9.5 is derived from computer output based on the specimen values of $Y$, *RC* and *RLA* given above and it will be seen that the standard error of a predicted observation is greater than the standard error of a predicted fitted value. Confidence intervals for the two kinds of prediction can be readily calculated based upon the *t*-distribution. In both cases, the appropriate degrees of freedom parameter is $\nu = T - k - 1$ (where, as before, $T$ is the total number of data values).

There are problems in the use of regression analysis for prediction. The most important is that such an exercise assumes that the regression model is still valid for the predicted observation. If the model is being used for prediction into the future, it may be bold to assume that there will be no change in the model relation. For example, the parameter values of both the systematic and the random component may change or other variables, so far excluded, may become important in explaining the dependent variable. Furthermore, it will have been noticed that only conditional prediction has been considered and therefore it has been assumed that the given values of the explanatory variables are not open to question. In time series prediction, these will have arisen from some separate procedure and any errors will carry through to errors in predicting the value of the dependent variable.

The whole discussion in this section has been limited to predicting a particular variable simply by means of regression analysis. *Forecasting*, as distinct from prediction as we have used the term, is a more widely based activity drawing upon a number of other sorts of evidence beyond the results of single prediction. In spite of the reservations expressed above concerning prediction based upon a regression model, the latter will often be a useful contribution to a wider forecasting exercise.

## 9.7 THE RESET AND LAGRANGE MULTIPLIER TESTS OF MISSPECIFICATION

In this section we shall introduce some procedures for testing possible misspecifications of the basic regression model. The test statistics we shall discuss are all easy to compute since they are based on regression computations, usually involving the residuals or fitted values calculated from the particular model being evaluated. These tests provide a more formal approach to the informal checking based on residuals discussed in earlier sections.

We first describe a simple version of the modified RESET test.[1] This is a general test of the assumption that, for any set of explanatory variable values, the probability distribution of the random component has a mean of zero. The point of testing this assumption is that it is likely to be violated in the presence of a number of types of misspecification. Consider, for example, the modelling of an essentially curvilinear relation by a linear systematic component. As is shown in Figure 8.1(a), the residuals exhibit a pattern in which their mean changes systematically as the fitted values (the $\hat{y}$'s) increase, although the

1. RESET stands for 'regressor specification error test'; regressor is an alternative name for explanatory variable.

overall mean is necessarily zero. In this situation, one might expect that, if the $\hat{y}$'s were introduced in some way into the existing regression model this would result in an increase in the value of $R^2$ which, as we shall see, can be tested for statistical significance.

A convenient form of the RESET test procedure is as follows. First, the fitted values of the dependent variable are calculated for the model under examination. A second regression is then carried out as before, using two additional explanatory variables: $\hat{y}_t^2$ (the squares of the fitted values) and $\hat{y}_t^3$ (their cubes). An $F$-test is then carried out to assess the collective significance of these two added variables (see Section 8.6). This procedure may be justified along the following lines. Any curvilinear pattern in a plot such as that in Figure 8.1(a) reflects a quadratic (or higher order) relationship between the residuals and the fitted values and this explains why square and cube terms are used in the test.

The RESET test, and subsequent tests in this section, will be illustrated by the model of Table 9.3, Case A, involving the regression of *M1* on *Y*, *RC* and *RLA*. The fitted values of *M1* are shown in Table 9.4. The fitted values are first standardised (to avoid the multicollinearity which would arise from the high correlation between the squares and cubes of the fitted values). The squared (standardised) fitted values (variable name *YHAT*2) and the cubed (standardised) fitted values (*YHAT*3) are then computed. The second regression with *M1* as dependent variable and *Y*, *RC*, *RLA*, *YHAT*2 and *YHAT*3 as explanatory variables gives $R^2 = 0.850$ compared to 0.753 for the original regression. The test statistic for the significance of this increase (see result (8.6.1)) is $[15(0.850 - 0.753)]/[2(1 - 0.850)] = 4.9$. The critical value for a 5% level of significance test is $F_{0.95}(2,20) = 3.5$. The basic regression model for the money data is therefore judged to be misspecified.

The RESET test has the advantage that it does not require the formulation of an alternative model. However, this has the consequential disadvantage that if, as here, the conclusion is that the model is misspecified, there is no specific indication of a better alternative.

Lagrange multiplier (LM) tests derive their name from a calculus technique, the method of Lagrange multipliers, which is used in their theoretical construction. LM tests can be derived for a number of different possible sources of misspecification. We illustrate the method for just one case: testing second order autocorrelation in the random component.

The LM test approach is similar to that of the RESET test in that a second regression model is constructed with additional variables, derived now from the residuals of the first regression. If the contribution of these additional variables is significant, this will again indicate misspecification. For the autocorrelation test, the additional explanatory variables are $e_{t-1}$ and $e_{t-2}$ and may be thought of as approximations to $u_{t-1}$ and $u_{t-2}$ in the basic model. If the lagged residuals make a significant additional contribution to the $R^2$ value, it would suggest that the basic model should have included $u_{t-1}$ and $u_{t-2}$ as part of the random component. In other words, the basic model was misspecified as regards independence of the random component.

The LM test procedure is as follows. First $F$, the $F$-ratio for the increase in $R^2$, is calculated. The LM test statistic equals $2F$ where 2 is the number of added lagged residual variables. If the basic regression model is valid, the LM test statistic has the approximate distribution $\chi^2(2)$, for which the 5% critical value is $\chi^2_{0.95}(2)$.

## 9.8 RIDGE REGRESSION

In Chapter 6 (Section 6.7) we noted that sometimes there is a choice of estimators for a particular model parameter, that is, more than one formula for estimating a parameter from the data. Two procedures for estimating the systematic parameters of the basic regression model have already been noted — least squares (LS) and maximum likelihood (ML). For the basic regression model, both give the same results, the usual regression coefficients, which are unbiased (Section 8.2).

Ridge regression is one of a large number of alternative methods for estimating the parameters of the systematic component of a regression model. We discuss this one in particular because its use has been illustrated in a number of model building exercises in economics and the social sciences generally.

The ordinary LS regression coefficients are calculated by solving the normal equations (see (8.2.2) for the case of 3 explanatory variables). It is usual in ridge regression to work in terms of standardised variables, with the dependent and explanatory variables transformed to variables with zero mean and unit variance. In this situation, since $\bar{y} = \bar{x}_1 = \bar{x}_2 = \bar{x}_3 = 0$, the equations (8.2.2) reduce to:

$$
\begin{aligned}
b_1\Sigma x_1^2 + b_2\Sigma x_1x_2 + b_3\Sigma x_1x_3 &= \Sigma x_1y \\
b_1\Sigma x_1x_2 + b_2\Sigma x_2^2 + b_3\Sigma x_2x_3 &= \Sigma x_2y \\
b_1\Sigma x_1x_3 + b_2\Sigma x_2x_3 + b_3\Sigma x_3^2 &= \Sigma x_3y
\end{aligned}
\qquad \textbf{(9.8.1)}
$$

In ridge regression, the ridge regression coefficients $b_1^*$, $b_2^*$ and $b_3^*$ are defined to be the solutions to the following equations:

$$b_1(\Sigma x_1^2 + \lambda) + b_2\Sigma x_1x_2 + b_3\Sigma x_1x_3 = \Sigma x_1 y$$
$$b_1\Sigma x_1x_2 + b_2(\Sigma x_2^2 + \lambda) + b_3\Sigma x_2x_3 = \Sigma x_2 y$$
$$b_1\Sigma x_1x_3 + b_2\Sigma x_2x_3 + b_3(\Sigma x_3^2 + \lambda) = \Sigma x_3 y \qquad (9.8.2)$$

In these equations, $\lambda$ (which is positive) is an extra parameter for which a value has to be determined as part of the analysis.[2] Comparing equations (9.8.2) with (9.8.1), the parameter $\lambda$ has been added to the first term in the first equation, the second term in the second equation and so on. The generalisation to any number of explanatory variables is thus straightforward. When $\lambda = 0$, the two sets of equations are identical.

Two questions now need to be considered: how is the value of $\lambda$ determined and in what sense might the resulting ridge coefficients be 'better' estimators than the ordinary regression coefficients? We consider the latter question first.

Although the LS regression coefficients are unbiased they may be rather poor estimators in the sense of having large standard errors. As we saw in Section 8.7, this happens if multicollinearity is significantly present. The purpose of ridge regression is to provide estimators of the systematic parameters which, although they are biased, will tend to have values closer to the systematic parameter value. In the language of Section 6.7, such a ridge estimator would have a smaller mean square error than the LS regression coefficient. In the presence of multicollinearity, a fitted ridge regression equation might be expected to be more stable and, when used in prediction, more reliable.

It is possible to show theoretically that there is some optimal value of $\lambda$ such that the resulting ridge coefficients are better than the LS regression coefficients in the above sense. The problem is that the theory tells us what the optimal $\lambda$ value should be in terms of the unknown model parameters. The $\lambda$ value therefore has to be estimated and the difficulties stem from this fact.

The first problem is that there is no clear-cut procedure for estimating $\lambda$. One procedure is based upon first plotting a *ridge trace*, one for each coefficient. A ridge trace is a plot showing how a ridge coefficient value changes for different values of $\lambda$ starting with $\lambda$ equal to zero (which gives the ordinary least squares coefficient). As $\lambda$ increases the numerical value of the ridge coefficient approaches zero, as is shown in the idealised diagram, Figure 9.3. The procedure is to superimpose the traces for all coefficients and choose that value of $\lambda$ beyond which

2. We use $\lambda$, rather than $k$ which is more commonly used in the literature of ridge regression, to avoid confusion with the number of explanatory variables, $k$.

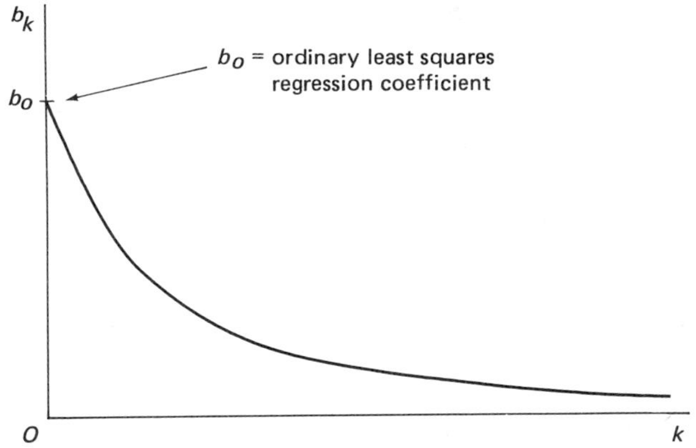


**Figure 9.3** Ridge Trace for a Single Regression Coefficient

all the ridge traces are relatively flat. Naturally, this procedure calls for some judgment and several other less subjective methods have been proposed.

The second problem is that there can be no guarantee that the ridge coefficients calculated from an *estimated* value of $\lambda$ will have the smaller mean square error property that is possible theoretically.

These problems have led many writers to discount completely the utility of ridge regression. Nevertheless, it appears, on occasions, to work in practice in the sense that better predictions are obtained from models whose coefficients are determined by ridge methods. Brown and Payne (1975) describe how ridge regression was used to predict the final outcome of the 1974 UK general election as successive constituency results were announced. They state (see the Discussion on the paper) that ridge regression with $\lambda = 4.0$ gave better predictions than ordinary regression. Perhaps the truth is that it can be a useful technique but only in experienced hands.

For any reader who wishes to explore the method empirically, we may finally note that ridge regression (for any prespecified value of $\lambda$) can easily be carried out within a normal regression program. This may be done by first standardising the variables and then adding $k$ artificial data points to the data matrix. Their values, in the order ($X_1$, . . . $X_k$, $Y$), are required to be ($\sqrt{[(n-1)\lambda]}$, 0,0 . . . 0,0) for the first, (0, $\sqrt{[(n-1)\lambda]}$, 0, . . . 0,0) for the second and so on to (0,0,0, . . . $\sqrt{[(n-1)\lambda]}$, 0) for the $k$th. The regression coefficients calculated by the program will be the ridge coefficients based on the selected value of $\lambda$.

## 9.9 WEIGHTED LEAST SQUARES

The proposal that individuals whose values are 'too large' or 'too small' should be dropped before calculating summary statistics is one which was first put forward in the discussion of outliers and trimming in Sections 2.13 and 2.14 and in particular, the notion of a trimmed mean was explained. A less drastic procedure would be to calculate a weighted mean (see (2.14.1)) in which extreme data values are given smaller weights than those near the centre of the distribution.

A similar procedure can be applied in regression analysis. We have already noted that outliers (those with excessively large residuals) can have an undue effect on the calculated coefficients and it may therefore be desirable to give a smaller weight to the set of values for any individual whose associated residual is relatively large.

The first stage in a weighted least squares procedure is to compute the residuals from the proposed model having fitted this by unweighted least squares. These residuals are then used to determine the weights to be applied to the values of the dependent and explanatory variables for each individual and a second regression is then carried out using the weighted least squares option which is available in some packages. The main problem is how to specify a formula for converting a residual value to a corresponding weight. There are several possibilities and some of these are described and illustrated in Mosteller and Tukey (1977).

## 9.10 NON-LINEAR REGRESSION

It will be recalled that the regression techniques discussed in Chapter 8 and the present chapter relate to linear models, in the sense that the *parameters* enter linearly into the model relation. It was pointed out in Section 8.11 that this does not preclude the variables themselves being non-linearly transformed.

A non-linear regression model is one where the systematic component is not a linear function of the parameters. Sometimes a non-linear systematic component may be transformed into a linear form as is the case with the constant elasticity model in economics (see Section 8.11). When such a transformation is not possible then the regression methods so far discussed cannot be applied.

Consider, for example, the growth function having the shape shown in Figure 9.4. A number of practical situations are represented by this shape. Thus, the variable $Y$ might be the number of colour television sets in use from the year of their introduction, to say, the present. The

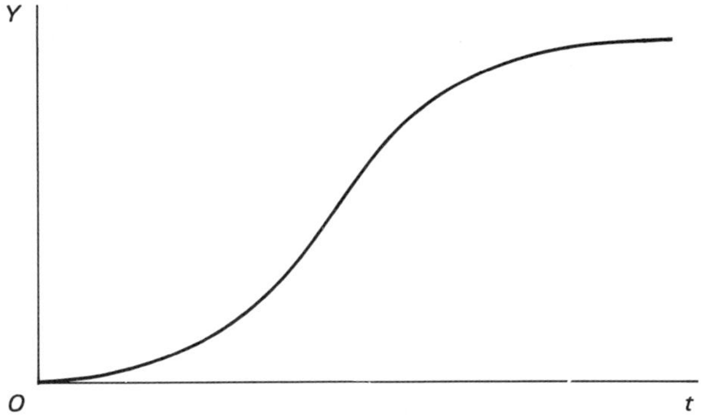


**Figure 9.4** Logistic Curve Growth Function

growth rate is relatively slow to start off with and then increases as colour television becomes more popular. Subsequently, the market becomes saturated and the curve flattens off, reflecting a slower growth as the number of colour TV sets approaches the number of households (ignoring the relatively few who will possess more than one set). Other examples arise in biology such as the growth in size of plants and the growth in numbers of biological populations.

It is often found that for phenomena such as these, their growth pattern is approximately fitted by the *logistic curve* which has the form $y = \gamma / [1 + \exp(\alpha - \beta t)]$ in which $\gamma$, $\alpha$ and $\beta$ are parameters ($\gamma$ and $\beta$ being positive) and $t$ is time. The role of the parameters is to accommodate different scales of measurement in the variable $Y$ whose growth is being described. In such circumstances, an appropriate regression model would be given by equating the logistic function to the systematic part of the model and adding the usual random component.

The parameters of the systematic component do not enter in a linear fashion in its expression, nor is it possible to make a transformation in order to achieve linearity. As a consequence, although the least squares procedure can be applied to give a set of normal equations, their form is too complex to allow a solution by the usual algebraic methods. For these reasons, the estimates of the parameters of the logistic function cannot be obtained by the use of commonly available computer regression programs.

Estimates of the parameters in the logistic function can in fact be obtained by the use of specially written programs which in effect solve the normal equations by an iterative method. Procedures of this kind are known as non-linear estimation and further discussion is beyond the scope of this text. There are of course a large number of other mathematical forms which give rise to similar estimation problems.

## 9.11 UNIVARIATE TIME SERIES ANALYSIS

At various points in this chapter we have noted that a typical feature of a time series variable is that the value at any particular point of time tends to be correlated with values at earlier points in times. Thus, in Section 9.1 the first order autocorrelation coefficient for $RC$ was calculated as $r_1 = 0.75$ and the second order autocorrelation coefficient as $r_2 = 0.59$. This suggests that it might be possible to formulate and fit a linear model of a time series variable $Y$, say, whose current value is modelled in terms of its lagged values. A model of this kind, where the only data values entering the analysis are those of a single time series variable, is known as a univariate time series model. Such a model has particular use in forecasting.

Before presenting some examples of univariate time series models we first consider an informative first step, namely to compute the *autocorrelation function* (acf) for the data. This is simply a calculation of the sequence of first order, second order, third order, ... autocorrelation coefficients; symbolically, the values $r_1, r_2, r_3, \ldots$. These are usually presented by a plot of $r_k$ against $k$, where $k$ is the lag (taking values 1, 2, 3 ...) and $r_k$ the corresponding autocorrelation coefficient.

Figure 9.5 shows a computer printout of the acf for the series of monthly gold prices in Table 9.6. The two left-hand columns give the successive values of $k$ and $r_k$ respectively and the asterisks are the points of the acf plot. The value of the acf decreases as the lag ($k$) increases. This reflects the fact that a value taken at one point of time is more closely dependent upon recent values than long past values. Many time series exhibit this pattern of decreasing autocorrelations with increasing lags although there are some instances for which this statement needs slight qualification. For example, quarterly time series which have a strong seasonal component will often show a 'jump' in the plot at lag 4 and, to a lesser extent, at lags which are multiples of 4.

Time series modelling is a complex subject and we can only hope in this section to give an introductory outline. The simplest possible time series model is one in which:

$$y_t = \mu + u_t, \text{ where } u_t \sim \text{NID}(0, \sigma^2) \qquad \textbf{(9.11.1)}$$

In this model, each value of $Y$ is, in effect, a random observation from the distribution $N(\mu, \sigma^2)$ and is uncorrelated with every other observation. In the light of earlier discussion, this will often be an implausible

**Table 9.6** London Monthly Gold Prices (and First Differences), 1974 Jan.—1980 Nov. US $ per fine oz.

| *1974* | | *1976* | | *1978* | | *1980* | |
|---|---|---|---|---|---|---|---|
| 132.50 | .. | 128.15 | −12.10 | 175.75 | 10.80 | 653.00 | 129.00 |
| 162.50 | 30.00 | 132.30 | 4.15 | 182.25 | 6.50 | 637.00 | −16.00 |
| 173.00 | 10.50 | 129.60 | −2.70 | 181.60 | −0.65 | 494.50 | −142.50 |
| 169.25 | −3.75 | 128.40 | −1.20 | 170.00 | −11.60 | 518.00 | 23.50 |
| 156.75 | −12.50 | 125.50 | −2.90 | 184.15 | 14.15 | 535.50 | 17.50 |
| 144.25 | −12.50 | 123.80 | −1.70 | 183.05 | −1.10 | 653.50 | 118.00 |
| 156.00 | 11.75 | 112.50 | −11.30 | 200.25 | 17.20 | 622.00 | −31.50 |
| 156.00 | 0.00 | 104.00 | −8.50 | 208.70 | 8.45 | 631.25 | 9.25 |
| 151.25 | −4.75 | 116.00 | 12.00 | 217.10 | 8.40 | 666.75 | 35.50 |
| 167.00 | 15.75 | 123.15 | 7.15 | 242.60 | 25.50 | 629.00 | −37.75 |
| 184.00 | 17.00 | 130.25 | 7.10 | 193.40 | −49.20 | 619.75 | −9.25 |
| 186.50 | 2.50 | 134.50 | 4.25 | 226.00 | 32.60 | | |
| *1975* | | *1977* | | *1979* | | | |
| 175.80 | −10.70 | 132.30 | −2.20 | 233.70 | 7.70 | | |
| 181.75 | 5.95 | 142.75 | 10.45 | 251.30 | 17.60 | | |
| 177.25 | −4.50 | 148.90 | 6.15 | 240.10 | −11.20 | | |
| 167.00 | −10.25 | 147.25 | −1.65 | 245.30 | 5.20 | | |
| 167.00 | 0.00 | 142.95 | −4.30 | 274.60 | 29.30 | | |
| 166.25 | −0.75 | 143.00 | 0.05 | 277.50 | 2.90 | | |
| 166.70 | 0.45 | 144.10 | 1.10 | 296.45 | 18.95 | | |
| 159.80 | −6.90 | 146.00 | 1.90 | 315.10 | 18.65 | | |
| 141.25 | −18.55 | 154.05 | 8.05 | 397.25 | 82.15 | | |
| 142.90 | 1.65 | 161.50 | 7.45 | 382.00 | −15.25 | | |
| 138.15 | −4.75 | 160.05 | −1.45 | 415.65 | 33.65 | | |
| 140.25 | 2.10 | 164.95 | 4.90 | 524.00 | 108.35 | | |

*Source:* Central Statistical Office, *Financial Statistics*, London.

```
                 S.E.
       AUTO-   RANDOM
ORDER  CORR.   MODEL   -1  -.75 -.50 -.25   0   .25  .50  .75  +1   ADJ.B-P
                       :----:----:----:----:----:----:----:----:
  1    0.944   0.108                   +   :   +              *     75.84
  2    0.882   0.108                   +   :   +             *      142.9
  3    0.820   0.107                   +   :   +           *        201.5
  4    0.772   0.106                   +   :   +          *         254.1
  5    0.717   0.106                   +   :   +         *          300.0
  6    0.645   0.105                   +   :   +        *           337.8
  7    0.599   0.104                   +   :   +       *            370.8
  8    0.550   0.104                   +   :   +      *             398.9
  9    0.501   0.103                   +   :   +     *              422.6
 10    0.422   0.102                   +   :   +   *                439.6
 11    0.345   0.102                   +   :   +  *                 451.2
 12    0.287   0.101                   +   :   + *                  459.3
 13    0.245   0.100                   +   :   +*                   465.3
 14    0.208   0.099                   +   :   *                    469.7
 15    0.172   0.099                   +   :  *+                    472.7
                       :----:----:----:----:----:----:----:----:
                       -1  -.75 -.50 -.25   0   .25  .50  .75  +1

                       * : AUTOCORRELATIONS
                       + : 2 STANDARD ERROR LIMITS (APPROX.)
```

**Figure 9.5** Autocorrelation Function of Monthly Gold Prices

time series model for two reasons: first, time series often possess a trend and therefore the assumption of a constant mean, $\mu$, is unrealistic and secondly, this model assumes the absence of autocorrelation. There is, however, one area in which the model (9.11.1) is appropriate, and this is where the time series follows what is known as a *random walk*. This feature is indeed exhibited by the gold price data to which we now return.

Gold prices, like other prices in speculative markets, such as commodity and stock exchanges, tend to exhibit random changes. That is to say, their movements from one time period to another cannot be predicted. Suppose that $x_t$ represents the price of gold at time $t$ and $x_{t-1}$ its price in the previous time period. Then a suitable model is:

$$\text{Change in gold price} = x_t - x_{t-1} = u_t \quad \text{where } u_t \sim \text{NID}(\mu, \sigma^2) \qquad \textbf{(9.11.2)}$$

This is equivalent to the model (9.11.1) in which the variable $y_t$ stands for the change in gold price between one observation and the next. In other words, we are modelling the price of gold in terms of the changes between successive time periods; these changes are known as first differences. It should be noted that an implication of this model is that the mean of the first difference is constant, its variance is also constant and its values are independent, that is, there is no autocorrelation.

```
                S.E.
        AUTO-  RANDOM
ORDER   CORR.  MODEL   -1   -.75 -.50 -.25   0    .25  .50  .75  +1   ADJ.B-P
                       :----:----:----:----:----:----:----:----:
  1    0.099  0.108                   +    : * +                     0.8362
  2   -0.200  0.108                   *    :   +                     4.289
  3   -0.246  0.107                  *+    :   +                     9.576
  4    0.160  0.106                   +    :  *+                     11.84
  5    0.242  0.106                   +    :   +*                    17.09
  6   -0.131  0.105                  +*    :   +                     18.65
  7    0.132  0.104                   +    :  *+                     20.26
  8    0.077  0.104                   +    : * +                     20.81
  9    0.115  0.103                   +    : * +                     22.05
 10   -0.173  0.102                  +*    :   +                     24.93
 11    0.063  0.102                   +    :*  +                     25.31
 12    0.100  0.101                   +    : * +                     26.29
 13   -0.044  0.100                   +   *:   +                     26.49
 14   -0.085  0.099                   +  * :   +                     27.21
 15   -0.022  0.099                   +    *   +                     27.26
                       :----:----:----:----:----:----:----:----:
                       -1   -.75 -.50 -.25   0    .25  .50  .75  +1

                       * : AUTOCORRELATIONS
                       + : 2 STANDARD ERROR LIMITS (APPROX.)
```

**Figure 9.6** Autocorrelation Function of Changes in Monthly Gold Prices

The plausibility of this model is a matter for empirical checking, which we now consider.

An informal procedure is to return to the autocorrelation function and compute this for the first differences, which are also given in Table 9.6. The computer output is exhibited in Figure 9.6. The autocorrelations are nearly all small and insignificant suggesting that the first differences are indeed uncorrelated for all lags.

The model (9.11.2) is a simple example of a model of a *stationary time series*. A stationary time series is one which exhibits no change in its mean, variance or autocorrelation pattern as time proceeds. By contrast, the time series of monthly gold prices (without any differencing) is not stationary. For it can be shown that the model (9.11.2) for the first differences implies the following model for the gold prices themselves:

$$z_t = \alpha + \nu_t$$
$$\text{where } \nu_t \sim (t\mu, t\sigma^2) \text{ and } \alpha \text{ is a constant} \qquad \textbf{(9.11.3)}$$

The random component $\nu_t$ now has both a mean and variance which shift over time and such a time series is said to be *non-stationary*.

The theory of time series analysis is mostly founded upon models of stationary time series. At first thought, this might seem unduly restrictive since empirical time series typically contain a trend of varying degrees of complexity with a shifting mean and so they could not be modelled as stationary time series. Fortunately, it is usually possible to transform a time series so that it can be modelled as a stationary time series. We have just applied one such procedure, namely taking first differences, but there are many others and selection of the appropriate transformation is part of the art of time series analysis.

One important kind of model is known as the *autoregressive model*. Indicating an individual value of the time series variable $Y$ by $y_t$, an autoregressive model takes the following form:

$$y_t = \alpha_1 y_{t-1} + \alpha_2 y_{t-2} + \ldots + \alpha_p y_{t-p} + u_t$$
$$\text{where } u_t \sim \text{NID}(0, \sigma^2) \qquad \textbf{(9.11.4)}$$

Stated in this general form, this regression type model specifies the systematic component in terms of lagged values up to and including that of order $p$ and is known as an autoregressive model or, alternatively, autoregressive process, of order $p$, denoted by AR($p$). Two simple versions are the AR(1) model, $y_t = \alpha_1 y_{t-1} + u_t$ and the AR(2) model, $y_t = \alpha_1 y_{t-1} + \alpha_2 y_{t-2} + u_t$. As long as the parameters satisfy certain conditions, an autoregressive process will be stationary: the values of $Y$ will fluctuate over time around a constant mean level. For example, an AR(1) process is stationary so long as $-1 < \alpha_1 < 1$.

Another class of stationary time series models is the *moving average model*. To take the simplest case first, suppose that a particular value of the time series, $y_t$, can be modelled as:

$$y_t = u_t + \beta_1 u_{t-1}$$
$$\text{where } u_t \text{ (and } u_{t-1}) \sim \text{NID}(0, \sigma^2) \qquad \textbf{(9.11.5)}$$

Model (9.11.5) expresses $y_t$ as a weighted average of the current and previous observation of independent drawings from N(0, $\sigma^2$). It follows that the value of $Y$ in time period $(t+1)$ is modelled as $y_{t+1} = u_{t+1} + \beta_1 u_t$ which accounts for the name moving average given to this kind of model. Model (9.11.5) is known as a first order moving average model or process, denoted by MA(1). More generally an MA($q$) process is defined as:

$$y_t = u_t + \beta_1 u_{t-1} + \beta_2 u_{t-2} + \ldots + \beta_q u_{t-q}$$
$$\text{where } u_t \sim \text{NID}(0, \sigma^2) \qquad \textbf{(9.11.6)}$$

The autoregressive and moving average models may be combined into what is known as an ARMA model by replacing the $u_t$ random component term in (9.11.4) by the moving average expression on the right-hand side of (9.11.6). Notationally such a model is referred to as ARMA($p$, $q$). For example an ARMA(2,1) model would be

$$y_t = \alpha_1 y_{t-1} + \alpha_2 y_{t-2} + u_t + \beta_1 u_{t-1}$$
$$\text{where } u_t \sim \text{NID}(0, \sigma^2) \qquad \textbf{(9.11.7)}$$

In practical time series analysis, a number of difficult problems have to be faced. These include:

(i) determining whether the assumption of stationarity is plausible for either the given series or some transformation of it;
(ii) determining the number of autoregressive, $p$, and/or moving average terms, $q$, to be used in the model;
(iii) estimating the parameter values of the selected model;
(iv) checking the fitted model.

Clearly, the subject is complex and further discussion would take us beyond the scope of this text, both as regards space and technical level.

## 9.12 CONCLUSION

The last two chapters have dealt with a number of methodological and practical issues and the reader may well be left feeling somewhat overwhelmed. To put these issues into perspective, it will help to first recall what was said at the end of Chapter 8, briefly that there are a number of reasons why the analyst cannot or need not apply the whole battery of analytical devices which we have reviewed. In addition, there are some more specific points which can be made on the subject of testing. In Sections 9.3, 9.4 and 9.7, for example, a number of tests were presented and some possible ways of reformulating the model were considered. This can prove onerous. However, whether such a task is desirable or potentially fruitful will depend upon two things: the size of the data set and the severity of the misspecification. With small data sets, it is unlikely that informative indications of misspecification can be obtained and so there may be little point in applying many tests[3] and, even with large data sets, the misspecification may not be severe enough to warrant the small gains which might result from respecification. ('Large' and 'small' are, of course, not capable of precise definition but, very roughly, one can think of 'large' as meaning $(n - k) > 50$ and 'small' as $(n - k) < 20$.

In any case, tests often do not provide clear guidance as to the nature of the respecification required and cannot replace informed judgment.

3. For example, the data sets used in this and the previous chapter may be considered small for some of the illustrative uses to which they have been put.

This is particularly true with time series since typically they exhibit trends and, as a result, it is common for high correlations to exist between variables. Such correlations may have little or no explanation in terms of causality, but can lead to high values of $R^2$ in time series regression analysis. For example, the annual output of pig iron in the UK is statistically well explained by the number of pigs, with $r^2 = 0.92$ for the period 1971–79. Clearly, high values of $R^2$ and statistical testing are insufficient, by themselves, as means for gauging success. What is missing is the theory of the subject matter under analysis, which must play an important role in model formulation particularly as regards the choice of variables. We reject the use of 'number of pigs' as an explanatory variable for the output of pig iron because we know how pig iron is produced.

A rather different, and perhaps more telling, approach to validation involves asking how well the relationship implied by a fitted model is reproducible for a further data set. In fact, it can be argued that this is the only satisfactory validation. At the very least this means that, for example, a positive marginal relationship established in one data set is reproduced, possibly with a different magnitude, in another.

Finally, we must not lose sight of the fact that a model is only a model. Perfection, even if it could be achieved, would be so complex as to be useless.

## EXAMPLES

**E 9.1** Determine whether the following calculated Durbin–Watson statistics are significant, non-significant or inconclusive (at the 5% level of significance):

(i) $T = 20, k = 4, d = 1.1$;
(ii) $T = 25, k = 3, d = 1.7$;
(iii) $T = 30, k = 5, d = 1.06$.

**E 9.2** The maximum possible value of the Durbin–Watson statistic is 4.0 and values greater than 2 tend to support the alternative hypothesis of *negative* first order autocorrelation; that is to say, random component values tend to be followed by ones of opposite sign and $\rho$ in (9.3.1) is negative. To test significance of $d$ for this case, one first computes $4 - d$ and then uses the standard table. Check the following results for negative autocorrelation:

(i) $T = 20, k = 4, d = 2.9$;
(ii) $T = 25, k = 3, d = 2.3$;
(iii) $T = 30, k = 5, d = 2.94$.

**E 9.3** In carrying out a Goldfeld–Quandt test for heteroscedasticity, the following regression results were obtained:

(a) First 25 ordered data points

| | *SS* | *DF* |
|---|---|---|
| RESIDUAL | 45 | 20 |

(b) Last 25 ordered data points

| | *SS* | *DF* |
|---|---|---|
| RESIDUAL | 120 | 20 |

Calculate the Goldfeld–Quandt statistic and test for significance.

**E 9.4** The following model combines the dummy variable models (9.5.2) and (9.5.4):

$$M1_t = \beta_0 + \beta_1 Y_t + \beta_2 RC_t + \beta_2^* (DREG_t) \times (RC_t) + \beta_3 RLA_t + \beta_4 DREG_t + u_t$$

(i) Verify this;
(ii) write down the model for the periods in which $DREG = 0$ and in which $DREG = 1$.

**E 9.5** Using the information in Table 9.5, calculate a 95% confidence interval for:

(i) the actual future observation $y_0^*$ and
(ii) the future expected value of $y_0$.

## FURTHER WORK

**F 9.1** In the original source of the data in Table 9.1, the value of *M1* for 1973II is 10,668. On first using the original data, we identified 1973II as an outlier and, to avoid problems in analysis, 10,668 was replaced by 9,501.

(i) Reinstate this value 10,668 and refit the regression model of Case A (Table 9.3).
(ii) Note from the residual plot how the outlier shows up.
(iii) Note also that the value of the Durbin–Watson statistic is substantially reduced. Check its value for significance. (This suggests that this statistic may be sensitive to the presence of outliers.)

**F 9.2** Obtain a series of daily (or weekly) stock market prices and compute the autocorrelation coefficients up to order 5, both for:

(i) the actual prices and
(ii) their first differences.

Compare the outcome with that for gold prices in Section 9.11.

*This chapter depends upon: Sections 0.9 and 1.4*

# 10

# Multivariate Analysis

## 10.1 INTRODUCTION

Many statistical investigations confront the analyst with data on several variables for each of a number of individuals. Sometimes, the set of values of one variable will be found to be closely related to the set of values of one or more other variables; sometimes there will be no such relationship. The study of such joint variation (or lack of it) is one aspect of multivariate analysis and we have already encountered some of the relevant techniques in the last two chapters on multiple regression.

Another aspect concerns the relationship between individuals, rather than variables. We might here ask such questions as: which individuals are alike, with respect to the variable values?—to what extent do groups of similar individuals exist and how should such a group be defined?—what individuals are outliers?

The multivariate methods to be discussed in subsequent chapters will primarily be concerned with numerical variables. They can be broadly categorised by whether the focus of attention is on the variables or on the individuals (or groups of individuals). For example, principal components analysis (Chapter 11) examines which variables are closely related (or correlated) and which are not. On the other hand, discriminant analysis (Chapter 12) investigates how far groups of individuals within the data matrix can be distinguished one from another in terms of their variable values.

Multivariate data sets can be very large. Even one with information on 10 variables and 50 individuals—not big by some standards—contains 500 data values. So clearly there can be a need for condensation. This may be achieved by the use of appropriate summary statistics in much the same way as in univariate analysis. Another method is to condense the data by reducing the number of variables entering the analysis. This may be done by identifying and discarding redundant variables—variables which are less effective than others for whatever is the purpose on hand. Or it may be achieved by calculating a small number of 'derived variables' (a term to be explained below in Section 10.4) which sum up the essence of the data set in the context of the aims of the analysis. Principal components analysis adopts this latter approach.

In earlier chapters, we have drawn a distinction between analysis by purely descriptive statistical procedures and analysis founded upon a statistical probability model of the data. A similar distinction runs throughout multivariate analysis. However, in the chapters which follow, the emphasis will almost exclusively be on descriptive procedures. There are two reasons for this. The first is that there is a more important role for description simply in sorting out the inevitable complexities of multivariate data. The second is that a discussion of statistical models of multivariate data would lead us into theory beyond the level set for this text. However, as in regression analysis, descriptive procedures and model estimation procedures often lead to consistent results, so the omission is not so serious as may appear.

The reader may have wondered why multiple regression analysis (which is certainly a multivariate method) has been considered separately in the previous two chapters. One explanation is that the regression model is based upon a *univariate* probability distribution, that of the random component. As a consequence, the estimation of the model requires the use only of univariate theoretical distributions such as the normal or $t$-distributions. By contrast, statistical modelling of the multivariate situations we shall now be considering would involve the use of, for example, the *multivariate normal distribution* which we do not attempt. There is further distinction between multiple regression analysis and most other multivariate methods. In the former, one variable in the data matrix is identified as the dependent variable and is therefore treated asymmetrically from the others. In the latter, variables are typically treated symmetrically.

The mathematics of multivariate analysis makes use of linear algebra, particularly matrix algebra. Our discussion will not assume this knowledge on the part of the reader and therefore we shall rely from time to time on an intuitive approach.

Multivariate methods, in practice, require the use of appropriate computer programs. One objective of the next few chapters is to relate the theoretical principles to computer output. The computational work in this chapter and in chapters 11, 12 and 13 has been carried out using two packages: SPSS (Statistical Package for the Social Sciences) and GENSTAT. A brief comparative discussion of these and other packages will be found in Appendix B.

## 10.2 THE DATA MATRIX AND THE VECTOR OF MEANS

The remainder of this chapter will be concerned with establishing some basic terminology and ideas, starting with the data matrix. The concept of a data matrix was introduced in Section 1.4; it exhibits the data in a table in which each row sets out the variable values for each individual and each column shows the values for a single variable across all individuals.

Table 10.1 shows a data set which will be used to illustrate the methods of this chapter. It sets out information of an economic and social nature relating to member countries of OECD. The individuals are 14 countries and the 7 variables are mainly related to aspects of standards of living, such as national income, infant mortality rates, television sets in use and so on. One might ask questions such as: which countries are the more like each other as regards their variable values? do the variables tell the same story, in the sense of being highly correlated with each other? is it possible to summarise the data matrix by calculating a meaningful overall index of living standards?

As regards general notation, $n$ will denote the number of individuals and $p$ the number of variables. Thus there are $np$ variable values in the whole matrix. The single symbol $\mathbf{X}$ will stand for the numerical information in the data matrix as a whole. Following earlier convention, individual variables will be denoted by capital letters with subscripts: $X_1, X_2, \ldots X_p$.

The notation is exhibited in Table 10.2 which also shows the notation for an individual variable value. The symbol $x_{rj}$ denotes, for the $r$th individual, its value for the $j$th variable. Thus, in Table 10.1, $x_{25} = 10$; it is the value the second country, France, has for the fifth variable, *INFMOR* (infant mortality rate).

Some simple mathematical terminology will be useful. Mathematically, a set of values (whether expressed numerically or algebraically) arranged in rows and columns is known as a *matrix* and

**Table 10.1** Country Data: Economic and Social Characteristics

| | $X_1$ *POPDEN* | $X_2$ *AGEMP* | $X_3$ *NATINC* | $X_4$ *CAPINV* | $X_5$ *INFMOR* | $X_6$ *ENERGY* | $X_7$ *TVSETS* |
|---|---|---|---|---|---|---|---|
| 1 Australia | 2 | 6 | 8.4 | 10.1 | 12 | 5.2 | 36 |
| 2 France | 97 | 9 | 10.7 | 9.2 | 10 | 3.7 | 28 |
| 3 W. Germany | 247 | 6 | 12.4 | 9.1 | 15 | 4.6 | 33 |
| 4 Greece | 72 | 31 | 4.1 | 8.1 | 19 | 1.7 | 12 |
| 5 Iceland | 2 | 13 | 11.0 | 6.6 | 11 | 5.8 | 25 |
| 6 Italy | 189 | 15 | 5.7 | 7.9 | 15 | 2.5 | 22 |
| 7 Japan | 311 | 11 | 8.7 | 10.9 | 8 | 3.3 | 24 |
| 8 New Zld | 12 | 10 | 6.8 | 8.0 | 14 | 3.4 | 26 |
| 9 Portugal | 107 | 31 | 2.1 | 5.5 | 39 | 1.1 | 9 |
| 10 Spain | 74 | 19 | 5.3 | 6.9 | 15 | 2.0 | 21 |
| 11 Sweden | 18 | 6 | 12.8 | 7.2 | 7 | 6.3 | 37 |
| 12 Turkey | 56 | 61 | 1.6 | 8.8 | 153 | 0.7 | 5 |
| 13 UK | 229 | 3 | 7.2 | 9.3 | 13 | 3.9 | 39 |
| 14 US | 24 | 4 | 10.6 | 7.3 | 13 | 8.7 | 62 |
| Means | 103 | 16 | 7.7 | 8.2 | 25 | 3.8 | 27 |
| SD's | 101 | 16 | 3.6 | 1.5 | 38 | 2.2 | 14 |

*Key:* The variables are measures of the following: *POPDEN* = population density, *AGEMP* = proportion in agricultural employment, *NATINC* = national income per head, *CAPINV* = investment in machinery etc., *INFMOR* = infant mortality rate, *ENERGY* = energy consumption per head, *TVSETS* = no. per hundred population.

*Source:* *OECD Observer*, no. 109, March 1981 which gives exact definitions of variables.

**Table 10.2** Data Matrix Notation

| | | *Variables* | | | | | | | |
|---|---|---|---|---|---|---|---|---|---|
| | | $X_1$ | $X_2$ | .. | $X_j$ | .. | $X_k$ | .. | $X_p$ |
| Individuals | | | | | | | | | |
| | 1 | $x_{11}$ | $x_{12}$ | .. | $x_{1j}$ | .. | $x_{1k}$ | .. | $x_{1p}$ |
| | 2 | $x_{21}$ | $x_{22}$ | .. | $x_{2j}$ | .. | $x_{2k}$ | .. | $x_{2p}$ |
| | .. | .. | .. | .. | .. | .. | .. | .. | .. |
| | $r$ | $x_{r1}$ | $x_{r2}$ | .. | $x_{rj}$ | .. | $x_{rk}$ | .. | $x_{rp}$ |
| | .. | .. | .. | .. | .. | .. | .. | .. | .. |
| | $s$ | $x_{s1}$ | $x_{s2}$ | .. | $x_{sj}$ | .. | $x_{sk}$ | .. | $x_{sp}$ |
| | .. | .. | .. | .. | .. | .. | .. | .. | .. |
| | $n$ | $x_{n1}$ | $x_{n2}$ | .. | $x_{nj}$ | .. | $x_{nk}$ | .. | $x_{np}$ |
| Means | | $\bar{x}_1$ | $\bar{x}_2$ | .. | $\bar{x}_j$ | .. | $\bar{x}_k$ | .. | $\bar{x}_p$ |
| SD's | | $s_1$ | $s_2$ | .. | $s_j$ | .. | $s_k$ | .. | $s_p$ |

an individual value, such as 10 or $x_{25}$, is known as an *element* of the matrix. In a data matrix, the elements in any one row relate to one individual and the elements in any one column relate to one variable.

A preliminary summarisation of the data is provided by calculating the mean of each variable. There are of course as many means as variables and the 7 means for the country data are shown in the last row of Table 10.1. As an obvious development of the univariate notation, the means will be denoted algebraically by the symbols $\bar{x}_1$, $\bar{x}_2$, ... $\bar{x}_p$. This set of all $p$ mean values, arranged in the order of the variables in the data matrix, is called the *mean vector* or alternatively the *vector of means.*[1] Another name for it is the *centroid*. It is conventional to enclose the values in brackets, so that the mean vector or centroid for the country data is (103, 16, 7.7, 8.2, 25, 3.8, 27). The single symbol which will stand for the mean vector is $\bar{\mathbf{x}}$.

In Section 2.5 we introduced the notion of standardising univariate data by first subtracting the mean from each data value and then dividing by the standard deviation. This standardisation can similarly be carried out separately on each variable (that is, the values in each column) of the data matrix. Thus, for the country data, we would at the first stage subtract $\bar{x}_1 = 103$ from the first column of the data

1. In mathematics, a vector is an ordered list of values, as in the present case.

matrix, $\bar{x}_2 = 16$ from the second column and so on. The resulting matrix is known as the *centred data matrix* and it will be indicated symbolically by $\mathbf{X}_0$. At the second stage, each column would be divided by the standard deviation of that column of variable values. The result is called the *standardised data matrix* which will be denoted by $\mathbf{X}_s$. The matrix $\mathbf{X}_s$ then has the property that for any single column, the mean of the values is zero and the standard deviation (and also the variance) is one. The standardised data matrix of Table 10.1 is given at Table 10.3.

In many multivariate techniques, there is a choice between basing the analysis upon the data matrix itself or upon the standardised data matrix. It is usually more appropriate to choose the standardised form since then the values of each variable do not depend upon possibly arbitrary units of measurement (see Section 2.5). Moreover, some computer programs do not allow any choice and employ only the standardised data matrix.

## 10.3 THE CORRELATION MATRIX AND THE DISPERSION MATRIX

It is a fortunate, and perhaps surprising, feature of multivariate analysis that the only other summary statistics which are usually needed are ones which have already been met in bivariate analysis. These are: (i) the variance of each variable, (ii) the correlation coefficient between each pair of variables.

Table 10.4 shows the correlation coefficients for each pair of variables in the country data. Its structure is fairly obvious. For example, the number $-0.18$ in the fourth row, second column is the correlation coefficient between the variable $X_4$ (*CAPINV*) and $X_2$ (*AGEMP*).

Table 10.4 is known as a *correlation matrix* and is denoted by the single symbol **R**. It has a number of features which are common to all correlation matrices. First, the elements (that is the individual values) in the diagonal running from top left-hand to bottom right-hand, the *leading diagonal* as it is known, are all one. This is because the correlation coefficient between a variable and itself is inevitably one. Secondly, each correlation coefficient appears twice in the table: $-0.18$ is both in the fourth row, second column and the second row, fourth column because the correlation coefficient between $X_4$ and $X_2$ is the same as that between $X_2$ and $X_4$. It follows that the upper triangle of the correlation matrix, indicated by putting the figures in italics, is redundant and this upper triangle (together with the leading diagonal

**Table 10.3** Standardised Country Data Matrix

| | $X_1$ *POPDEN* | $X_2$ *AGEMP* | $X_3$ *NATINC* | $X_4$ *CAPINV* | $X_5$ *INFMOR* | $X_6$ *ENERGY* | $X_7$ *TVSETS* |
|---|---|---|---|---|---|---|---|
| 1 Australia | −1.00 | −0.64 | 0.20 | 1.30 | −0.33 | 0.64 | 0.62 |
| 2 France | −0.06 | −0.45 | 0.84 | 0.68 | −0.39 | −0.04 | 0.06 |
| 3 W. Germany | 1.42 | −0.64 | 1.31 | 0.61 | −0.25 | 0.37 | 0.41 |
| 4 Greece | −0.30 | 0.95 | −0.99 | −0.07 | −0.15 | −0.94 | −1.05 |
| 5 Iceland | −1.00 | −0.20 | 0.92 | −1.10 | −0.36 | 0.91 | −0.14 |
| 6 Italy | 0.85 | −0.07 | −0.54 | −0.21 | −0.25 | −0.58 | −0.35 |
| 7 Japan | 2.05 | −0.32 | 0.28 | 1.84 | −0.44 | −0.22 | −0.21 |
| 8 New Zld | −0.90 | −0.39 | −0.24 | −0.14 | −0.28 | −0.17 | −0.07 |
| 9 Portugal | 0.04 | 0.95 | −1.54 | −1.85 | 0.38 | −1.21 | −1.26 |
| 10 Spain | −0.28 | 0.19 | −0.66 | −0.89 | −0.25 | −0.80 | −0.42 |
| 11 Sweden | −0.84 | −0.64 | 1.42 | −0.69 | −0.47 | 1.14 | 0.69 |
| 12 Turkey | −0.46 | 2.86 | −1.68 | 0.41 | 3.40 | −1.39 | −1.53 |
| 13 UK | 1.24 | −0.83 | −0.13 | 0.75 | −0.31 | 0.05 | 0.83 |
| 14 US | −0.78 | −0.77 | 0.81 | −0.62 | −0.31 | 2.23 | 2.43 |
| Means | 0 | 0 | 0 | 0 | 0 | 0 | 0 |
| SD's | 1 | 1 | 1 | 1 | 1 | 1 | 1 |

*Source:* Table 10.1

**Table 10.4** Correlation Matrix of Country Data

| | POPDEN $X_1$ | AGEMP $X_2$ | NATINC $X_3$ | CAPINV $X_4$ | INFMOR $X_5$ | ENERGY $X_6$ | TVSETS $X_7$ |
|---|---|---|---|---|---|---|---|
| $X_1$ | 1.00 | *−0.15* | *0.02* | *0.49* | *−0.13* | *−0.25* | *−0.07* |
| $X_2$ | −0.15 | 1.00 | *−0.79* | *−0.18* | *0.89* | *−0.72* | *−0.78* |
| $X_3$ | 0.02 | −0.79 | 1.00 | *0.19* | *−0.60* | *0.83* | *0.72* |
| $X_4$ | 0.49 | −0.18 | 0.19 | 1.00 | *0.00* | *0.01* | *0.13* |
| $X_5$ | −0.13 | 0.89 | −0.60 | 0.00 | 1.00 | *−0.49* | *−0.52* |
| $X_6$ | −0.25 | −0.72 | 0.83 | 0.01 | −0.49 | 1.00 | *0.92* |
| $X_7$ | −0.07 | −0.78 | 0.72 | 0.13 | −0.52 | 0.92 | 1.00 |

*Source:* from data of Table 10.1

of ones) is usually suppressed in computer printouts. The symbol $r_{jk}$ will indicate the correlation coefficient between $X_j$ and $X_k$ so $r_{24} = -0.18 = r_{42}$.

The *dispersion matrix,* alternatively called the *variance–covariance matrix,* is a table showing the variances of all the variables and the covariance for each pair of variables. Its symbol is **S**. Consistent with earlier work, $s_j^2$ will denote the variance of the variable $X_j$; $s_{jk}$ will denote the covariance between the two variables $X_j$ and $X_k$. Table 10.5 shows the structure of the dispersion matrix in terms of this notation. The variances appear in the leading diagonal of **S** and the covariances appear in the other 'off-diagonal' positions.

A feature of the standardised data matrix, $\mathbf{X}_s$ may be noted:

The dispersion matrix of $\mathbf{X}_s$ is identical to the correlation matrix, **R**. **(10.3.1)**

The reason is that the dispersion matrix of $\mathbf{X}_s$ has in its leading diagonal the variances of the variables of $\mathbf{X}_s$ which all equal one since these variables have been standardised. Also, the off-diagonal positions contain, by definition, the covariances of the standardised variables which equal the correlation coefficients of the original variables (see Section 0.9).

The correlation coefficients in **R** provide a first indication of the association between pairs of variables in the data matrix. For example, the most strongly correlated pairs of variables are $X_6$ and $X_7$ with a correlation coefficient of 0.92 and the weakest are $X_4$ and $X_5$ with a correlation coefficient of virtually zero. However, the reservation expressed in Section 4.11, concerning the limited utility of the correlation coefficient, remains.

**Table 10.5** Dispersion Matrix Notation

| | | *Variables* | | | | | | |
|---|---|---|---|---|---|---|---|---|
| | | $X_1$ | ... | $X_j$ | ... | $X_k$ | ... | $X_p$ |
| Variables | $X_1$ | $s_1^2$ | ... | $s_{1j}$ | ... | $s_{1k}$ | ... | $s_{1p}$ |
| | ... | ... | ... | ... | ... | ... | ... | ... |
| | $X_j$ | $s_{j1}$ | ... | $s_j^2$ | ... | $s_{jk}$ | ... | $s_{jp}$ |
| | ... | ... | ... | ... | ... | ... | ... | ... |
| | $X_k$ | $s_{k1}$ | ... | $s_{kj}$ | ... | $s_k^2$ | ... | $s_{kp}$ |
| | ... | ... | ... | ... | ... | ... | ... | ... |
| | $X_p$ | $s_{p1}$ | ... | $s_{pj}$ | ... | $s_{pk}$ | ... | $s_p^2$ |

We have now encountered two kinds of matrix. On the one hand, there is the data matrix with its centred and standardised versions. These have $n$ rows and $p$ columns. As we shall see, in multivariate analysis the number of individuals is typically greater than the number of variables so $n > p$. On the other hand, the correlation and dispersion matrices have identical numbers of rows and columns (equal to the number of variables, $p$). Such matrices are said to be *square matrices*. Furthermore, **R** and **S** are also *symmetric matrices*. A symmetric matrix is one whose elements in the upper right-hand triangle equal corresponding elements in the lower left-hand triangle in exactly the same fashion as was noted for the correlation matrix and is also the case for the dispersion matrix.

## 10.4 LINEAR COMBINATIONS OF VARIABLES

It was remarked in Section 10.1 that one method of summarisation in multivariate analysis is to calculate what was there called 'derived variables'. We can now be more precise.

The data matrix of Table 10.1 has $p = 7$ variables. Now, at a very simple level, we might think of summarising this data by calculating, for each individual, its average value across all variables. Thus, for Australia, this value would be $(1/7)(2) + (1/7)(6) \ldots + (1/7)(0.36) = 6.3$. Repeating this calculation for all countries produces $n = 14$ values of a derived variable, $Z_1$ say. Mathematically, the variable $Z_1 = (1/7)X_1 + (1/7)X_2 + \ldots + (1/7)X_7$, since the values of $Z_1$ arise from multiplying each corresponding variable value by the same number 1/7.

The derived variable $Z_1$ is known as a *linear combination* of the original variables. The numbers 1/7, 1/7 ... multiplying the variables are the *coefficients* of the linear combination.

In this example, the coefficients are all equal. This is not at all typical. Almost invariably, some variables will have more importance in the analysis than others and these will attract larger valued coefficients. Consider, for example, the derived variable determined by $Z_2 = 10X_1 + 2X_2 - X_3 + 0.5X_4 + X_5 - X_6 + X_7$. Here, one would expect the values of $Z_2$ to reflect a greater numerical influence of the data variables $X_1$, since its values are multiplied by the larger number.

To sum up, a particular linear combination is determined by its coefficients and there are as many coefficients as there are variables in the data matrix. Algebraically, they may be denoted by $(a_1, a_2, \ldots a_p)$. This list of the coefficients in order is called the *vector of coefficients* or the *coefficient vector* (corresponding to the terminology for

**Table 10.6** Variables Derived by Linear Combination of Original Variables

| | | $X_1$ | $X_2$ | $X_3$ | $X_4$ | $X_5$ | $X_6$ | $X_7$ | $Z_1$ | $Z_2$ |
|---|---|---|---|---|---|---|---|---|---|---|
| *Coefficient* | | | | | | | | | | |
| *Vectors:* | $Z_1$ | 0.14 | 0.14 | 0.14 | 0.14 | 0.14 | 0.14 | 0.14 | | |
| | $Z_2$ | 10.0 | 2.0 | −1.0 | 0.5 | 1.0 | −1.0 | 1.0 | | |
| Australia | | 2 | 6 | 8.4 | 10.1 | 12 | 5.2 | 0.36 | 6.2 | 36 |
| France | | 97 | 9 | 10.7 | 9.2 | 10 | 3.7 | 0.28 | 19.6 | 988 |
| ... | | .. | .. | .. | .. | .. | .. | .. | .. | .. |
| Turkey | | 56 | 61 | 1.6 | 8.8 | 153 | 0.7 | 0.05 | 39.4 | 837 |
| UK | | 229 | 3 | 7.2 | 9.3 | 13 | 3.9 | 0.39 | 37.2 | 2,303 |
| US | | 24 | 4 | 10.6 | 7.3 | 13 | 8.7 | 0.62 | 9.6 | 246 |

*Source:* data of Table 10.1

the mean vector). The single symbol **a** stands for the vector as a whole.

Having specified a derived variable in terms of its coefficient vector, its values can be calculated for each individual. Table 10.6 exhibits, in abbreviated form, the results for the derived variables $Z_1$ and $Z_2$, the coefficient vectors being shown at the head of the table. In effect, two further columns have been added to the original data matrix. The $Z_2$ values are obviously highly correlated with the $X_1$ values.

## 10.5 EIGENVALUES AND EIGENVECTORS

Many techniques of multivariate analysis lead to mathematical procedures for calculating appropriate linear combinations of variables, that is, appropriate coefficient vectors. As we have suggested, one purpose (among others) is to arrive at a derived variable which summarises in some relevant sense the information contained in the $p$ variables of the data set.

Some methods involve the calculation of just one linear combination. More often, several different linear combinations are calculated and the corresponding derived variables can always be put in order in terms of their effectiveness for the purpose of the analysis on hand. Having put them in order, the first linear combination corresponding to the most effective derived variable is determined by a certain coefficient vector, say, $\mathbf{a}_1$. The next linear combination, determined by its coefficient vector, $\mathbf{a}_2$ say, represents the second most effective derived variable, and so on.

These coefficient vectors arise from the mathematical solution of whatever problem is posed by the objectives of the analysis. As already stated, we shall not be going into mathematical detail of this kind. However, one technical aspect cannot be avoided since it concerns the interpretation of results and indeed the reading of computer output. Whatever the multivariate method, the different coefficient vectors are typically calculated from some symmetric square matrix derived from the data. The dispersion and correlation matrices are examples of such a matrix and indeed these matrices are the basis for principal components analysis (see Chapter 11). Mathematically, the coefficient vectors are known as the *eigenvectors* (or sometimes the *latent vectors*) of the square matrix. The number of rows (or columns) in this square matrix equals the number of eigenvectors.

Additionally, the mathematics provides a measure of the relative effectiveness of each linear combination for the purpose in hand. These measures are numerical values, usually positive, which are called the *eigenvalues* (or *latent roots*) of the square matrix. One eigenvalue is calculated for and associated with each eigenvector. So the eigenvector which is associated with the largest eigenvalue determines the most effective derived variable, the eigenvector which has the next highest eigenvalue determines the next most effective derived variable, and so on.

## 10.6 DISTANCE MATRIX FOR BIVARIATE DATA

In the previous sections, attention has been focused upon the analysis of the variables in the data matrix. We now take a different point of view and focus attention upon the individuals. We shall be particularly concerned with the question: which individuals are similar to each other and which individuals are different as regards their values for the variables in the data matrix?

We first consider how this question might be answered if there were just two variables in the data matrix. Figure 10.1 shows a scatterplot for the country data based on the two variables $X_3$ (*NATINC*) and $X_6$ (*ENERGY*). The *centroid* is also shown; this is defined as the point whose coordinates are respectively the means of $X_3$ and $X_6$. This scatterplot shows immediately which countries are alike and which are very different. For example New Zealand and the United Kingdom have similar values for both variables and their data points are close to each other. On the other hand the United States and Turkey could hardly be more different.

This geometrical approach suggests an obvious single way of measuring the degree of difference between two individuals, namely

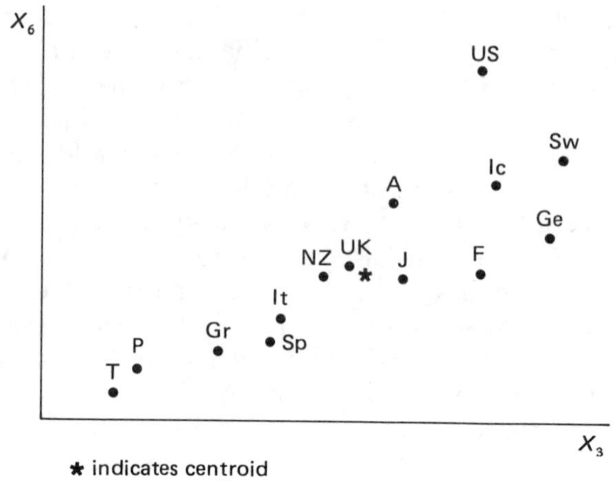


**Figure 10.1** Scatterplot of Variables $X_3$ and $X_6$

the distance between their data points on the scatterplot. This distance can be calculated by applying Pythagoras's Theorem. The method of calculation is demonstrated in Figure 10.2. The left-hand diagram shows the calculation of the distance between Italy and the United States in terms of each country's values for $X_3$ and $X_6$. The right-hand diagram shows, in more general algebraic terms, the calculation of the distance between individual *r* and individual *s* based upon the variables $X_j$ and $X_k$.

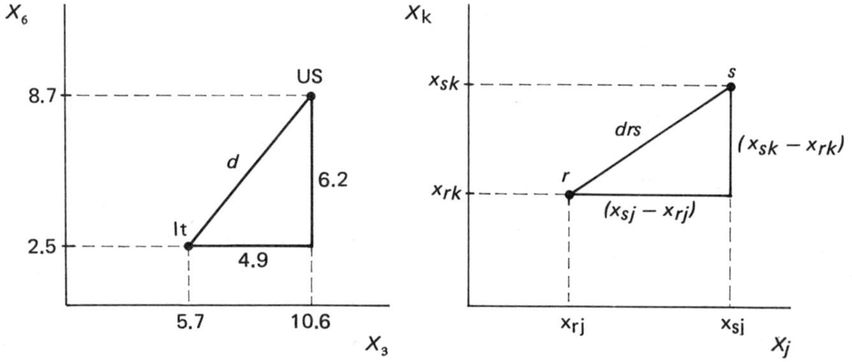


(a) Distance calculation between Italy and US:
$d^2 = 4.9^2 + 6.2^2 = 62.45$
so $d = 7.9$

(b) Distance calculation between individuals *r* and *s* based on variables $X_j$ and $X_k$:
$d_{rs}^2 = (x_{sj} - x_{rj})^2 + (x_{sk} - x_{rk})^2$

**Figure 10.2** Distance Calculation for a Pair of Individuals

Similarly, the distance between each pair of individuals in the (bivariate) data matrix can be calculated and the resulting distances can be shown in a table which is known as the *distance matrix*. A distance matrix has as many rows and columns as there are individuals so that the distance matrix for the country data would have $n = 14$ rows and $n = 14$ columns. A distance matrix can be and is usually very large. For reasons of space, Table 10.7 shows the distance matrix for five of the countries based on the variables $X_3$ and $X_6$.

Table 10.7 exhibits some obvious general properties of distance matrices. They are square matrices with as many rows as columns. They are symmetric, in the same sense as the correlation matrix for, just like a table of road distances between towns, the distance between individual $r$ and individual $s$ is the same as that between individual $s$ and individual $r$; the top right-hand triangle comprising the figures in italics is therefore redundant. The entries in the leading diagonal are inevitably zero.

Finally we note two important practical points. The first is that the distance between two individuals is defined in terms of the variable values and the Pythagorean formulae as shown in Figure 10.2. This distance will be equivalent to the geometrical distance on the scatterplot only if the scales on the two axes are the same. This will not usually be the case but even so the scatterplot by itself gives a useful approximate indication.

The second practical point is that the distance between two individuals may be relatively big even though the values of one of the two variables are very similar. For example New Zealand and France are far apart on the scatterplot because their values for $X_3$ are very different; their values for $X_6$ however are quite close.

## 10.7 GEOMETRY AND MULTIVARIATE ANALYSIS

From time to time, we have seen how geometrical diagrams are an aid both in understanding statistical methods and in the interpretation of data. This is particularly so for bivariate data analysis. For example the bivariate regression model is wholly and exactly specified in algebraic terms, but while a geometrical picture (such as that of Figure 4.2) does not add anything substantive, it is an aid to an intuitive understanding of the model. Also, a scatterplot of a bivariate data set offers to the eye an immediate appreciation of some of the essential features of the data which are far from obvious from the data matrix itself.

Another example has just been met in the preceding section. There,

**Table 10.7** Distance Matrix, for Selected Countries, for Variables $X_3$ and $X_6$

| | *Australia* | *Italy* | *Japan* | *United Kingdom* | *United States* |
|---|---|---|---|---|---|
| Australia | 0 | 3.8 | 1.9 | 1.8 | 4.1 |
| Italy | 3.8 | 0 | 3.1 | 2.1 | 7.9 |
| Japan | 1.9 | 3.1 | 0 | 1.6 | 5.7 |
| United Kingdom | 1.8 | 2.1 | 1.6 | 0 | 5.9 |
| United States | 4.1 | 7.9 | 5.7 | 5.9 | 0 |

*Source:* from data of Table 10.1

geometrical ideas were used to arrive at a purely algebraic definition of the distance between two individuals. Only the algebraic definition is strictly necessary for the specification of the distance matrix, but its geometrical interpretation is an aid both to an intuitive understanding of the algebra and its application to a particular data set.

Now in multivariate analysis, we are concerned not with two but with many variables. It might be thought that geometry has no part to play since we cannot for example draw scatterplots for seven variables. Even so we shall see in subsequent discussions that geometrical ideas can still be helpful, though in a less concrete fashion than is the case for bivariate data.

The bivariate scatterplot is a picture in two dimensions, that is to say, it is a graphical picture with two axes at right angles to each other, one for each of the two variables. Now suppose we have a trivariate data set — one with three variables. A scatterplot can still be constructed, at least in principle, in the following way. Three axes at right angles to each other are now needed, one for each of the three variables. We can think of these axes as analagous to the edges of a room: two of the axes are where the floor meets two adjacent walls and the third axis is where the two walls themselves meet.

Figure 10.3 shows how one might start to construct a scatterplot of trivariate data, based upon the three variables $X_3$, $X_6$ and $X_7$ of the country data matrix. Thus, Italy has, in the same order, the variable values (5.7, 2.5, 22) and so its data point position in the room is at a height of 22 above the point which is 5.7 from one floor edge and 2.5 from the other floor edge. The data values for other individuals may be plotted in a similar way. The centroid of the data set is, by analogy with the two variable case, the point defined by the vector of means. Figure 10.2 is of course a two dimensional picture of a three dimensional scatterplot, just as a photograph is a two dimensional picture of a three dimensional scene.

It need hardly be said that multivariate statistical analysis does not involve making three dimensional models of scatterplots even if the number of variables is only $p = 3$. The point is that the above discussion allows one to conceptualise the implications of different configurations of data points which might emerge from the information in the data matrix. The two right-hand diagrams of Figure 10.3 show two very different configurations. The upper one represents a situation where there is no relationship between any pair of variables: it is the three variable equivalent of the two variable scatterplot when the correlation coefficient is small. The lower diagram exhibits a pattern resulting from relatively large intercorrelations between the variables. In this latter case, the 'cloud' of points tends to be elongated along a

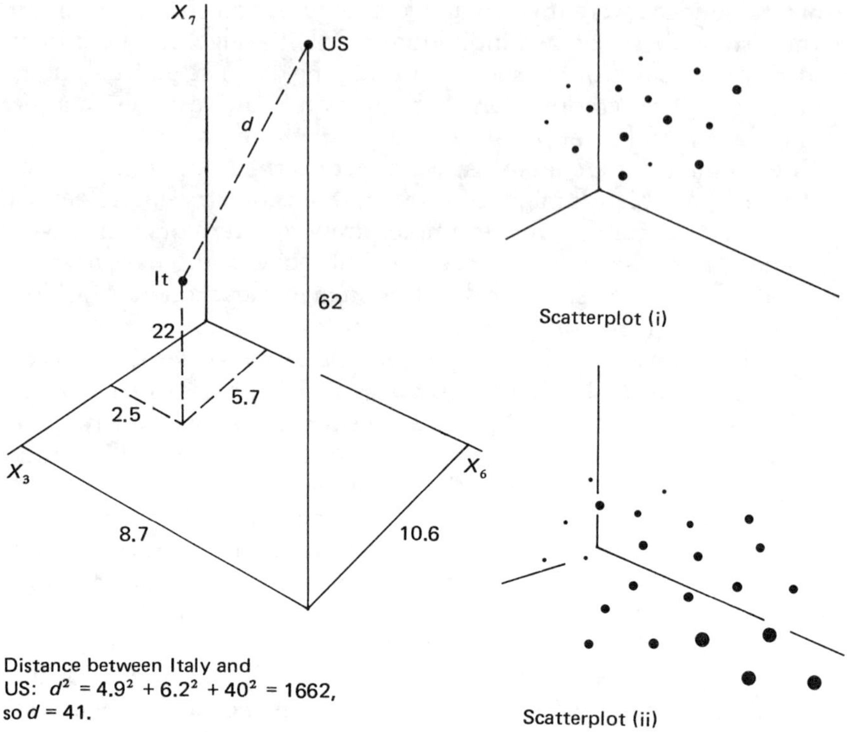


**Figure 10.3** 'Scatterplot' for Three Variables

particular direction. In both cases, the centroid would be in the middle of the cloud.

When one turns to more than three variables, it is of course true that diagramatic representation fails. Space is not four dimensional. Even so, multivariate procedures for more than three variables can still helpfully be interpreted by thinking of what essentially happens geometrically for the simpler cases of $p = 2$ and $p = 3$. That this is so will be clear from the following section where we consider how a distance matrix can be defined and interpreted when $p > 3$.

## 10.8 DISTANCE MATRIX FOR MULTIVARIATE DATA

In Section 10.6 we discussed the calculation of distances between pairs of individuals in terms of bivariate data. Figure 10.2 showed how these distances are calculated both in numerical and in algebraic terms. In verbal terms:

> The squared distance between two individuals equals the sum, over both variables, of the squared differences between the two individuals' corresponding variable values. **(10.8.1)**

The preceding Section 10.7 extended the idea of a bivariate scatterplot to a three dimensional scatterplot where the data set is trivariate. Figure 10.3 showed how one may 'plot' such data. Its left-hand diagram gives the construction of the data points for Italy and the United States as examples. Again, the distance between Italy and the United States can be defined in terms of the geometrical distance between their data points and this can be calculated by means of the three dimensional version of Pythagoras's Theorem. This is a straightforward extension of the two dimensional version. It amounts to adding on an extra term equal to the squared difference, for the third variable, of the two individual values.

The distance calculation between Italy and the United States, for the three variables $X_3$, $X_6$ and $X_7$, is given in Figure 10.3. The squared distance $d^2 = (10.6 - 5.7)^2 + (8.7 - 2.5)^2 + (62 - 22)^2$, which implies that $d = 41$. If the reader compares this calculation with that for the two variables $X_3$ and $X_6$ at Figure 10.2, it can be seen that the effect of adding a third variable is simply to add the term $(62 - 22)^2$ to the squared distance; it is the squared difference, for the third variable, of the two individual values.

The distance definition at (10.8.1) can now be restated for the trivariate case:

> The squared distance between two individuals equals the sum, over all three variables, of the squared differences between the two individuals' corresponding values. **(10.8.2)**

We cannot apply ordinary Euclidean geometry to defining and calculating the 'distance' between two individuals in terms of more than three variables. However, we can define such a distance by making an obvious extension of definitions (10.8.1) and (10.8.2). Specifically, for multivariate data:

> The squared distance between two individuals equals the sum, over *all* variables, of the squared differences between the two individuals' corresponding variable values. **(10.8.3)**

Taking into account all seven variables in the data matrix at Table 10.1, the squared distance between Italy and the United States is $d^2 = (24 - 189)^2 + (4 - 15)^2 + (10.6 - 5.7)^2 + (7.3 - 7.9)^2 + (13 - 15)^2 + (8.7 - 2.5)^2 + (62 - 22)^2$, and $d$ is the square root of this expression.

More formally, the multivariate distance can be defined algebraically in terms of the notation of Table 10.2:

The distance, $d_{rs}$ between individuals $r$ and $s$, is such that

$$d_{rs}^2 = (x_{r1} - x_{s1})^2 + (x_{r2} - x_{s2})^2 + \ldots + (x_{rp} - x_{sp})^2 \tag{10.8.4}$$

The multivariate distance matrix has exactly the same form as the bivariate distance matrix introduced in Section 10.6; the only difference is that the elements of the matrix, that is the individual numbers in it, are calculated from all the variables rather than just two of them.

## 10.9 CONCLUSION

In this chapter we have introduced the reader to some fundametal ideas in multivariate analysis. The earlier part of the chapter focused attention on the variables leading to the discussion in Section 10.4 of what we called derived variables. This discussion will be developed in Chapter 11 where we look at principal components analysis. This has as a principal aim the calculation of a small number of derived variables which sum up the essential features of the data set. In other words, principal components analysis is a method of 'data reduction' just as is the calculation of summary statistics.

An important strand in the present chapter has been the consideration of the role of geometrical and graphical methods in multivariate analysis. The ideas presented in Section 10.7 were there applied to the concept of distance between two individuals. In Chapter 11, we shall see how they can also help in the intuitive understanding of principal components analysis.

## EXAMPLES

**E 10.1** Using the infcrmation in Tables 10.1 and 10.4, calculate the following elements in the dispersion matrix (ref. Table 10.5): $s_{36}$, $s_3$, $s_6$, $s_{63}$.

**E 10.2** Calculate the centred data matrix based upon Table 10.1. What is the relation between any column of the centred data matrix and the corresponding column of the standardised data matrix at Table 10.3?

**E 10.3** Using the centred data matrix, calculate the values for each country of the following linear combinations:

(a) $Z_1 = 0.87X_3 + 0.49X_6$ and (b) $Z_2 = -0.49X_3 + 0.87X_6$ (the coefficients of the other variables being zero).

**E 10.4** For the countries Portugal, Turkey, US, calculate the following distance matrices:

(a) based upon the (unstandardised) variables $X_1$, $X_3$ and $X_6$ in Table 10.1
(b) based upon the standardised variables $X_1$, $X_3$ and $X_6$ in Table 10.3.

Compare the two distance matrices.

## FURTHER WORK

**F 10.1** Many multivariate methods are more effective or more reliable if the variables have approximately symmetrical distributions. Examine each variable in Table 10.1 for excessive skewness (see Section 2.10) and identify any outliers. Investigate the effect of a logarithmic or square root transformation on any skew variables.

*This chapter depends upon: Chapter 10*

# 11

# Principal Components Analysis

## 11.1 INTRODUCTION

The idea that linear combinations of variables could be analytically useful was briefly introduced in the last chapter and we now proceed to examine this possibility further. In particular, this chapter will outline the construction and discuss the use of a particular set of linear combinations which are known as the principal components. To introduce some basic ideas we will concentrate initially on the case of just two variables; the more general multivariate case will come later.

The appearance of a scatter plot of the values of two variables which possess a strong correlation will, by now, be familiar. If one were asked to make a quick verbal description of such a plot it would generally be sensible to remark that the points formed an elongated 'cloud' and if the correlation is positive, the elongation is upward from left to right across the graph. Figure 10.1 illustrated a scatter of points possessing high positive correlation for which this description would be reasonable. (Reference to the correlation matrix at Table 10.4 shows that the correlation coefficient between the two variables, $X_3$ and $X_6$, is 0.83.)

We shall see how it is possible to achieve some economy of description of the data for two correlated variables by relating the data points not to the 'natural' axes (those defined by the variables themselves) but to new axes, one of which passes through or close to the greatest elongation of the scatter, the second of which is at right angles to the first.

In the following section we shall work out the analytical implications of this approach and take a preliminary look at its utility. In later sections, the method will be extended to the multivariate case.

## 11.2 PRINCIPAL COMPONENTS ANALYSIS OF BIVARIATE DATA

Principal components analysis is based upon the *centred* data matrix (see Section 10.2), that is, the data matrix with the values in each column expressed in terms of deviations from the mean. Table 11.1

**Table 11.1** Centred Data Matrix and PC Scores for Variables $X_3$ (*NATINC*) and $X_6$ (*ENERGY*)

| | $X_3$ *NATINC* | $X_6$ *ENERGY* | $Z_1$ | $Z_2$ |
|---|---|---|---|---|
| | *a* | *b* | *c* | *d* |
| Australia | 0.7 | 1.4 | 1.3 | 0.9 |
| France | 3.0 | −0.1 | 2.6 | −1.5 |
| Germany | 4.7 | 0.8 | 4.5 | −1.6 |
| Greece | −3.6 | −2.1 | −4.1 | −0.1 |
| Iceland | 3.3 | 2.0 | 3.9 | 0.1 |
| Italy | −2.0 | −1.3 | −2.3 | −0.2 |
| Japan | 1.0 | −0.5 | 0.7 | −0.9 |
| New Zealand | −0.9 | −0.4 | −0.9 | 0.1 |
| Portugal | −5.6 | −2.7 | −6.2 | 0.4 |
| Spain | −2.4 | −1.8 | −2.9 | −0.4 |
| Sweden | 5.1 | 2.5 | 5.7 | −0.3 |
| Turkey | −6.1 | −3.1 | −6.8 | 0.3 |
| UK | −0.5 | 0.1 | −0.4 | 0.3 |
| US | 2.9 | 4.9 | 5.0 | 2.9 |
| Means | 0 | 0 | 0 | 0 |
| SD's | 3.6 | 2.2 | 4.1 | 1.1 |
| Variances | 13.1 | 4.9 | 16.8 | 1.2 |

$Z_1 = 0.87X_3 + 0.49X_6$ $\quad$ $Z_2 = -0.49X_3 + 0.87X_6$

*Source:* Table 10.1

shows in columns *a* and *b* the centred data matrix for the two variables $X_3$ and $X_6$ of the country data matrix at Table 10.1.

A scatterplot of this bivariate centred data matrix is shown in Figure 11.1; the sloping axes $Z_1$ and $Z_2$ should be ignored for the present. It is useful to compare this scatterplot with that of the original data at Figure 10.1. The arrangement of the data points in the two scatterplots is exactly the same and the only difference is that the axes in Figure 11.1 have been moved so that the new origin is at the centroid $(\bar{x}_3, \bar{x}_6)$, $= (7.7, 3.8)$, on the original scatterplot. In other words, Figure 11.1 could have been drawn by simply rubbing out the axes in Figure 10.1 and putting them in the new position. Incidentally, this demonstrates that the distances between the data points do not change as a result of centring the data matrix.

The axis labelled $Z_1$ on Figure 11.1 has been drawn in the direction of the greatest elongation of the data points and the axis $Z_2$ is at right angles to $Z_1$. Exactly what is the criterion for specifying the 'direction of greatest elongation' and how this leads to a calculation of the position of the axis $Z_1$ (and therefore $Z_2$) is something to which we shall return. At this stage, we take the positioning of $Z_1$ for granted and

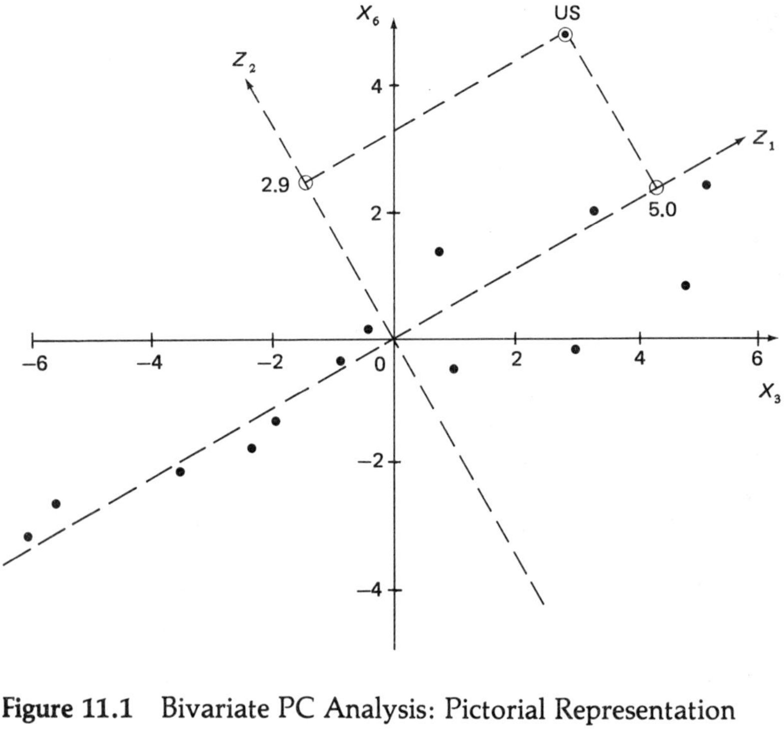


**Figure 11.1** Bivariate PC Analysis: Pictorial Representation

follow through its implications. In fact, the equation of the $Z_1$-axis is given by $-0.49X_3 + 0.87X_6 = 0$ and the $Z_2$-axis is $0.87X_3 + 0.49X_6 = 0$.

First let us see, by reference to Figure 11.1, how a point which is defined in terms of the first set of axes, the X-axes, may be related to the second set, the Z-axes. The US data point with X-coordinates (2.9, 4.9) will be used as an example. Geometrically, its $Z_1$-coordinate can be found—as shown in Figure 11.1—by dropping a perpendicular onto the $Z_1$-axis and measuring the distance from the foot of the perpendicular to the origin; it turns out to be 5.0. Similarly, the $Z_2$-coordinate of the US data point turns out to be 2.9.

The Z-coordinates may be precisely calculated algebraically (starting from the equations of the $Z_1$- and $Z_2$-axes already given). For a data point with X-coordinates $(x_3, x_6)$, its Z-coordinates $(z_1, z_2)$ are in fact given by:

$$\begin{aligned} z_1 &= 0.87x_3 + 0.49x_6 \text{ and} \\ z_2 &= -0.49x_3 + 0.87x_6 \end{aligned} \qquad (11.2.1)$$

The US Z-coordinates are calculated from:

$$\begin{aligned} (0.87)(2.9) + (0.49)(4.9) &= 5.0 \text{ and} \\ (-0.49)(2.9) + (0.87)(4.9) &= 2.9 \end{aligned}$$

Similar calculations may be made of the Z-coordinates of the other data points and the results of doing so are set out in columns *c* and *d* of Table 11.1.

In terms of the terminology of Section 10.4, each of the columns headed $Z_1$ and $Z_2$ gives values of a derived variable which is a linear combination of the original variables $X_3$ and $X_6$. $Z_1$ is the linear combination $0.87X_3 + 0.49X_6$ and $Z_2$ the linear combination $-0.49X_3 + 0.87X_6$. Note that $Z_1$ has a larger coefficient multiplying $X_3$ than $X_6$, so $Z_1$ will be more influenced by $X_3$ than by $X_6$. $Z_1$ is said to have a higher *loading* (or a higher weighting) on $X_3$. The contrary is the case for $Z_2$.

We have now demonstrated that the process of referring a set of values of two variables to a pair of new axes amounts to calculating a linear combination of those variables. The coefficients are determined by the orientation of the new axis system since they are calculated from the equations of the Z-axes.

## 11.3 THE SPECIFICATION AND PROPERTIES OF THE PRINCIPAL COMPONENTS

The $Z_1$-axis was described as having a direction along the 'greatest elongation' of the points in the scatterplot. We can now be more specific. It is in fact chosen so that the *variance of the* $Z_1$ *values of the individuals has a maximum possible value.* In Table 11.1, this is the variance of the values in column *c*, which, as shown there, equals 16.8.

The mathematics behind this specification, and how it leads to the equation specifying the direction of the *Z*-axis, need not detain us. The important point is that the $Z_1$-axis defines a derived variable which is a linear combination of the observed variables and allows the coefficient vector to be calculated from the data; in our example it is (0.87, 0.49). The coefficient vector for the second $Z_2$ variable is (−0.49, 0.87).

It is no coincidence that the sum of the squares of the coefficients of each vector is unity: $0.87^2 + 0.49^2 = 1 = (-0.49)^2 + (0.87)^2$. This is a mathematical consequence of making the *Z*-axes' scale the same as the *X*-axes' scale. (The repetition of the numbers 0.87 and 0.49 is peculiar to bivariate analysis.)

The derived variable $Z_1$ is known as the *first principal component* and the variable $Z_2$ is known as the *second principal component.* The term principal component is often abbreviated to PC. The values of the PC variables, such as the numbers in column *c* and *d* of Table 11.1, are called the PC *scores.*

We are now in a position to get some preliminary idea of the possible utility of PC analysis. Because our chosen variables are relatively highly correlated, and because the direction of the first PC axis $Z_1$ is directed along the elongation of the data points, we may, without losing too much information, describe each individual in terms of just one variable, the first PC. A comparison of the first PC scores of Table 11.1 with the scatterplot of Figure 11.1 shows that these scores very well sum up the relative positions of the data points since the second PC scores are generally quite small. (One obvious exception is the US which is a clear outlier and has a large second PC score.)

This is an example of an important aspect of PC analysis in general. Where there are correlations between the variables, one can expect that PC analysis will result in some degree of data condensation in that the essential features of the data can be summed up in terms of a smaller number of principal component variables.

## 11.4 THE PRINCIPAL COMPONENTS OF VARIANCE

The first PC variable has, as already stated, the property that it is the linear combination of the original variables which has a maximum variance (subject to the scaling condition that the sum of squares of the elements of the coefficient vector is one). The variance has a further central role in PC analysis as we shall now see.

We note from Table 11.1 that the variance of $X_3$, $\mathrm{Var}(X_3)$, $= 13.1$ and $\mathrm{Var}(X_6) = 4.9$. Their sum, which we shall call the *total variance*, $= 18.0$. Turning to the PC variables, $\mathrm{Var}(Z_1) = 16.8$ and $\mathrm{Var}(Z_2) = 1.2$ and their total variance is also 18.0. This is an example of a general property of PC variables:

> The total variance of the (original) variables equals the total variance of the principal components. **(11.4.1)**

The PC analysis can therefore be looked upon as a method of reallocating the total variance between two derived variables, the principal components, in such a way that the first PC takes on as large a proportion of the total variance as possible. This proportion for our data set was $16.8/18.0 = 93\%$. The larger this proportion, the better can the first PC by itself be said to represent the original bivariate data.

There is a further important property. If we were to calculate the correlation coefficient between the two principal components, that is the correlation coefficient between the values in columns *c* and *d* of Table 11.1, it would be found to be zero. Again this is a general property:

> Any pair of principal component variables is uncorrelated; their correlation coefficient (and covariance) is zero. **(11.4.2)**

## 11.5 COMPUTATION

Table 11.2 sums up the results of the bivariate PC analysis which we have been examining. The terms *factor* and *eigenvalues* are those which are typically found in computer output. Computer programs often use the word *factor* instead of principal component. The reason for this is explained in Section 11.11 below.

The use of the word *eigenvalues* for variances takes us back to the discussion of eigenvalues and eigenvectors in Section 10.5. Mathematically, the calculation of the principal components coefficients results from calculating the eigenvectors and eigenvalues of the *dispersion matrix* for the data. It turns out that each eigenvector of the dispersion matrix equals the coefficient vector of one of the principal components. And each associated eigenvalue equals the variance of that principal component. The principal component associated with

**Table 11.2** Summary of Bivariate PC Analysis Output

*Variances (Eigenvalues)*

| *PC (Factor)* | *Variance (Eigenvalue)* | *% of Var.* | *Cum. %* |
|---|---|---|---|
| 1 | 16.8 | 93 | 93 |
| 2 | 1.2 | 7 | 100 |

*PC coefficient vectors (Factor score coefficients)*

| | *PC 1 (Factor 1)* | *PC 2 Factor 2)* |
|---|---|---|
| *NATINC* | 0.87 | −0.49 |
| *ENERGY* | 0.49 | 0.87 |

*Source:* GENSTAT program analysis of Table 11.1 data

the largest eigenvalue therefore has the largest variance. So we have here an example of the general point made in Section 10.5: that the eigenvalues measure the relative effectiveness of a particular linear combination determined by the associated eigenvector.

## 11.6 PRINCIPAL COMPONENTS ANALYSIS OF MULTIVARIATE DATA

Bivariate principal components analysis has been presented mainly from a geometrical standpoint. We now use some of the geometrical ideas discussed in Section 10.7 to see how PC analysis may be generalised to multivariate data.

In Section 10.7 we focused attention upon trivariate data and Figure 10.3 showed two 'scatterplots'. Let us now see at an intuitive level how a PC analysis would work on the trivariate data matrix behind scatterplot (ii) of that figure. First, the data matrix would be centred, which implies that the origin of the scatterplot would be moved to the centroid of the data (keeping the same orientation of axes). Then it is necessary to identify the directions of *three* new axes, mutually perpendicular to each other. The first, $Z_1$, would as before be directed along the line of greatest elongation of the cloud of data points. The second $Z_2$, would be in the direction of next greatest elongation at right angles to $Z_1$. The direction of the third axis, $Z_3$, is then fixed as that at right angles to both $Z_1$ and $Z_2$. Having settled the directions of the three principal component axes, the three principal component variables are, as before, defined as the coordinates of the data points in relation to these axes.

We cannot start to talk about a geometrical approach to the PC analysis of a data set where $p > 3$, but the above discussion makes plausible the following basic features of multivariate analysis:

(i) there are as many principal components variables as there are variables in the data matrix to be analysed;
(ii) the PC variables can be placed in order, the first accounting for the highest proportion of the total variance;
(iii) each principal component is a linear combination of the original variables.

The mathematical procedure for the multivariate PC analysis has exactly the same form as that outlined above for the bivariate case. Again, the eigenvalues and eigenvectors of the dispersion matrix are calculated. If there are $p$ variables, the dispersion matrix (with $p$ rows and columns) has $p$ eigenvectors and $p$ associated eigenvalues. The largest eigenvalue equals the variance of the first PC, with the corresponding eigenvector yielding its coefficients. The second largest eigenvalue equals the variance of the second PC, and so on.

## 11.7 USE OF STANDARDISED DATA MATRIX

We must now face a basic problem in PC analysis which has so far been avoided. It is brought out by contrasting Figure 11.2 with the scatterplot of our bivariate data set at Figure 11.1.

In Figure 11.2, which is based upon hypothetical data, the variance of $X_1$ is very much greater than that of $X_2$ and consequentially the direction of greatest elongation of the data points, the $Z_1$-axis, lies close to the $X_1$-axis. This in turn would mean that the first PC would be dominated by (have a greater weighting on) the variable $X_1$ with the larger variance. And the first PC alone could account for a large proportion of total variance simply as a result of possibly arbitrary measurement scales used for one or both of the variables. No data condensation in any real sense would be achieved: the first PC would be closely equal to $X_1$ and the second PC to $X_2$.

Multivariate data matrices often contain a set of rather disparate variables both in terms of definition and in terms of their basic statistical properties such as their variances. In general terms, where the variances from one variable to another differ considerably, then

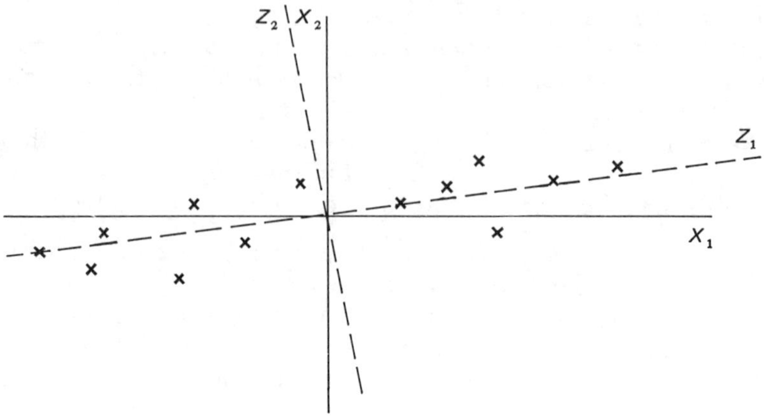


**Figure 11.2** Bivariate Data with Unequal Variances

each will in turn dominate the determination of the principal components successively from the first to the last. As in the above bivariate example, the PC scores thus derived represent no useful condensation of the original data set. Can anything be salvaged?

One solution is to standardise the centred data matrix so that each variable has a variance of one. This puts all the variables on an 'equal footing' so to speak.

A PC analysis carried out on standardised data yields different results from one based on unstandardised data. The geometrical reason for this is fairly clear. If the data behind the scatterplot of Figure 11.7 were standardised, the variances of $X_1$ and $X_2$ would each be one and the slope of $Z_1$ would be tipped away from the $X_1$-axis. (It can in fact be shown that the slope would be 45° exactly.) The coefficient vector of $Z_1$ would change, and also the variance of its scores.

Mathematically, the results differ for the following reason. The PC analysis of (unstandardised) data is based upon calculating the eigenvalues and eigenvectors of the dispersion matrix. On the other hand, the PC analysis of the standardised data matrix is based upon calculating eigenvalues and eigenvectors of the correlation matrix, since the latter equals the dispersion matrix of the standardised data (see result (10.3.1)).

Although in principle there is a choice between analysing unstandardised or standardised data, in practice there may not be any choice. The reason is that many computer programs based their PC analysis *only* upon standardised data. SPSS is a case in point.

Subsequent computations in this chapter will relate to standardised data.

## 11.8 PC ANALYSIS: AN EXAMPLE

We now present and discuss the results of a PC analysis on all 7 variables in the country data matrix of Table 10.1. SPSS was used and so the analysis was inevitably based upon standardised data. In any case, it would not have been sensible to do otherwise since the variances of some of the variables (notably $X_1$, *POPDEN*) are extremely large in relation to the rest and would inevitably have dominated the analysis.

Table 11.3 is based upon the SPSS output, though the form is typical of most packages. It has the same structure as the bivariate analysis Table 11.2 and the reader will recall that 'Factor' means 'Principal component'.

The upper part of Table 11.3 shows that the analysis has had some degree of success in condensing the data. The first three PC's together

**Table 11.3** Multivariate PC Analysis

| *Factor* | *Eigenvalue* | *% of Var.* | *Cum. %* |
|---|---|---|---|
| 1 | 3.94 | 56.2 | 56.2 |
| 2 | 1.56 | 22.3 | 78.6 |
| 3 | 0.81 | 11.6 | 90.2 |
| 4 | 0.36 | 5.1 | 95.3 |
| 5 | 0.27 | 3.9 | 99.1 |
| 6 | 0.05 | 0.6 | 99.8 |
| 7 | 0.02 | 0.2 | 100.0 |

Factor Score Coefficients

| | *Factor 1* | *Factor 2* | *Factor 3* | ...... |
|---|---|---|---|---|
| *POPDEN* | 0.00 | 0.57 | −0.28 | ...... |
| *AGEMP* | −0.24 | −0.09 | 0.29 | ...... |
| *NATINC* | 0.23 | 0.01 | 0.16 | ...... |
| *CAPINV* | 0.04 | 0.51 | 0.60 | ...... |
| *INFMOR* | −0.20 | −0.05 | 0.70 | ...... |
| *ENERGY* | 0.23 | −0.19 | 0.35 | ...... |
| *TVSETS* | 0.23 | −0.07 | 0.30 | ...... |

*Source:* SPSS FACTOR program analysis of Table 10.1 data

account for as much as 90.2% of the total variance (which is measured by the eigenvalues) and even the first two PC's account for nearly 80%. Incidentally, this part of the table offers a check on result (11.4.1) which stated that the total variance of the principal components equals the total variance of the original variables. For, in the case of standardised data, each variable has a variance of unity, and so—with seven variables—the total variance is 7. Also, the total variance of the principal components equals the total of their eigenvalues and the total of the eigenvalues 3.94 + 1.56 + ... + 0.02 does indeed equal 7.

It can therefore be expected that the earlier principal components will well describe the data set and the differences and similarities between individual countries. The question to which we now turn is: do the PC variables themselves have any interpretative meaning? The answer lies in the information in the bottom part of Table 11.3.

Each principal component is a linear combination of the variables and the successive *columns Factor 1* (i.e. PC 1), *Factor 2*, ... give the coefficients of these linear combinations. (For reasons of space, only the first three columns are reproduced.) PC 1 can be interpreted as a variable which indicates the level of economic development. It has positive coefficients attached to those variables for which advanced countries will have relatively large numerical values (*NATINC, ENERGY, TV SETS*) and negative coefficients to variables working the opposite way (*INFMOR, AGEMP*).

PC 2 offers no such simple interpretation and appears to be little more than a mathematical artefact. It has high loadings (i.e. large coefficients) on the two variables *POPDEN* and *CAPINV* which were virtually excluded from PC 1. The principal components analysis has put into PC 2 the sources of variablility not taken account of in PC 1.

The PC scores can also be computed and those for the first three principal components are shown in Table 11.4. Because the first two principal components account for a large proportion of the total variance, one can be fairly confident that a bivariate scatterplot of the PC 1 and PC 2 scores will well sum up the information in the original multivariate data matrix. This plot is reproduced in Figure 11.3 and we leave the reader to assess its plausibility.

## 11.9 USES AND INTERPRETATIONS OF PC ANALYSIS

We can now make some general comment in the light of the case study presented in the last section.

The obvious use is by now clear. If it is the case that the first few

**Table 11.4** Scores for PC 1, PC 2 and PC 3

| | *PC 1* | *PC 2* | *PC 3* |
|---|---|---|---|
| Australia | 1.20 | 0.01 | 0.97 |
| France | 0.81 | 0.48 | 0.15 |
| Germany | 1.41 | 1.39 | 0.06 |
| Greece | −1.75 | −0.06 | −0.53 |
| Iceland | 0.90 | −1.56 | −0.24 |
| Italy | −0.54 | 0.66 | −0.86 |
| Japan | 0.43 | 2.79 | 0.03 |
| New Zealand | 0.05 | −0.63 | −0.23 |
| Portugal | −2.56 | −0.91 | −1.46 |
| Spain | −0.91 | −0.56 | −0.98 |
| Sweden | 1.89 | −1.25 | 0.13 |
| Turkey | −4.75 | −0.18 | 2.11 |
| UK | 0.93 | 1.39 | −0.10 |
| US | 2.89 | −1.57 | 0.93 |
| Means | 0 | 0 | 0 |
| Variances | 3.94 | 1.56 | 0.81 |

*Source:* GENSTAT analysis of Table 10.1 data. Note that the SPSS output standardises each set of PC scores to have unit variance.

principal components account for a large proportion of total variance, then a worthwhile reduction in 'dimensionality' (the number of variables required to describe the set-up) will have been achieved. In such a case, the analysis is a useful exploratory device: a means of looking for 'structure'—groupings of individuals, gaps, outliers, etc.—in an explicitly *multivariate* setting.

In this context, plots of individuals as points according to their scores on each of two components are a valuable technique. A plot of individuals' scores on the first two PC components (such as was presented in Figure 11.3) might be thought of as a two dimensional 'window' into the several dimensional space in which the individuals exist as points. Through this window, some structure of the multivariate data as a whole can be seen. Plots of this kind may conveniently be called 'plots in the space of the first and second PC components'. Plots in the space of other components may also be useful in directing attention to persistent outlying individuals or groupings of individuals.

In interpreting these plots, it should be borne in mind that the set of

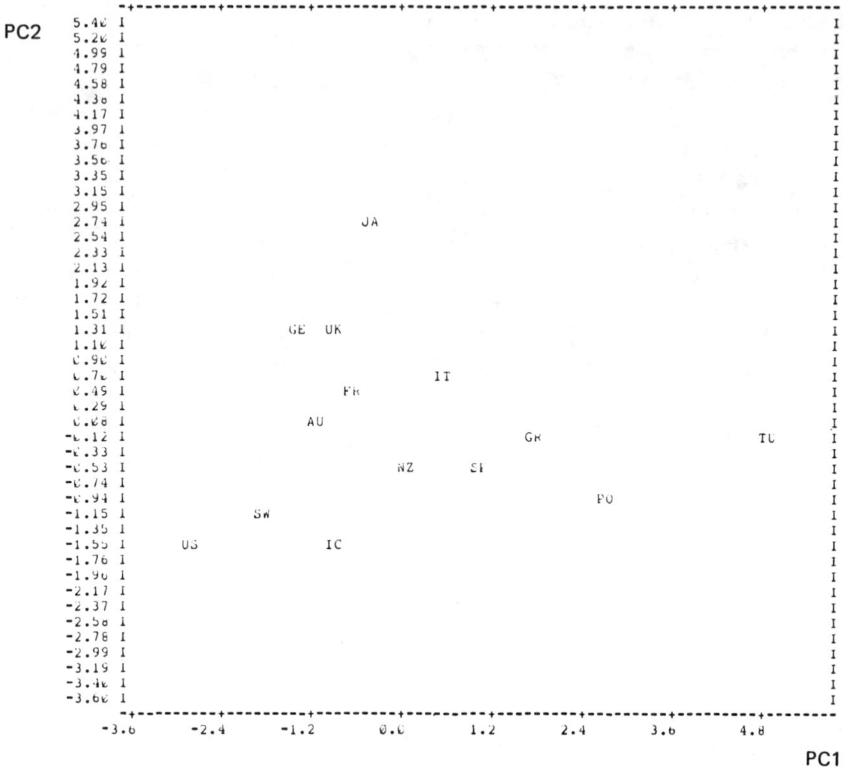


*Source*: SPSS analysis at Table 11.3

**Figure 11.3** Plot of PC1 and PC2 Scores

scores on one principal component is uncorrelated with the set of scores on any other (see result (11.4.2)). Therefore, the plots *might* present a completely unstructured appearance but often they will not as in Figure 11.3.

PC analysis offers a bonus if in addition some interpretative meaning can be given to the principal components. In our country data example, we were able to make some sense of the first PC. This is fairly typical, simply because some or all the variables in a data matrix usually have some contextual connections which are reflected in mutual correlations. However, interpretation of later PC's is rarely straightforward or convincing.

In Section 11.8, our interpretation of PC 1 depended not so much upon the absolute but on the relative magnitudes of the coefficients. This is fortunate because coefficients are often rescaled in computer printouts and in published work for one reason or another. Table 11.5 shows five equivalent presentations of the coefficients of PC 1. Column *a* gives the coefficients which arise from the standard PC analysis; they have the property that the sum of their squared values is unity as for the bivariate coefficients in Section 11.3. Column *b* gives the SPSS coefficients which we reproduced in Table 11.3; SPSS divides each coefficient by the standard deviation of PC 1 ($\sqrt{3.94} = 1.98$) in order that the PC scores should have unit SD and variance. Column *c*, simply to aid interpretation, reproportions the coefficients so that the largest, that of *AGEMP*, is numerically unity. Column *d* is the same as column *a* except that the signs are all reversed. This reflects an important theoretical point that the PC analysis does not *and cannot* distinguish between columns *a* and *d* as acceptable solutions. The reason is clear from the geometrical bivariate analysis diagram at Figure 11.1. If the direction of $Z_1$ were reversed, the $Z_1$-values would still meet the PC condition of maximising the variance but the PC coefficients and scores would each take an opposite sign.

A second approach to an interpretation of a particular principal component is based upon an examination of the correlation coefficients between each of the original variables and the particular principal component whose interpretation is sought. However, this is equivalent to looking at the magnitudes of the coefficients for there is a simple relation between the coefficients of a particular PC and its correlation with each of the variables. Taking Table 11.5, column *a* (the PC solution coefficients) as an example, the correlation coefficients between PC 1 and each variable are given by multiplying each coefficient by the square root of the eigenvalue, $\sqrt{3.94}$. Alternatively, the SPSS coefficients in column *b* would be multiplied by the eigenvalue, 3.94. These correlation coefficients are shown in column *e*.

**Table 11.5** Alternative Presentations of PC Coefficients

| | *a* | *b* | *c* | *d* | *e* |
|---|---|---|---|---|---|
| *POPDEN* | 0.00 | 0.00 | 0.02 | 0.00 | 0.02 |
| *AGEMP* | −0.48 | −0.24 | −1.00 | 0.48 | −0.94 |
| *NATINC* | 0.45 | 0.23 | 0.95 | −0.45 | 0.90 |
| *CAPINV* | 0.08 | 0.04 | 0.17 | −0.08 | 0.16 |
| *INFMOR* | −0.40 | −0.20 | −0.83 | 0.40 | −0.78 |
| *ENERGY* | 0.45 | 0.23 | 0.95 | −0.45 | 0.89 |
| *TVSETS* | 0.45 | 0.23 | 0.95 | −0.45 | 0.90 |

*Source:* coefficients of PC 1 (Factor 1) in Table 11.3

The reader will have noticed that we have not offered any prescription about how many of the later PC's can be disregarded. However, it often turns out in practice that the percentage variance accounted for by successive components beyond the third or fourth is usually small, suggesting that most of the variation in the multivariate data set can be adequately described by the first few components.

## 11.10 PRINCIPAL COMPONENTS IN REGRESSION ANALYSIS

In Chapter 8, we considered the problem of multicollinearity in regression analysis. Briefly, if there are intercorrelations between the explanatory variables, it becomes difficult or impossible to disentangle their separate effects on the dependent variable and regression coefficients tend to have large standard errors.

It might be thought that one simple way out of this would be to perform a PC analysis on the explanatory variables and then regress the dependent variable on some or all of the principal components, since the latter are uncorrelated. Unfortunately this procedure rarely helps. Unless there is a satisfactory interpretation to the principal components, such a regression would have little contextual significance. And in any case it would throw no additional light on the nature of the relation between the dependent variable and the original explanatory variables.

## 11.11 FACTOR ANALYSIS

Principal components analysis is a descriptive statistics procedure in which each of the PC variables is defined as a linear combination of the original variables. Mathematically, the analysis sets up, in the $p$ variable case, equations of the form:

$$\begin{aligned} \mathrm{PC}(1) &= a_{11}X_1 + \dots + a_{1p}X_p \\ \mathrm{PC}(2) &= a_{21}X_1 + \dots + a_{2p}X_p \\ &\dots\dots \end{aligned} \tag{11.11.1}$$

The original $X$-variables can similarly be expressed as linear combinations of the PC variables. It is a matter of algebra to solve equations (11.11.1) and arrive at equations of the form:

$$\begin{aligned} X_1 &= b_{11}\mathrm{PC}(1) + \dots + b_{1p}\mathrm{PC}(p) \\ X_2 &= b_{21}\mathrm{PC}(1) + \dots + b_{2p}\mathrm{PC}(p) \\ &\dots\dots \end{aligned} \tag{11.11.2}$$

(If the PC coefficients of (11.11.1) are scaled so that their sum of squares is unity, as in column *a* of Table 11.3, there is a very simple connection between the *a*-coefficients of equations (11.11.1) and the *b*-coefficients of equations (11.11.2). The *columns* of the *a*'s equal the *rows* of the *b*'s.)

Factor analysis on the other hand is based upon setting up a statistical model of the data. Each variable is modelled as a linear combination of a small number of 'factors' with the addition of a random component term, a little like regression analysis. So the factor analysis model has the form:

$$\begin{aligned} X_1 &= k_{11}F_1 + \ldots + k_{1m}F_m + u_1 \\ X_2 &= k_{21}F_1 + \ldots + k_{2m}F_m + u_2 \\ &\ldots\ldots \end{aligned} \tag{11.11.3}$$

where $m$ denotes the number of factors in the model and $m \leqslant p$. (Unlike regression analysis, the factors $F_1$, $F_2$, ... are hypothetical variables.)

In the limiting case when the number of factors equals the number of variables, the random component terms disappear and the factor analysis model equations (11.11.3) reduce to the PC equations (11.11.2). PC analysis can therefore be looked upon as a special case of factor analysis. This is the reason why computer packages often incorporate PC analysis within a factor analysis framework and why, as we noted in Section 11.5, program output adopts factor analysis terminology. In particular, the term *factor* is used instead of *principal component* and sometimes *loading* instead of *coefficient.* The reference to *communality* in the SPSS and other output is redundant for principal component analysis.

Factor analysis is a complex subject which we do not develop in this text. In any case we share the view which has been expressed by others that the very many different procedural possibilities in factor analysis make interpretation difficult and conclusions of doubtful foundation (see Chatfield and Collins (1980)).

## EXAMPLES

E 11.1 Using Table 10.3, plot the scatterplot of the standardised values of $X_3$ and $X_6$. Draw in the two principal component axes, which are the 45° lines $x_3 = x_6$ and $x_3 = -x_6$ for bivariate standardised variables. Calculate the first and second principal component scores, which are respectively given by linear combinations:

$$Z_1 = (1/\sqrt{2})X_1 + (1/\sqrt{2})X_2 \text{ and}$$
$$Z_2 = -(1/\sqrt{2})X_1 + (1/\sqrt{2})X_2$$

Check a few of the calculations graphically.

**E 11.2** Repeat E 11.1 using the standardised variables $X_4$ and $X_6$.

**E 11.3** Calculate the variances of the first and second principal components for: (i) case E 11.1 and; (ii) case E 11.2. Check that in each case the total variance = 2.

**E 11.4** Why does the first PC account for a greater proportion of the total variance in case E 11.1 than in case E 11.2?

**E 11.5** Two multivariate distance matrices are calculated. The first, $\mathbf{D}_1$, is based upon all the PC variables (calculated from standardised data). The second, $\mathbf{D}_2$, is based upon the original standardised variables themselves. What is the geometrical justification for the theoretical result that $\mathbf{D}_1 = \mathbf{D}_2$?

**E 11.6** A distance matrix $\mathbf{D}_3$, is calculated using only the PC 1 and PC 2 values. Why would one expect $\mathbf{D}_3$ to closely approximate $\mathbf{D}_1$ and $\mathbf{D}_2$?

## FURTHER WORK

**F 11.1** Carry out a PC analysis on the explanatory variables of Table 8.3 (i.e. all variables except *IHDMR*). Review the results, particularly considering the effectiveness of the first few principal components and whether or not they have any plausible interpretation.

**F 11.2** Carry out a regression analysis with *IHDMR* as the dependent variable and the first two principal components as explanatory variables. Can the results be interpreted?

*This chapter depends upon: Chapters 10 and 11*

# 12

# Discriminant Analysis

## 12.1 INTRODUCTION

Discriminant analysis is concerned with distinguishing (that is, discriminating) between two or more groups of individuals on the basis of a common set of variable values for each of the individuals in each group. For example, we may have data for a number of economic and social variables for each of a number of countries where each country is categorised as being developed or less-developed. Discriminant analysis examines how far the statistical data in itself serves to discriminate between the two groups. Or again, suppose a set of students has sat a number of examination papers, the variables being the marks on each paper, and suppose that two groups of students with differing educational backgrounds sat the examination. Discriminant analysis could be used to answer the question whether there is any difference in performance—as measured by the marks—between the two groups.

As these two examples demonstrate, one purpose of discriminant analysis is exploration and explanation: how far, if at all, does the multivariate data confirm the prior categorisation of the individual into the various groups?

Table 12.1 shows the archaeological data set which will be used to illustrate the methods of this chapter. The individuals are Greek pottery fragments and, for each fragment, various chemical metallic

**Table 12.1** Greek Pottery: Metallic Oxide Content

| | | *SI* | *AL* | *FE* | *MG* | *CA* | *TI* |
|---|---|---|---|---|---|---|---|
| Attic | | 55.8 | 14.0 | 10.2 | 4.9 | 5.0 | 0.88 |
| | | 51.2 | 12.5 | 10.1 | 4.4 | 4.8 | 0.86 |
| | | 57.1 | 14.0 | 8.3 | 6.4 | 11.2 | 0.75 |
| | | 53.8 | 13.1 | 9.3 | 4.9 | 6.6 | 0.81 |
| | | 59.4 | 14.8 | 9.8 | 5.5 | 5.4 | 0.89 |
| | | 56.2 | 14.0 | 9.9 | 4.9 | 5.4 | 0.89 |
| | | 54.3 | 13.7 | 10.3 | 6.3 | 5.1 | 0.89 |
| | | 52.9 | 13.4 | 9.9 | 7.5 | 8.5 | 0.89 |
| | | 58.5 | 17.0 | 9.3 | 2.9 | 5.2 | 0.90 |
| | | 56.7 | 14.7 | 9.8 | 4.6 | 4.6 | 0.87 |
| | | 57.9 | 14.7 | 10.4 | 5.3 | 4.4 | 0.88 |
| | | 69.9 | 13.1 | 9.8 | 4.4 | 4.4 | 0.89 |
| | | 55.0 | 13.1 | 10.0 | 4.4 | 3.9 | 0.89 |
| | Means | 56.8 | 14.0 | 9.8 | 5.1 | 5.7 | 0.87 |
| Eritrean | | 50.9 | 15.7 | 9.5 | 2.5 | 5.8 | 0.86 |
| | | 50.8 | 15.5 | 9.5 | 2.3 | 4.9 | 0.85 |
| | | 51.5 | 17.2 | 9.3 | 5.5 | 8.7 | 0.78 |
| | | 53.2 | 14.3 | 7.4 | 2.4 | 7.0 | 0.76 |
| | | 58.8 | 17.7 | 8.1 | 2.5 | 5.7 | 0.86 |
| | | 53.9 | 16.5 | 9.5 | 2.5 | 4.7 | 0.85 |
| | | 63.1 | 17.6 | 9.3 | 2.7 | 4.2 | 0.87 |
| | | 54.5 | 17.3 | 9.8 | 2.6 | 4.9 | 0.85 |
| | | 48.1 | 15.8 | 9.8 | 2.0 | 5.6 | 0.83 |
| | | 52.7 | 16.3 | 8.4 | 5.1 | 7.7 | 0.76 |
| | | 54.1 | 13.6 | 8.1 | 4.4 | 12.8 | 0.71 |
| | | 51.0 | 13.0 | 8.5 | 4.2 | 9.4 | 0.71 |
| | | 57.2 | 17.3 | 9.3 | 2.7 | 5.6 | 0.83 |
| | | 52.4 | 16.4 | 8.8 | 3.3 | 7.7 | 0.77 |
| | Means | 53.7 | 16.0 | 8.9 | 3.2 | 6.7 | 0.81 |

oxide constituents were measured, giving information on $p = 6$ variables. From prior archaeological knowledge, the fragments have been categorised into four groups according to their origin: (i) Attic, (ii) Eritrean, (iii) probably Eritrean, and (iv) probably Attic. The archaeological interest in this data set is how far the variable values, that is the chemical constituents of the pottery, can be used to distinguish between their origins.

| | *SI* | *AL* | *FE* | *MG* | *CA* | *TI* |
|---|---|---|---|---|---|---|
| Probably Eritrean | 52.8 | 16.5 | 9.9 | 2.2 | 3.9 | 0.87 |
| | 63.4 | 13.6 | 8.8 | 5.3 | 5.1 | 0.90 |
| | 59.6 | 18.1 | 9.3 | 2.7 | 4.0 | 0.88 |
| | 62.0 | 18.3 | 9.8 | 2.7 | 3.4 | 0.89 |
| | 51.5 | 16.1 | 10.1 | 2.4 | 4.9 | 0.87 |
| | 48.4 | 14.4 | 8.2 | 3.4 | 6.7 | 0.76 |
| | 51.3 | 14.3 | 8.1 | 4.0 | 8.6 | 0.74 |
| | 52.4 | 15.5 | 8.9 | 3.0 | 6.2 | 0.80 |
| | 53.4 | 11.2 | 8.2 | 4.8 | 6.8 | 0.79 |
| Means | 55.0 | 15.3 | 9.0 | 3.4 | 5.5 | 0.83 |
| Probably Attic | 55.2 | 12.2 | 8.7 | 5.5 | 8.8 | 0.82 |
| | 48.4 | 10.8 | 8.1 | 5.2 | 10.3 | 0.75 |
| | 49.7 | 10.9 | 8.1 | 4.4 | 9.5 | 0.75 |
| | 49.3 | 11.7 | 9.7 | 4.4 | 5.8 | 0.82 |
| Means | 50.6 | 11.4 | 8.6 | 4.9 | 8.6 | 0.78 |
| Overall Means | 54.7 | 14.7 | 9.2 | 4.0 | 6.3 | 0.83 |

*Source:* W.B. Stern and J.-P. Descoeudres, 'X-ray fluorescence analysis of archaic Greek pottery', *Archaeometry*, Vol. 19 (1977)
Variable labels are chemical symbols for metals concerned

We have chosen this archaeological data set because it well illustrates another important use of discriminant analysis, that of *classification*. Some of the fragments in the data set are of doubtful origin. Given the data, can such fragments be shown to have Attic (or Eritrean) characteristics? If so, the analysis could be used to classify the doubtful cases accordingly.

The plan of this chapter is as follows. Discriminant analysis for the simple case of bivariate data will be considered first. This will allow the introduction of some basic terminology and procedures. The discussion will then be extended to multivariate data. In all these earlier sections, discussion will be restricted to the two group case, that is, to just the (known) Attic and Eritrean pottery fragments. It will then be extended to the multiple group case where there are more than two groups of individuals in the data matrix. The use of discriminant analysis for classifying individuals to particular groups will then follow.

## 12.2 BIVARIATE DISCRIMINANT ANALYSIS FOR TWO GROUPS

In the present section, we shall make use of just two columns of the data matrix of Table 12.1 (the variables *MG* and *CA*) and only the individuals in the first two groups Attic and Eritrean. Table 12.2 reproduces these data; the remaining columns headed *d* can be ignored for the present.

The essential methodology of discriminant analysis can be approached geometrically through a scatterplot of the two variables (see Figure 12.1). For this particular data set, the data points of the Attic jars on the whole lie mainly in one region of the graph, the 'north west', and the set for the Eritrean jars in another region, the 'south east', though there is some overlapping. This kind of picture is typical of situations where the two variables in the data matrix achieve some degree of discrimination between the groups. Where there is no discrimination in terms of the variables, the data points would be completely intermingled; where there is very good discrimination the two clusters of data points would be completely separated.

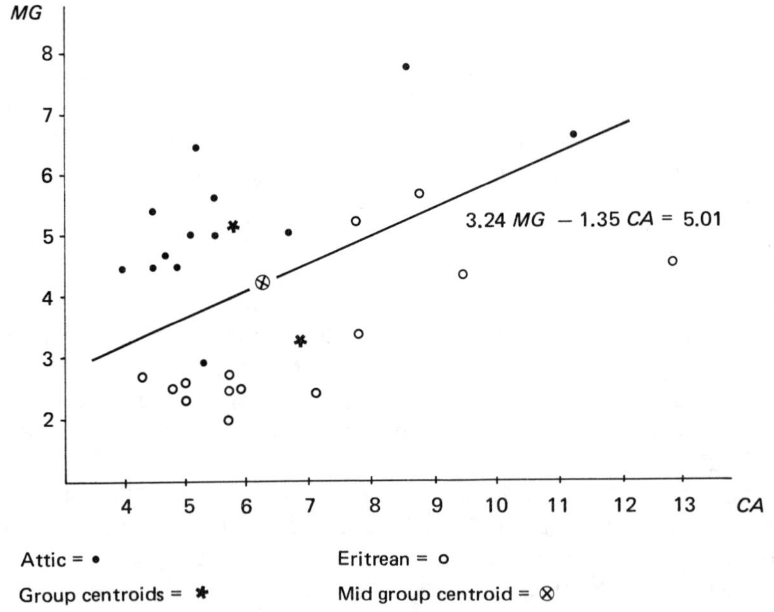


**Figure 12.1** Scatterplot of Attic and Eritrean Pottery Groups (variables *MG* and *CA*)

The line which has been drawn on the scatterplot is that straight line which 'best' divides the two groups of data points. This line is known as the *linear discriminant boundary.* How its equation is calculated (and what is meant by 'best') is something we shall come back to. Meanwhile, we note that its equation can be shown to be $3.24MG - 1.35\,CA = 5.01$ which, collecting all the terms on one side of the equation, can be written as

$$3.24\,MG - 1.35\,CA - 5.01 = 0 \qquad \textbf{(12.2.1)}$$

The left-hand side of this equation is known as the linear discriminant function, or briefly, the *discriminant function.*

The discriminant function, $3.24\ MG - 1.35\ CA - 5.01$ for the present data, plays a central role in discriminant analysis. We can calculate the value of the discriminant function for any point on the graph—and, in particular, any data point—by substituting the coordinates of the point in the discriminant function. This discriminant function value is called the *discriminant score.* Thus, the first of

**Table 12.2** Bivariate Discriminant Analysis. Fisher Discriminant Scores

| | *Attic* | | | *Eritrean* | | |
|---|---|---|---|---|---|---|
| | *MG* | *CA* | *d* | *MG* | *CA* | *d* |
| | 4.9 | 5.0 | 4.1 | 2.5 | 5.8 | −4.7 |
| | 4.4 | 4.8 | 2.8 | 2.3 | 4.9 | −4.2 |
| | 6.4 | 11.2 | 0.6 | 5.5 | 8.7 | 1.1 |
| | 4.9 | 6.6 | 1.9 | 2.4 | 7.0 | −6.7 |
| | 5.5 | 5.4 | 5.5 | 2.5 | 5.7 | −4.6 |
| | 4.9 | 5.4 | 3.6 | 2.5 | 4.7 | −3.3 |
| | 6.3 | 5.1 | 8.5 | 2.7 | 4.2 | −1.9 |
| | 7.5 | 8.5 | 7.8 | 2.6 | 4.9 | −3.2 |
| | 2.9 | 5.2 | −2.6 | 2.0 | 5.6 | −6.1 |
| | 4.6 | 4.6 | 3.7 | 5.1 | 7.7 | 1.1 |
| | 5.3 | 4.4 | 6.2 | 4.4 | 12.8 | −8.0 |
| | 4.4 | 4.4 | 3.3 | 4.2 | 9.4 | −4.1 |
| | 4.4 | 3.9 | 4.0 | 2.7 | 5.6 | −3.8 |
| | | | | 3.3 | 7.7 | −4.7 |
| Means | 5.1 | 5.7 | 3.8 | 3.2 | 6.7 | −3.8 |

*Source:* SPSS analysis of Table 12.1 data

the Attic fragments has coordinates $MG = 4.9$, $CA = 5.0$ and so its discriminant score is

$$(3.24)(4.9) - (1.35)(5.0) - 5.01 = 4.1$$

Similarly the first of the Eritrean jars has coordinates $MG = 2.5$, $CA = 5.8$ and so its discriminant score is

$$(3.24)(2.5) - (1.35)(5.8) - 5.01 = -4.7$$

In a similar manner, we can calculate the discriminant scores for each of the data points—each Attic and each Eritrean fragment—and they are shown in the columns headed $d$ of Table 12.2.

The discriminant function, and the associated discriminant scores, play a central role in discriminant analysis. The reason stems from two simple general algebraic properties:

(i) The discriminant score of any point lying on the discriminant boundary is zero, for its values would satisfy the discriminant boundary equation.
(ii) The discriminant scores of points lying on one side of the discriminant boundary have an opposite sign to those of points lying on the opposite side. **(12.2.2)**

Reference to Table 12.2 shows that, on the Attic side of the discriminant boundary, the values of the discriminant function are all positive. Most of the Attic data points exhibit positive discriminant scores. Those which do not, which have negative discriminant scores, correspond to data points on the Eritrean side of the discriminant boundary.

As a consequence, a glance down the columns of discriminant scores quickly shows how effective the discriminant line is in separating the two groups of jars. The more inconsistency of signs there is within each group, the less satisfactory is the discrimination and the more mixed up are the data points of the two groups.

Furthermore, the numerical magnitude of the discriminant score of a data point tells us how far the data point is away from the discriminant boundary. For example, the Attic fragment with values $MG = 7.5$, $CA = 8.5$ has a discriminant score of 7.8, whilst the Attic fragment with values $MG = 6.4$, $CA = 11.2$ has a discriminant score of 0.6. By looking at Figure 12.1, the reader will see that the corresponding data points are respectively far from and near to the discriminant boundary. (If the discriminant function values were rescaled by dividing them by $\sqrt{[(3.24^2 + (-1.35)^2]}$, these numbers being the coefficients in the discriminant function, the discriminant scores would then exactly equal the perpendicular distances of the data points from the discriminant boundary.)

## 12.3 SPECIFICATION OF BIVARIATE DISCRIMINANT FUNCTION

We shall not go into the detail of the specification and calculation of the bivariate discriminant function since in any case the procedures for the more useful multivariate situation would take us beyond the mathematical limits set for this text. Nevertheless, it is useful to have some intuitive understanding if only to appreciate the limitations of the methods and where they might be inappropriate for particular kinds of data.

In order to calculate a discriminant boundary, and hence a discriminant function, we first need acceptable criteria which the calculation procedure should satisfy, just as in regression analysis, the least squares principle is the criterion which determines the calculation of the line of best fit.

The first point is that, as in the previous section, the discriminant boundary is defined to be a straight line. As a consequence, the discriminant function is a simple linear function of the two variables. For the pottery data, the linear discriminant function was $3.24\ MG - 1.35\ CA - 5.01$. In general, for two variables denoted by $X_1$ and $X_2$, the linear discriminant function can be written as $a_1X_1 + a_2X_2 + c$ where the parameters $a_1$, $a_2$ and $c$ are calculated from the data. Although discriminant analysis can be extended to the case where the discriminant boundary is curved, we shall limit ourselves to the linear case and henceforth 'discriminant function' will stand for 'linear discriminant function'.

We now come to the criteria for calculating the 'best' discriminant boundary and the associated discriminant function. The values of the coefficients $a_1$ and $a_2$ are chosen so that the discriminant scores of individuals in any one group have the smallest possible variance and, at the same time, the difference between the average discriminant scores of the two groups is as big as possible. The value of the constant $c$ is chosen to make the discriminant boundary pass through the point midway between the centroids of the two groups, as shown on Figure 12.1.

In Section 10.4 we introduced the notion of deriving new variables by setting up linear combinations of the data variables and we saw in Chapter 11 that the principal components are particular linear combinations having certain properties. The same is true of the discriminant function. Apart from the addition of the constant, $c$, the discriminant function in this bivariate case is a linear combination of the two variables $X_1$ and $X_2$. The coefficients $a_1$ and $a_2$ form the coefficient vector for this linear combination; the coefficient vector for the pottery data was $(3.24, -1.35)$. The discriminant *scores* are the values for each individual of this derived variable. The effect of adding the constant, $c$, to this derived variable simply ensures that the discriminant scores of one group tend to have an opposite sign to those of the other group.

## 12.4 COMPUTATION—BIVARIATE DATA

Computer output in this area can be extremely complex and this is particularly true of SPSS, which will be used for illustrative purposes in the present chapter.

In SPSS, the discriminant function discussed above can be derived from that part of the output reproduced in Table 12.3. The output in fact shows the coefficients for two functions:

$4.41\,MG - 0.36\,CA - 10.94$ and $1.17\,MG + 1.00\,CA - 5.93$. The *difference* between them gives the discriminant function for the pottery data $3.24\,MG - 1.36\,CA - 5.01$ at (12.2.1). (This decomposition of the discriminant function is connected with the use of discriminant analysis for classification purposes and we shall return to this point in Section 12.7.)

The discriminant function specified up to now is often known as the Fisher linear discriminant function and indeed is so named in the SPSS output of Table 12.3. For some purposes it is convenient to rescale this discriminant function by a simple linear transformation which alters the mean and the variance of the discriminant scores. Unfortunately, SPSS does not give the discriminant scores arising from the Fisher linear discriminant function and only prints out transformed values. Table 12.4 shows, for each pottery fragment, the (Fisher) discriminant scores from Table 12.2 together with the SPSS transformed scores.

As we have already noted, the Fisher discriminant scores have positive values in the Attic region of the scatterplot and negative

**Table 12.3** Fisher Discriminant Function Coefficient

| | *Classification function coefficients (Fisher's linear discriminant functions)* | |
|---|---|---|
| | *AT* | *ER* |
| MG | 4.41 | 1.17 |
| CA | −0.36 | 1.00 |
| (Constant) | −10.94 | −5.93 |

*Source:* SPSS DISCRIMINANT output (Option 12)

**Table 12.4** Fisher and SPSS Transformed Discriminant Scores

| | | | | | | | | | |
|---|---|---|---|---|---|---|---|---|---|
| *Attic* | Fisher | 4.1 | 2.8 | 0.6 | 1.9 | 5.5 | 3.6 | 8.5 | 7.8 |
| | SPSS | 1.5 | 1.1 | 0.3 | 0.8 | 2.1 | 1.3 | 3.1 | 2.9 |
| | Fisher | −2.6* | 3.7 | 6.2 | 3.3 | 4.0 | | | |
| | SPSS | −0.9* | 1.4 | 2.3 | 1.3 | 1.5 | | | |
| *Eritrean* | Fisher | −4.7 | −4.2 | 1.1* | −6.7 | −4.6 | −3.3 | −1.9 | −3.2 |
| | SPSS | −1.7 | −1.5 | 0.4* | −2.4 | −1.6 | −1.1 | −0.6 | −1.1 |
| | Fisher | −6.1 | 1.1* | −8.0 | −4.1 | −3.8 | −4.7 | | |
| | SPSS | −2.2 | 0.5* | −2.9 | −1.4 | −1.3 | −1.7 | | |

*Source:* Derived from Tables 12.2 and 12.3

values in the Eritrean. So the Attic fragments lying in the Eritrean region and the Eritrean fragments lying in the Attic region, can readily be identified by the change of sign; such inconsistencies are identified by * in Table 12.4. The SPSS transformation, which ensures that the overall average discriminant score is zero, does not in general have this simple property, but the asterisks in the printout identify the same inconsistencies.

Table 12.5 shows an extract from that part of the SPSS output which gives the transformed scores. The final column gives the SPSS discriminant scores already shown in Table 12.4 and the symbol *** identifies those individuals which do not lie in their own group region. (The middle columns and the later rows labelled UNGRPD should be ignored for the present.) The SPSS discriminant function behind these scores is $1.16\ MG - 0.49\ CA + 1.77$.

## 12.5 MULTIVARIATE DISCRIMINANT ANALYSIS FOR TWO GROUPS

In practical terms, it is a simple matter to extend bivariate analysis to multivariate analysis. In the case of two groups of individuals, again there is one discriminant function which is a linear combination of many variables rather than just two. In general, the discriminant function takes the form $a_1X_1 + a_2X_2 + \ldots + a_pX_p + c$. The Fisher discriminant function for the pottery data, including all six variables in the analysis, is

$$-2.6\ AL - 2.2\ FE + 3.4\ MG - 0.5\ CA + 103.7\ TI - 37.7 \qquad \textbf{(12.5.1)}$$

**Table 12.5** Bivariate Discriminant Analysis Computer Output

| SUBFILE | *Case* SEQNUM | Actual group | Highest probability group | $p(x/g)$ | $p(g/x)$ | 2nd Highest group | $p(g/x)$ | Discriminant scores |
|---|---|---|---|---|---|---|---|---|
| AT | 1 | 1 | 1 | 0.9093 | 0.9840 | 2 | 0.0160 | 1.5440 |
| AT | 2 | 1 | 1 | 0.7069 | 0.9408 | 2 | 0.0592 | 1.0541 |
| AT | 3 | 1 | 1 | 0.2457 | 0.6461 | 2 | 0.3539 | 0.2693 |
| AT | 4 | 1 | 1 | 0.5027 | 0.8760 | 2 | 0.1240 | 0.7599 |
| AT | 5 | 1 | 1 | 0.5330 | 0.9960 | 2 | 0.0040 | 2.0535 |
| AT | 6 | 1 | 1 | 0.9346 | 0.9728 | 2 | 0.0272 | 1.3480 |
| AT | 7 | 1 | 1 | 0.0871 | 0.9998 | 2 | 0.0002 | 3.1412 |
| AT | 8 | 1 | 1 | 0.1454 | 0.9996 | 2 | 0.0004 | 2.8859 |
| AT | 9 | 1*** | 2 | 0.6728 | 0.9333 | 1 | 0.0667 | −0.9057 |
| AT | 10 | 1 | 1 | 0.9659 | 0.9755 | 2 | 0.0245 | 1.3873 |
| AT | 11 | 1 | 1 | 0.3798 | 0.9980 | 2 | 0.0020 | 2.3084 |
| AT | 12 | 1 | 1 | 0.8572 | 0.9647 | 2 | 0.0353 | 1.2502 |
| AT | 13 | 1 | 1 | 0.9481 | 0.9817 | 2 | 0.0183 | 1.4952 |
| ER | 1 | 2 | 2 | 0.7323 | 0.9914 | 1 | 0.0086 | −1.6701 |
| ER | 2 | 2 | 2 | 0.8917 | 0.9849 | 1 | 0.0151 | −1.4642 |
| ER | 3 | 2*** | 1 | 0.3203 | 0.7431 | 2 | 0.2569 | 0.4362 |
| ER | 4 | 2 | 2 | 0.2947 | 0.9988 | 1 | 0.0012 | −2.3758 |
| ER | 5 | 2 | 2 | 0.7694 | 0.9902 | 1 | 0.0098 | −1.6211 |
| ER | 6 | 2 | 2 | 0.8438 | 0.9630 | 1 | 0.0370 | −1.1310 |

| | | | | | | | | |
|---|---|---|---|---|---|---|---|---|
| ER | 7 | 2 | 2 | 0.4983 | 0.8739 | 1 | 0.1261 | −0.6508 |
| ER | 8 | 2 | 2 | 0.8286 | 0.9611 | 1 | 0.0389 | −1.1114 |
| ER | 9 | 2 | 2 | 0.4054 | 0.9978 | 1 | 0.0022 | −2.1600 |
| ER | 10 | 2*** | 1 | 0.3300 | 0.7534 | 2 | 0.2466 | 0.4560 |
| ER | 11 | 2 | 2 | 0.1239 | 0.9997 | 1 | 0.0003 | −2.8666 |
| ER | 12 | 2 | 2 | 0.9144 | 0.9837 | 1 | 0.0163 | −1.4354 |
| ER | 13 | 2 | 2 | 0.9929 | 0.9787 | 1 | 0.0213 | −1.3369 |
| ER | 14 | 2 | 2 | 0.7394 | 0.9912 | 1 | 0.0088 | −1.6606 |
| PE | 1 | UNGRPD | 2 | 0.8132 | 0.9590 | 1 | 0.0410 | −1.0917 |
| PE | 2 | UNGRPD | 1 | 0.5925 | 0.9949 | 2 | 0.0051 | 1.9654 |
| PE | 3 | UNGRPD | 2 | 0.4382 | 0.8410 | 1 | 0.1590 | −0.5527 |
| PE | 4 | UNGRPD | 2 | 0.2849 | 0.7015 | 1 | 0.2985 | −0.2587 |
| PE | 5 | UNGRPD | 2 | 0.9851 | 0.9793 | 1 | 0.0207 | −1.3466 |
| PE | 6 | UNGRPD | 2 | 0.7833 | 0.9546 | 1 | 0.0454 | −1.0529 |
| PE | 7 | UNGRPD | 2 | 0.9606 | 0.9751 | 1 | 0.0249 | −1.2785 |
| PE | 8 | UNGRPD | 2 | 0.9603 | 0.9751 | 1 | 0.0249 | −1.2782 |
| PE | 9 | UNGRPD | 1 | 0.3757 | 0.7958 | 2 | 0.2042 | 0.5443 |
| PA | 1 | UNGRPD | 1 | 0.2970 | 0.7165 | 2 | 0.2835 | 0.3872 |
| PA | 2 | UNGRPD | 2 | 0.5305 | 0.8883 | 1 | 0.1117 | −0.7007 |
| PA | 3 | UNGRPD | 2 | 0.9373 | 0.9731 | 1 | 0.0269 | −1.2493 |
| PA | 4 | UNGRPD | 1 | 0.3865 | 0.8045 | 2 | 0.1955 | 0.5640 |

*Source:* SPSS output based upon Table 12.2 data

The SPSS transformed discriminant function is

$$-0.64\,AL - 0.34\,\mathbf{FE} + 0.84\,MG - 0.12\,CA + 25.17\,TI - 9.07 \tag{12.5.2}$$

(Note that the coefficients of (12.5.2) are proportional to those of (12.5.1).)

The computer printout of the discriminant scores for this multivariate case has exactly the same form as shown in Table 12.5. For space reasons we do not reproduce it, but Table 12.6 compares the (SPSS) discriminant scores for the bivariate analysis with those from the multivariate analysis. It can be seen that only one pottery fragment (no.9 in the Attic group) has a discriminant score appropriate to the other group compared to three such cases in the bivariate analysis. In other words, the multivariate analysis has achieved a more effective discrimination of the two groups than did the bivariate analysis.

**Table 12.6** Discriminant Scores: Bivariate and Multivariate Two Group Analysis

| | | | | | | | | | |
|---|---|---|---|---|---|---|---|---|---|
| Attic | $p = 2$ | 1.5 | 1.1 | 0.3 | 0.8 | 2.1 | 1.3 | 3.1 | 2.9 |
| | $p = 6$ | 2.1 | 2.2 | 0.4 | 1.2 | 2.5 | 2.5 | 3.7 | 4.6 |
| | $p = 2$ | −0.9* | 1.4 | 2.3 | 1.3 | 1.5 | | | |
| | $p = 6$ | −0.5* | 1.4 | 2.0 | 2.8 | 2.8 | | | |
| Eritrean | $p = 2$ | −1.7 | −1.5 | 0.4* | −2.4 | −1.6 | −1.1 | −0.6 | −1.1 |
| | $p = 6$ | −1.2 | −1.4 | −1.9 | −1.9 | −1.7 | −1.8 | −1.7 | −2.4 |
| | $p = 2$ | −2.2 | 0.5* | −2.9 | −1.4 | −1.3 | −1.7 | | |
| | $p = 6$ | −2.6 | −1.6 | −2.2 | −1.7 | −2.7 | −3.1 | | |

$p = 2$ gives bivariate scores, $p = 6$ gives multivariate scores
*Source:* Table 12.4 ($p = 2$) and SPSS DISCRIMINANT output ($p = 6$)

## 12.6 MULTIPLE GROUP DISCRIMINANT ANALYSIS

So far, we have considered the problem of discriminating between two groups of individuals. We have seen how a single discriminant function can be calculated which has the property that the discriminant scores of individuals in one group tend to be closer to one another than the discriminant scores of individuals in different groups. This of course will only happen if indeed there are overall differences between the two groups as regards the variable values.

We now turn to the case where there are more than two identified groups of individuals as in the full set of pottery data in Table 12.1. There, four groups are identified: in addition to Attic and Eritrean there are the two groups 'probably Eritrean' and 'probably Attic'. For the purposes of exposition, we shall for the present treat them as four separately identified groups, although there are obvious contextual links between them.

There are two main approaches to multiple group discriminant analysis. The first, which we shall only briefly consider, is essentially based upon applying the two group procedure to each pair of groups in the data matrix. In our case of four groups, there are as many as six pairs which implies six separate two group discriminant analyses. In practice, however, the situation is not as complex as may appear, particularly if the main interest is in deciding in which group region an individual data point lies.

It will be recalled (Section 12.4), that the two group Fisher discriminant function, $D$ say, was derived by taking the difference between two component functions, $D_{AT}$ and $D_{ER}$.[1] That is, $D = D_{AT} - D_{ER}$. Now an individual is in the Attic region if its $D$ value is greater than 0, that is if $D_{AT} > D_{ER}$. So the rule for deciding in which group region an individual lies can be reformulated by saying that it lies in the Attic region if $D_{AT}$ is the greatest and in the Eritrean region if $D_{ER}$ is the greatest. In the case of four groups, four consistent functions can be calculated, one connected with each of the four groups, and SPSS prints out the coefficients of these four functions. In order to decide in which of the four group regions an individual lies, its data values are successively substituted in each of the functions and the function with the greatest value determines the appropriate region.

The second approach to multiple group discrimination involves the calculation of a sequence of *canonical discriminant functions*. In many ways, it provides a more convenient method of evaluating a complex data set with many groups and it forms the basis of the standard SPSS printout.

This canonical method adopts the approach outlined in general terms in Section 10.5, and one which was used in principal components analysis. It calculates an ordered sequence of discriminant functions, each being a different linear combination of the variables in the data matrix. The first discriminant function is that single derived variable whose values best separate out the individuals into the groups. The second discriminant function, together with the first, provides the

1. The subscripts $AT$ and $ER$ relate these constituent functions to the two groups for reasons which we do not elaborate here.

**Table 12.7** Multiple Group Discriminant Analysis Output

| *Actual group* | *Predicted group* | *Discriminant scores* | | |
|---|---|---|---|---|
| AT | AT | 2.0 | 0.7 | 0.3 |
| AT | AT | 2.1 | −0.3 | −0.3 |
| AT*** | ER | 0.3 | −0.5 | 1.6 |
| AT | AT | 1.0 | −0.2 | −0.2 |
| AT | AT | 1.9 | 1.2 | 0.0 |
| ........... | | | | |
| ER | ER | −0.8 | −0.5 | 0.6 |
| ER*** | PE | −1.0 | −0.3 | −0.1 |
| ER | ER | −1.0 | 1.7 | 2.0 |
| ER*** | PE | −2.5 | −1.5 | −2.1 |
| ER | ER | −2.4 | 0.3 | −1.2 |
| ........... | | | | |
| PE*** | ER | −1.1 | 0.5 | 0.0 |
| PE*** | AT | 2.1 | 0.3 | −1.9 |
| PE | PE | −1.7 | 1.5 | −0.6 |
| PE | PE | −1.5 | 1.9 | −0.3 |
| PE*** | ER | −0.6 | 0.1 | 0.9 |

| *Function* | *Eigenvalue* | *Percent of variance* |
|---|---|---|
| 1 | 2.34 | 78.0 |
| 2 | 0.51 | 17.0 |
| 3 | 0.15 | 5.0 |

*Source:* Abstract from SPSS DISCRIMINANT output based upon Table 12.1 data

best discrimination in terms of two derived variables and so on. If there are $g$ groups, there are $(g - 1)$ discriminant functions (unless $p < (g - 1)$ in which case there are $p$ discriminant functions). So for the pottery data, there are 3 discriminant functions.

For each individual, three discriminant scores can now be calculated and an abstract from the SPSS printout is given at Table 12.7. The symbol *** again indicates those individuals whose data points do not lie in their own group region; the column headed 'predicted group' gives the group region in which such data points lie. The full output shows that as many as 5 in the Attic and Eritrean groups are 'misclassified' by the analysis, that is to say, the discriminant function values put them in a group region other than that to which they actually belong. Reference back to Table 12.6 (the two group analysis) shows that only one individual was misclassified in this way. What has happened is that the discriminant analysis in this multigroup case is trying to distinguish as best it may between four groups at the same time and so it is not as efficient as the two group analysis.

Similar to the case of principal components analysis, the coefficient vectors of the discriminant functions are calculated from the eigenvectors of a particular matrix based upon the information in the data matrix. An eigenvalue is calculated for each discriminant function and measures the effectiveness of that function. SPSS gives the three eigenvalues in order as 2.34, 0.51 and 0.15, having a total of 3.0. So the first discriminant function has an effectiveness measure, in percentage terms of $(100)\cdot(2.34/3.0) = 78\%$ and the first two functions together an effectiveness of $(100)\cdot(2.85/3.0) = 95\%$. The discriminant scores on the first two functions for this particular data set therefore account for a very large proportion of the discrimination power of the whole analysis.

The canonical discriminant analysis has thus achieved a considerable degree of economy in its presentation of the results. We started with a data matrix with six variables and four groups of individuals. The six variables have been reduced to two derived variables, the first two discriminant functions, whose values best separate the four groups. This has the advantage that a bivariate

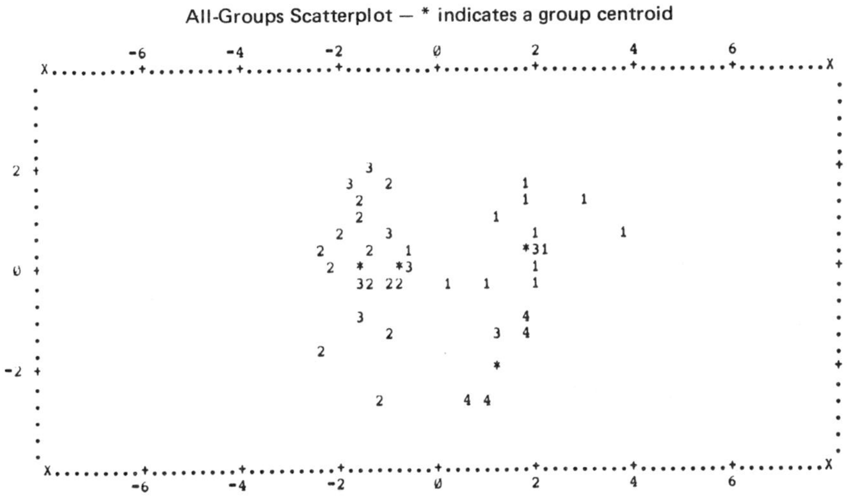


*Source*: SPSS output based upon Table 12.1 data

**Figure 12.2** Plot of First Two Discriminant Function Scores scatterplot of the two sets of discriminant function scores gives a visual picture of the discriminant analysis, as in Figure 12.2. It can be seen that groups 1 (Attic) and 2 (Eritrean) are well separated. The individuals in group 3 (probably Eritrean) are mainly mingled with the Eritrean group. On the other hand, group 4 (probably Attic) occupies its own separate region of the graph.

Finally, we note that in the case of two groups only, this multiple group canonical discriminant analysis gives the same result as the earlier presentation based upon the Fisher discriminant function.

## 12.7 DISCRIMINANT ANALYSIS AS A METHOD OF CLASSIFICATION

So far in this chapter, discriminant analysis has been presented as a method of data exploration, a method which tells us whether, for example, the Eritrean pottery is essentially different from the Attic pottery as regards the available data—the metallic oxide contents. Now it may, with some justice, be argued that this is not particularly significant. After all, if it is known that particular fragments are in one group or another, the archaeologist may not care very much whether the groups are distinguishable on the basis of their chemical constituents.

What is perhaps a more useful aspect of a discriminant analysis is that it offers a method of classifying an individual whose group membership is not known, or not known with any confidence. This is the case with the pottery data. Most of the individuals are classified as of Attic or Eritrean origin but a number are doubtful. We shall now see how the analysis can be used to allocate or classify the doubtful cases.

The scatterplot of figure 12.2 has already offered a provisional answer, where we saw that most of the 'probably Eritrean' fragments did indeed have Eritrean characteristics as measured by the first two discriminant scores. However, in the present context, a discriminant analysis which treats the four groups as separate is not a sensible one since, as was remarked in the last section, the analysis of the individuals of known origin is contaminated by those in the two groups of doubtful origin. The classification procedure is therefore based upon the discriminant analysis of only those individuals whose classification is certain and predetermined—in the present example upon the Attic and Eritrean groups.

We therefore return to the two group discriminant analysis of the Attic and Eritrean pottery fragments described in the earlier sections of this chapter. Our discussion will be based upon the bivariate analysis for the variables $MG$ and $CA$. (The reader will appreciate from the earlier discussion how similar are bivariate and multivariate analysis and that both lead to a single discriminant function.)

In general terms, the classification method is very simple. First the discriminant analysis is carried out for those individuals whose classification is known, leading to a computed discriminant function. Second, we take the variable values of any individual of unknown or doutbful classification and substitute those values in the discriminant function, which gives the individual's discriminant score. The value of this discriminant score then determines the group in which the individual is classified.

For example, the (bivariate) Fisher discriminant function was $3.24\ MG - 1.35\ CA - 5.01$ with the property that positive scores were typical of Attic fragments and negative scores of Eritrean fragments. Now from Table 12.1 the first individual of doubtful origin (under the head 'probably Eritrean') has variable values $MG = 2.2$, $CA = 3.9$ and so its discriminant score is $(3.24)\ (2.2) - (1.35)(3.9) - 5.01 = -3.15$. Since this is negative, this individual is indeed to be classified as Eritrean. On the other hand, the next individual, with $MG = 5.3$, $CA = 5.1$ has a discriminant score of $+5.2$ and would be classified as Attic.

This mathematical classification procedure can also be applied to the individuals of *known group membership*. If this is done, it is equivalent to checking the effectiveness of the discriminant function for the original data. The fewer the number of individuals classified by their discriminant scores in the wrong group, the better the discrimination.

## 12.8 DISCRIMINATION, CLASSIFICATION AND PROBABILITY

In the last section we offered what was in effect a classification rule: classify an individual of doubtful origin in one group (Attic) if its discriminant score is positive and in the other (Eritrean) if its discriminant score is negative. This is obviously too simple a rule since it does not invariably work even for those individuals whose classification is known. A look back at Table 12.5 will remind the reader that there were three 'starred' individuals whose discriminant scores had the wrong sign. If the classification rule had been used for these individuals, it would have wrongly classified them.

For the reasons explained in Chapter 10, our discussion of multivariate analysis is being limited to a descriptive statistics framework. This has its limitations. As in univariate and bivariate analysis, descriptive statistics cannot offer a satisfactory answer to questions such as: is the calculated discriminant function 'reliable'? what are the possible ranges of error of its coefficients? what is the probability that the classification rule is incorrect? To answer questions of this kind we need a statistical model of the data. Just as the normal distribution, *where valid*, is a useful model of univariate data, so is the multivariate normal distribution for multivariate data. Computer packages include in their output results based upon a multivariate distribution model for the data and we shall now see how the SPSS output can be interpreted to resolve the question of reliability of classification.

The relevant information is given in columns of the printout of Table 12.5 which have not yet received any comment. Taking this table column by column, the first three are self-explanatory. They show, respectively, the code letters for the groups, integers to identify each individual, and code numbers giving the prespecified group membership (where 1 = *AT* and 2 = *ER*). The fourth column, below the words 'Highest group', shows the group region in which the individual data point lies and we have already observed that three individuals, identified by ***, do not lie in their own group region; they have a different group number in column 3 from column 4.

The classification probabilities are given in the two columns headed $p(g/x)$ (which is a notation for 'probability that individual lies in a particular group $g$ given its variable values $x$).[2] In the final 13 rows, the program has used the discriminant function to classify the doubtful cases. Thus, the first doubtful individual, PE 1, has variable values which give an SPSS discriminant score of −1.0917. This is well within the negative Eritrean region and its group classification, 2 = *ET*, has a high probablity of 0.9590; an individual with the same variable values, *could* be Attic but the probability of this is only 0.0410 (= 1 − 0.9590).

In the first 27 rows of the table, the classification procedure has also been applied to the 13 Attic and 14 Eritrean individuals who are in fact already classified by prior knowledge. In effect, the calculations answer the question: how would we have classified these individuals without this prior knowledge but given the discriminant function and the variable values? As we have seen, the result is that three individuals are 'misclassified'; individuals with their variable values would have the greater probability of being in the other group than their actual group.

Classification procedures of this kind have been much employed in medical diagnosis. The results of a set of medical tests are used to decide whether someone is to be classified, and therefore treated, as

2. We do not consider the column headed $p(x/g)$.

having a particular disease. The procedure would be as follows. First, the tests would be applied to two groups of people, one group known to have the disease and the other known not to have the disease. This would yield a two group multivariate data matrix just as we had for the Attic and Eritrean pottery. A discriminant analysis would then calculate a discriminant function (which could be checked for effectiveness). This function could be used to classify the person being diagnosed as being in the 'disease' group or in the 'non-disease' group depending upon his 'variable values'—his medical test results. The doctor could make use of the probability information in the following fashion. If it were the case that failure to treat someone with the disease was potentially dangerous, he might decide to classify a person as being in the disease group whenever the corresponding probability was greater than, say, 0.3. On the other hand, if failure to treat was not dangerous and if treatment itself was hazardous, he might set a high probability of say 0.8 before classifying and treating someone as if they had the disease.

## 12.9 SOME MISCELLANEOUS ISSUES

Finally, we touch on some issues which have not been mentioned and which merit some brief comment even in an introductory account such as the present one.

The limitations of a purely descriptive statistics approach have already been noted in the context of classification procedures. It also has its limitations at an earlier stage in the analysis when the discriminant function itself is calculated. Returning to the medical diagnosis example, a discriminant function may be calculated based upon the test results of, say, 50 people with and 50 people without the disease. But how reliable might it be for use in classification? It would be unlikely, for instance, to be as effective a discriminant function as one based upon a much larger data set.

The setting up of a statistical model would of course be one way of tackling questions of reliability and error. Another simple practical procedure is to follow through the idea in the preceding paragraph by dividing the basic data set (if it is large enough) into two parts, the first being called the *calibrating set* (or alternatively the *training set*) and the second the *validation set* (or the *test set*). The discriminant analysis would be carried out on the calibration set and then the effectiveness of the discriminant function and the reliability of any classification method could be checked out on the validation set.

A second issue concerns the choice of effective discriminating variables. Some variables may be very good in discriminating between groups of individuals, but others may be poor or even useless. When this is the case, it means that a discriminant function may be based upon a smaller number of variables than the total number of variables in the data matrix and this can result in economies of a practical nature. For example, in medical diagnosis, the classification of a patient would require fewer medical tests (that is, fewer variable values). Computer packages typically allow the identification of 'best' variables, usually by a stepwise procedure as for variable selection in multiple regression analysis.

Finally, we return to the beginning as it were, and stress the importance of preliminary examination of the data set before embarking upon discriminant analysis, particularly if the ultimate purpose is one of classification. This implies the removal of outliers, at least on the basis of a univariate examination of the data. Excessively skewly distributed variables can cause problems in two ways. First, the skewness can lead to the variable receiving an undue and unreliable weighting in the discriminant function. Secondly, the assignment of probabilities in classification is based upon a multivariate normal distribution model which itself assumes that each variable is consistent with a (univariate) normal distribution. So classification can be unreliable in the presence of skewly distributed variables.

## EXAMPLES

**E 12.1** Using the data in Table 12.1, draw the bivariate scatterplot based on *CA* and *TI*, distinguishing the individuals in the Attic and Eritrean groups. Plot the two group centroids and the point, *P* say, midway between them. Draw by eye a discriminant boundary, passing through *P*, which appears best to separate the two groups. Identify the individuals which are misclassified.

**E 12.2** Repeat the discriminant analysis of E 12.1 using a computer program. Compare the results with your own geometrical analysis of E 12.1.

**E 12.3** Compare the discriminant analysis of E 12.2 with the bivariate analysis in the text based upon the variables *MG* and *CA*. What conclusion can be drawn about the relative discriminatory effectiveness of the variables *MG* and *TI*?

**E 12.4** The multiple group discriminant analysis of Section 12.6 and the discriminant function plot of Figure 12.2 suggested that the two groups, Eritrean and Probably Eritrean, are well mingled. Apply a check of this conclusion by drawing a scatterplot of the two groups based upon, say, the variables *MG* and *CA*.

## FURTHER WORK

**F 12.1** Table 12.8 shows certain measured properties of three kinds of brandy. Examine how far the variable values discriminate between the brandies.

**Table 12.8** Amounts of Flavour Compounds in Grape Brandies

| | Sample No. | *Amount* ($\mu g$ *litre*$^{-1}$) *of flavour compounds* *A* | *B* | *C* | *D* |
|---|---|---|---|---|---|
| German grape brandies | 1 | 1962 | 4968 | 29 | 139 |
| | 2 | 3244 | 6710 | 31 | 85 |
| | 3 | 2551 | 6895 | 41 | 121 |
| | 4 | 2363 | 7164 | 28 | 100 |
| | 5 | 1762 | 6734 | 14 | 58 |
| | 6 | 1376 | 5241 | 16 | 80 |
| | 7 | 739 | 3087 | 20 | 61 |
| | 8 | 1323 | 4418 | 3 | 60 |
| Means | | 1915 | 5652 | 23 | 88 |
| SD's | | 794 | 1457 | 12 | 30 |
| French cognacs | 1 | 1002 | 13270 | 77 | 210 |
| | 2 | 1038 | 11245 | 83 | 154 |
| | 3 | 623 | 12338 | 93 | 122 |
| | 4 | 903 | 11987 | 112 | 146 |
| | 5 | 1068 | 11583 | 87 | 103 |
| | 6 | 810 | 11691 | 85 | 92 |
| | 7 | 1994 | 7569 | 55 | 133 |
| | 8 | 604 | 13614 | 119 | 131 |
| Means | | 1005 | 11662 | 89 | 136 |
| SD's | | 437 | 1847 | 20 | 36 |
| French grape brandies | 1 | 1828 | 9769 | 26 | 60 |
| | 2 | 822 | 9283 | 13 | 139 |
| | 3 | 962 | 6368 | 18 | 88 |
| | 4 | 1708 | 10896 | 25 | 71 |
| | 5 | 1247 | 8040 | 21 | 76 |
| | 6 | 1450 | 6760 | 10 | 121 |
| | 7 | 1085 | 8110 | 19 | 77 |
| | 8 | 1300 | 8461 | 19 | 90 |
| Means | | 1300 | 8461 | 19 | 90 |
| SD's | | 350 | 1507 | 5 | 27 |

*Notation:* $A$ = ethyl butanoate $B$ = ethyl octanoate
$C$ = ethyl 2-furanoate $D$ = ethyl myristate

*Source:* Extract from Schreier, P. and Reiner, L., 'Characterisation of grape brandies', *Journal of the Science of Food and Agriculture*, 1979, 30

*This chapter depends upon: Chapter 10*

# 13

# Distances and Similarities

## 13.1 INTRODUCTION

The idea of a distance between a pair of individuals was first introduced in Chapter 10; there it was shown that, given $n$ individuals, a symmetric distance matrix, of order $n$, could be built up which set out the distances between all pairs of individuals. This chapter will be concerned with the further analysis of distance matrices and will also introduce a related idea, that of a similarity matrix.

One possible objective of further analysis is to identify, if possible, groups of individuals which, ideally, should be compact and well separated from each other. Such groups may be called *clusters* and the techniques for finding them constitute the subject matter of *cluster analysis*. Finding such clusters may constitute an informal check on the existence of some previously suspected 'structure' in the data matrix; alternatively, it may allow some reduction of the data matrix, from $n$ individuals to a smaller number of homogeneous groups. In this mode, the analysis proceeds from a data matrix to the grouping of individuals sometimes, but not always, with the calculation of a distance or similarity matrix as an intermediate stage.

It is sometimes desirable to proceed in the other direction, that is, given a distance matrix, we may ask what configuration of data points is consistent with it, in the sense that the latter will closely reproduce the former. The necessary techniques here constitute the subject matter of multidimensional scaling, which is the other main topic of this chapter.

We regard the techniques of this chapter as being mainly exploratory. By using them, we are seeking insights into the nature of the variation present in a multivariate data set. What use may be made of any insights gained will depend upon the context of the analysis.

Most of the computational work for this Chapter has been carried out using GENSTAT (see Appendix B). It will be clear that to make use of the methods presented here will depend very much on having access to appropriate programs.

## 13.2 DISTANCES AND SIMILARITIES

It will be recalled, from Section 10.8 that, given the values of $p$ numerical variables for two individuals, $r$ and $s$, say, then the square of the Euclidean distance between them is:

$$d_{rs}^2 = \sum_{j=1}^{p} (x_{rj} - x_{sj})^2$$

and that a set of the distances between all pairs of individuals may be arranged in the form of a symmetric matrix of order $n$. There are other ways of defining distances, but we will not survey them here. Everitt (1980) presents some further possibilities and the reader is referred there for details. Two practical points which will be needed in the discussion which follows are these:

(i) in general, the computation of distances should, we feel, be based upon standardised rather than raw data values. The arguments behind this proposal have been presented in the earlier chapters on multivariate analysis and need not detain us here. In Section 13.6 we illustrate the consequence of working with raw data and compare the results obtained with those obtained when the data have been first standardised;

(ii) the notion of a distance between two individuals is easily extended to admit the idea of a distance from an individual to a centroid (or, indeed, to any arbitrary, non-individual, point). Thus, if there are $p$ numerical variables whose means are denoted by $\bar{x}_1, \bar{x}_2, \ldots, \bar{x}_p$, then the distance squared from individual $s$ to this point is, by analogy with (10.8.4),

$$d_{sc}^2 = \sum_{j=1}^{p} (x_{sj} - \bar{x}_j)^2 \qquad \textbf{(13.2.1)}$$

where the subscript $c$ denotes the centroid.

Related to distances are similarities. If the distance between two individuals is relatively large they may be said to be dissimilar and this suggests that we may define a measure with respect to two individuals which is large if they are similar and small if they are dissimilar or distant. Such a measure is called a similarity coefficient.

To be more precise, a similarity coefficient is a number whose value lies between 0 and 1 with the upper limit implying that the two individuals are identical in terms of the respective values of the data variables. To facilitate visual inspection, similarity coefficients may also be expressed in percentage terms, thereby giving perfect similarity a value of 100. The set of coefficients between all pairs of individuals may, as in the case of distances, be assembled into a similarity matrix, which will be symmetrical.

**Table 13.1** Votes of 19 Selected Nations on 12 Selected Issues, UN General Assembly 1968

| | | *1* | *2* | *3* | *4* | *5* | *6* | *7* | *8* | *9* | *10* | *11* | *12* |
|---|---|---|---|---|---|---|---|---|---|---|---|---|---|
| USSR | (RU) | 1 | 1 | 1 | 1 | 3 | 1 | 2 | 3 | 1 | 3 | 2 | 2 |
| Bulgaria | (BU) | 1 | 1 | 1 | 1 | 3 | 1 | 1 | 3 | 1 | 3 | 2 | 2 |
| Yugoslavia | (YU) | 1 | 3 | 3 | 3 | 3 | 1 | 1 | 3 | 1 | 2 | 3 | 1 |
| Syria | (SY) | 1 | 2 | 2 | 2 | 3 | 1 | 1 | 3 | 1 | 2 | 3 | 1 |
| UAR | (UA) | 1 | 3 | 3 | 3 | 3 | 1 | 1 | 3 | 2 | 2 | 3 | 1 |
| Kenya | (KE) | 1 | 3 | 3 | 3 | 3 | 1 | 1 | 3 | 2 | 4 | 3 | 1 |
| Tanzania | (TA) | 1 | 2 | 2 | 2 | 3 | 1 | 1 | 3 | 2 | 4 | 3 | 1 |
| Senegal | (SE) | 1 | 3 | 3 | 3 | 1 | 2 | 2 | 2 | 2 | 1 | 3 | 1 |
| Dahomey | (DA) | 1 | 3 | 3 | 3 | 1 | 3 | 1 | 3 | 4 | 1 | 3 | 1 |
| USA | (US) | 1 | 3 | 3 | 3 | 1 | 3 | 3 | 1 | 3 | 1 | 1 | 3 |
| UK | (UK) | 1 | 3 | 3 | 3 | 1 | 1 | 3 | 2 | 3 | 1 | 1 | 3 |
| France | (FR) | 1 | 3 | 3 | 3 | 3 | 1 | 2 | 3 | 3 | 1 | 1 | 3 |
| Sweden | (SW) | 1 | 3 | 3 | 3 | 3 | 1 | 2 | 3 | 3 | 1 | 1 | 3 |
| Norway | (NO) | 1 | 3 | 3 | 3 | 3 | 1 | 3 | 2 | 3 | 1 | 1 | 3 |
| Australia | (AL) | 1 | 3 | 3 | 3 | 1 | 3 | 1 | 3 | 3 | 1 | 1 | 3 |
| NZ | (NZ) | 1 | 3 | 3 | 3 | 1 | 3 | 1 | 1 | 3 | 1 | 1 | 3 |
| Mexico | (MX) | 1 | 2 | 2 | 2 | 1 | 3 | 3 | 1 | 3 | 1 | 1 | 1 |
| Venezuala | (VE) | 1 | 2 | 2 | 2 | 1 | 3 | 3 | 1 | 3 | 1 | 2 | 1 |
| Brazil | (BR) | 1 | 2 | 2 | 2 | 1 | 3 | 3 | 1 | 3 | 1 | 1 | 3 |

*Source:* Dixon, W.J. and Nicholson, W.L., *Exploring Data Analysis* (1974), Ch. 2 by Hartigan, J.A.

*Key:* for issues, see Source
for votes: 1 = yes, 2 = abstain, 3 = no, 4 = absent

A useful coefficient of similarity is that devised by Gower (1971). This allows similarities to be calculated from both numerical and categorical variables and from mixtures of the two. We illustrate the calculations by using a subset of the data of Table 10.1 (for the numerical case) and the data of Table 13.1 (for the categorical case). This latter data set is a record of the voting of several member nations of the United Nations General Assembly on various 'issues' in the 1968 session. Although the values are coded to numbers 1, 2, 3 and 4, these represent the categorical responses YES, ABSTAIN, NO and ABSENT.

To start with the numerical case, we have selected six countries—France, Germany, Italy, Sweden, UK and USA—from Table 10.1 and five variables—$X_2$, $X_3$, $X_5$, $X_6$ and $X_7$. This constitutes the basic data set. To compute Gower's similarity coefficient between, say, France and Germany we first lay out their data values:

| | $X_2$ | $X_3$ | $X_5$ | $X_6$ | $X_7$ |
|---|---|---|---|---|---|
| France | 9 | 10.7 | 10 | 3.7 | 28 |
| Germany | 6 | 12.4 | 15 | 4.6 | 33 |

and then calculate the differences, ignoring the sign;

| | | | | | |
|---|---|---|---|---|---|
| Difference: | 3 | 1.7 | 5 | 0.9 | 5 |

Each difference is then divided by the range of that variable in the whole data set, giving the following 'weighted' differences:

| | | | | |
|---|---|---|---|---|
| 0.25 | 0.24 | 0.63 | 0.15 | 0.13 |

The sum of the weighted differences is 1.4; dividing this by 5 (the number of variables) gives 0.28 and, finally, subtracting this from 1 gives 0.72 as the similarity. The complete similarity matrix is shown in Table 13.2, with the similarity coefficients expressed as percentages.

**Table 13.2** Similarity Matrix of Country Data

| | | | | | | | |
|---|---|---|---|---|---|---|---|
| France | 1 | — | | | | | |
| Germany | 2 | 72.3 | — | | | | |
| Italy | 3 | 56.5 | 53.9 | — | | | |
| Sweden | 4 | 68.7 | 71.4 | 25.2 | — | | |
| UK | 5 | 66.5 | 70.1 | 57.8 | 55.5 | — | |
| US | 6 | 50.8 | 58.9 | 22.9 | 55.2 | 61.8 | — |
| | | 1 | 2 | 3 | 4 | 5 | 6 |

*Source:* Table 10.1, using variables $X_2$, $X_3$, $X_5$, $X_6$, $X_7$

To illustrate the calculations where the variables are categorical, let us take the voting record for the first two countries, USSR and Bulgaria. The values are noted as letters, Y for yes, N for no and A for abstain.

| | | | | | | | | | | | | |
|---|---|---|---|---|---|---|---|---|---|---|---|---|
| USSR | Y | Y | Y | Y | N | Y | A | N | Y | N | A | A |
| Bulgaria | Y | Y | Y | Y | N | Y | Y | N | Y | N | A | A |

Scoring 1 for an agreement and 0 otherwise, we have

1 1 1 1 1 1 0 1 1 1 1 1

which gives an average score of 11/12 = 0.917. This is the similarity coefficient and a sample of the complete matrix is given in Table 13.3.

Standardisation of the data set prior to calculation of the similarity coefficient is far less important than is standardisation as a preliminary to the calculation of distances. This is because the similarity coefficient formula incorporates the range of each set of variable values. Prior standardisation would, of course, alter the range as well as the data values.

Other definitions of similarity are possible and to add richness to the range of possibilities it may be noted that, in turn, similarity coefficients may be converted into distance like measures by computing, for example, $d_{ij} = (1 - s_{ij})$, where the subscripts refer to individuals $i$ and $j$.

To sum up, therefore, a distance matrix and a similarity matrix are alternative ways of representing the degree of difference between each pair of individuals in a data matrix. The term *proximity matrix* will be used to mean either a distance matrix or a similarity matrix.

The computation of proximity matrices is not, as we remarked in the introduction, an end in itself however. Indeed, starting with a matrix of $n \times p$ data values, the calculation of a proximity matrix leaves us with $\frac{1}{2}n(n - 1)$ new values to be examined which, if $n$ is

**Table 13.3** Extract from the Full Similarity Matrix based on the Data of Table 13.1

| | | | | | | |
|---|---|---|---|---|---|---|
| USSR | | | | | | |
| Bulgaria | 91.7 | | | | | |
| Kenya | 33.3 | 41.7 | | | | |
| Tanzania | 33.3 | 41.7 | 75.0 | | | |
| US | 8.3 | 8.3 | 33.0 | 8.3 | | |
| France | 41.7 | 33.3 | 58.3 | 33.3 | 66.7 | |
| | USSR | Bulgaria | Kenya | Tanzania | US | France |

much more than $p$, means that no simplification has been achieved. One possibility is to list, for each individual or for a given set of individuals, the three or four others who are nearest or most similar to it. While this may be useful with small data sets, it will not be pursued here. A second possibility involves linking individuals who are close to each other in a chain-like arrangement, and one way of doing this will now be discussed.

## 13.3 THE MINIMUM SPANNING TREE

In Chapter 10 it was pointed out that given $p$ numerical variables, the $n$ individuals will form a set of $n$ points in $p$ dimensional space. Suppose now we envisage joining these points together by straight lines such that:

(i) None of the lines forms a closed loop.
(ii) Each point is visited by at least one line.
(iii) The resulting collection of lines is *connected*. That is, there is a path, which possibly passes through a number of intermediate points, from one point to any other point.

Such a collection of lines defines what is called a *spanning tree* and if we envisage the spanning tree being constructed in such a way that the sum of the lengths of the lines which comprise it is minimised then we have what is called a *minimum spanning tree.*

For those who find physical analogies helpful, one might think of the $n$ points as being cities in a country which are to be connected by telephone line. The problem of finding the minimum length of telephone line is exactly analogous to finding the minimum spanning tree.

Selecting the individuals which are to be linked to form the minimum spanning tree is one which is laborious and requires computer methods. The presence of several variables adds no new conceptual problems, although a two dimensional representation of the minimum spanning tree becomes a matter of convention only; there is no way in which it can be drawn exactly to scale. It should be noted that, if we are working with similarities, rather than distances, minimising the length of the set of line segments implies linking together those individuals between which there is the *greatest* similarity.

Graphical representation may be achieved either by a layout such as that shown in Figure 13.1 which is derived from the similarity matrix of Table 13.2, or by superimposing the minimum spanning tree

```
FRANCE      GERMANY  SWEDEN

1......     2.......  4

 72.3       ( 71.4

            (

            (          UK            ITALY

            (.......... 5........... 3

              70.1     (     57.8

                       (

                       (             US

                       (.......... 6

                             61.8
```

*Source*: Copied from GENSTAT output.

**Figure 13.1** Minimum Spanning Tree from Similarity Matrix in Table 13.2

linkages onto some bivariate plot of the data. This latter possibility is demonstrated in Figure 13.2 which is the PC plot first shown in Figure 11.3 with the points linked by the minimum spanning tree.

The minimum spanning tree provides some further evidence about the nature of the (many dimensional) scatter; two individuals who appear close together in the (necessarily) two dimensional representation of a principal component analysis plot, may in fact, be some distance from each other. This will be indicated by the fact that the minimum spanning tree does not link them directly. Thus Australia and France are close together in terms of their first two PC values, but are not directly linked by the tree since their other PC values are more discrepant.

The minimum spanning tree imposes a simplification by focusing attention upon a chain of similar individuals. The nature of the linkage is determined ultimately by the requirement that the length of the spanning tree should be minimised. For example, Figure 13.1 shows that France is similar to Germany and Germany is similar to the UK, but this, by itself, tells us nothing about the similarity of France to the UK; only the similarity matrix contains that information. Consideration of the minimum spanning tree leads to one possible method for defining groups or clusters of individuals. This method

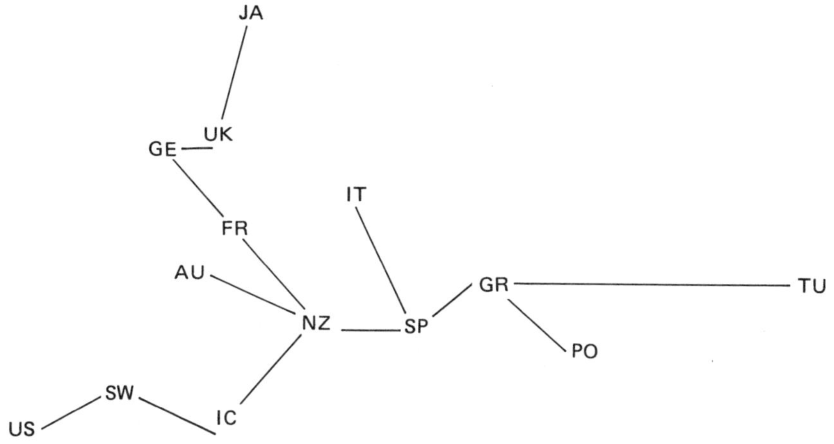


**Figure 13.2** Minimum Spanning Tree Superimposed on PC Plot of Figure 11.3

bases itself upon the notion that each individual in the cluster is linked to some— possibly only one—other individual in the same cluster by a similarity link which is not less than a prespecified value. There is no guarantee, therefore, that *all* individuals are within the same distance of each other; the clusters are not necessarily 'tight', in other words. We sketch the steps in this method in the next section.

## 13.4 HIERARCHICAL CLUSTER ANALYSIS

Keeping the minimum spanning tree in Figure 13.1 in view, let us select a level of similarity, say 70.0. We can now easily see that France, Germany, Sweden and UK are connected by links of at least that level and we may say that these individuals form a *cluster.* Italy and the US, on the other hand, have no such connection with each other or with any member of this cluster and are said to be ungrouped at the selected level of similarity. Let us now lower the level to 60. The US has a link slightly higher than this (61.8) and so now joins the others. Italy remains 'outside' until the level at which individuals are to be considered similar falls to less than 57.8. At that point all six individuals may be deemed 'similar'; all are in a single cluster.

This discussion illustrates what is call 'fusion' in cluster analysis. We have to begin with, six individuals, and at the first stage four of them fuse to form a cluster. This is then joined, as the level of similarity

```
                  LEVELS 75.0  65.0  55.0

FRANCE            1..

GERMANY           2..)

SWEDEN            4..)

UK                5..).....

US                6.........)..

ITALY             3............)............
```

*Source*: Copied from GENSTAT output.

**Figure 13.3** Dendrogram of Single Linkage Cluster Analysis Derived from Minimum Spanning Tree (Figure 13.1)

becomes smaller, by each of the remaining individuals in turn. The whole process may be summarised by means of a *dendrogram* as shown in Figure 13.3. The horizontal scale of this diagram shows decreasing levels of similarity as one moves from left to right. The right-hand parenthesis character is used by the program to show how the individuals link together as the fusion proceeds.

The clustering strategy illustrated here is known as *nearest*

**Table 13.4** Linking of Countries According to Furthest Neighbour Cluster Analysis

| *At similarity level ...* | *... are linked* | *Ungrouped, are (for example)* |
|---|---|---|
| 70.0 | (i) France, Germany | Sweden, since its link to France is only 68.7 |
| 65.0 | (i) France, Germany Sweden | UK, since its link to France is only 66.5 |
| 60.0 | (i) France, Germany Sweden<br>(ii) UK, US | Italy, since its link to France is only 56.6 and a link to the second emergent cluster cannot be made because its similarity to US is only 22.9. |

*Source:* Derived from Table 13.2

```
         LEVELS   75.0  65.0  55.0  45.0  35.0  25.0

FRANCE            1 . .

GERMANY           2 . . ) . .

SWEDEN            4 . . . . . ) . . . . . . . .

UK                5 . . . . . . . .         )

US                6 . . . . . . . ) . . . . . . ) . . . . . . . . . . . . . . . .

ITALY             3 . . . . . . . . . . . . . . . . . . . . . . . . . . . . . ). . . . . . . .
```

*Source*: Copied from GENSTAT output.

**Figure 13.4** Dendrogram of Furthest Neighbour Cluster Analysis

*neighbour* or *single linkage cluster analysis.* A little reflection will suggest other possible strategies however. For example, we might ask that a cluster should only be defined if there is a similarity link of at *least* a given level from any individual to *all* other individuals who are candidates for grouping. (Compare this carefully with the single linkage criterion outline above). Table 13.4 sets out a few steps in the fusion of the six individuals according to this latter rule. It will aid the reading of this table if the similarity matrix of Table 13.2 is kept in view.

Notice now that, at the level of similarity 60.0, *two* clusters emerge, but these cannot be fused together because the similarity link of US to France is only 50.8. At level 50.0, however, all countries except Italy merge into a single cluster. The progress of the fusion may, as before, be mapped by means of a dendrogram which is shown in Figure 13.4.

This strategy for fusing individuals is an example of *furthest neighbour* or *complete linkage cluster analysis.* Both methods illustrated so far (and there are several others) are known as *hierarchical methods* since the individuals form clusters which are then, in turn, clustered. Once an individual is fused with another or admitted to an existing cluster, it remains in that group as further fusion takes place at lower levels of similarity.

The above outlines two strategies for obtaining clusters of like individuals but there are other possibilities. The key to these lies in the different methods used to redefine similarity or distance between a cluster which has been formed by the fusion of two or more individuals and the remaining individuals or the other existing clusters. Without discussing these, however, it will be apparent that we have:

(i) a multitude of choices of similarity/dissimilarity measures and, when this is resolved,
(ii) a further rich selection of procedures for defining hierarchical clusters.

To add yet more to the problem of choice, there are other strategies for defining clusters; one of these leads to a set of procedures known as non-hierarchical methods and the next section illustrates one of these.

## 13.5 A NON-HIERARCHICAL CLUSTERING PROCEDURE

As we noted above, the distinguishing feature of hierarchical clustering methods is that the grouping proceeds from single individuals to clusters and then to clusters of clusters and so on. Once two individuals are linked together they stay together as the clustering proceeds. By contrast, the distinguishing feature of non-hierarchical methods is that this property does not necessarily hold as the number of clusters decreases—or increases, since we are not necessarily tied to moving only in one direction.

One system for non-hierarchical clustering depends upon the relationship between the *total sum of squares*, the *within group sum of squares* and the *between group sum of squares* and we therefore begin by explaining these terms and their relationship. To keep the discussion simple, we will start by considering just one variable—$X_5$—from the data matrix of Table 10.1.

Suppose we divide the subset of countries used in Section 13.4 into three *arbitrary* clusters, each with two individuals, thus:

| | *Cluster 1* | *Cluster 2* | *Cluster 3* |
|---|---|---|---|
| Data values: | 10 | 15 | 13 |
| | 15 | 7 | 13 |

The cluster means are 12.5, 11 and 13 respectively. If we now compute the sum of the squared differences between each individual value and the corresponding cluster mean we have:

$$\begin{aligned}
&\text{Cluster 1:}\quad (10-12.5)^2+(15-12.5)^2 &&= 12.5\\
&\text{Cluster 2:}\quad (15-11)^2+(7-11)^2 &&= 32\\
&\text{Cluster 3:}\quad (13-13)^2+(13-13)^2 &&= 0
\end{aligned}$$

and adding these up gives us the *within group sum of squares* whose value is 44.5.

The *total sum of squares* is simply the sum of the squared differences between each value and the mean of the whole univariate data set ignoring, that is to say, the arbitrary grouping. This is easily shown to be 48.84. The difference between the total and the within group sum of squares is:

$$48.84 - 44.5 = 4.34$$

and this is called the *between group sum of squares*, a quantity whose calculation may alternatively be approached by taking the sum of the squared differences between each cluster mean and the mean of all the values, weighted so as to reflect the number of values in each group. Thus we have:

$$2(12.5 - 12.2)^2 + 2(11 - 12.2)^2 + 2(13 - 12.2)^2 = 4.34$$

which is the figure obtained earlier. This demonstrates a relationship reminiscent of (indeed, formally identical to) that used in regression, namely:

| | | | | |
|---|---|---|---|---|
| Total sum of squares | = | Between group sum of squares | + | Within group sum of squares |
| (TSS | = | BGSS | + | WGSS) |

except that, in regression, the terms on the right-hand side are usually called the 'explained' and the 'residual' sums of squares respectively. If we bring two or more variables into the calculations, we have:

TSS for two (several) variables = BGSS for two (several) variables
+ WGSS for two (several) variables

which may be easily verified with a suitable small data set.

We now have the first approach to another method of clustering. Clusters, we might say, 'ought' to be such that the WGSS is as small as possible relative to the BGSS and individuals should be shuffled round between clusters so as to achieve this. The following steps are therefore suggested:

(i) Compute the total sum of squares for all variables entering the analysis.
(ii) Choose the number of clusters to be formed and assign the individuals either randomly or on the basis of prior information to these clusters.
(iii) Compute the mean of each variable for each cluster. These may be conveniently called the cluster centroids.
(iv) Take the first individual and compute its Euclidean distance from its own cluster centroid and from each of the other cluster centroids. Then re-assign this individual to the cluster for which this distance is smallest.

(v) Recompute the cluster centroids for the cluster which has lost and the cluster which has gained the individual.
(vi) Repeat steps (iv) and (v) for each of the remaining individuals in turn.

Having passed through the data set once in this fashion, the process may then be repeated a number of times since individuals whose re-allocation was done early in the process may now be nearer to a cluster centroid adjusted by some later re-allocation. Ideally, there will come a point at which no re-allocations are made during a particular pass through the data. In practice, the number of passes is preset and any subsequent minor adjustments thereby disregarded. At the conclusion, we have a clustering of data values, and therefore individuals, which has as small a *within group* sum of squares, and therefore as large a *between group* sum of squares as can be achieved.

We may, in practice, start with all individuals in one cluster and then move successively to two, three and more by splitting the cluster whose contribution to the within group sum of squares is largest; this split may be made by dividing the individuals into two groups, those which are below and those which are above the mean of the variable which has the largest variance within that cluster. The re-allocation outlined above then proceeds.

## 13.6 NON-HIERARCHICAL CLUSTERING—A PRACTICAL EXAMPLE

Many similar methods have been discussed in the literature and implemented as computer programs. We do not attempt to describe the many possibilities, even less embark upon a comparative discussion of computing methods. Nevertheless, in order to give the reader some flavour of the kind of output which is typically produced, we present the results of a non-hierarchical clustering applied to the data of known origin (Attic and Eritrean) in Table 12.1. The interest here is to see how far the clustering procedure, which takes no account of the information on group membership, succeeds in clustering the two known groups of individuals.

The clustering was done using a procedure and programs set out in Hartigan (1975). It will be seen that the analysis based upon the standardised data is a better reflection of what is known about this data set.

**Table 13.5** Results of Applying a Non-Hierarchical Clustering Procedure to the Data of Table 12.1

*Data values unstandardised*

| *Cluster no.* | *Total number of individuals* | *Number known to be Attic* | *Number known to be Eritrean* |
|---|---|---|---|
| 1 | 19 | 8 | 11 |
| 2 | 8 | 5 | 3 |

*Data values standardised*

| *Cluster no.* | *Total number of individuals* | *Number known to be Attic* | *Number known to be Eritrean* |
|---|---|---|---|
| 1 | 14 | 12 | 2* |
| 2 | 13 | 1** | 12 |

*Individuals 11, 12 } as annotated on Table 12.5
**Individual 9 }

## 13.7 PRACTICAL ISSUES IN CLUSTER ANALYSIS

It would be pleasing at this point to be able to provide some general recommendations as to how to proceed in practice. This, however, is not possible. What can be done is to point out some of the reasons why such advice cannot be given.

The root of the problem is that, generally speaking, it is not clear what constitutes a cluster. Intuitively, a cluster sounds as though it should consist of individuals whose data values appear as well defined 'swarms' or clouds of points, distinctly separated from other such swarms. However, such swarms could conceivably have any number of shapes; they will not necessarily be circular or elliptical. Figure 13.5 illustrates, in the context of two variables, some possibilities. When several variables are being used, the possible 'shapes' which may satisfy intuition become even more diverse. All that can be said is that we are trying to identify the existence of multidimensional shapes which, if they exist at all, will be compact and separated from each other.

To add to our problems, we may have no information as to *how many* clusters might exist. We are fishing around, in other words, in a multidimensional 'sea' for an unknown number of usefully

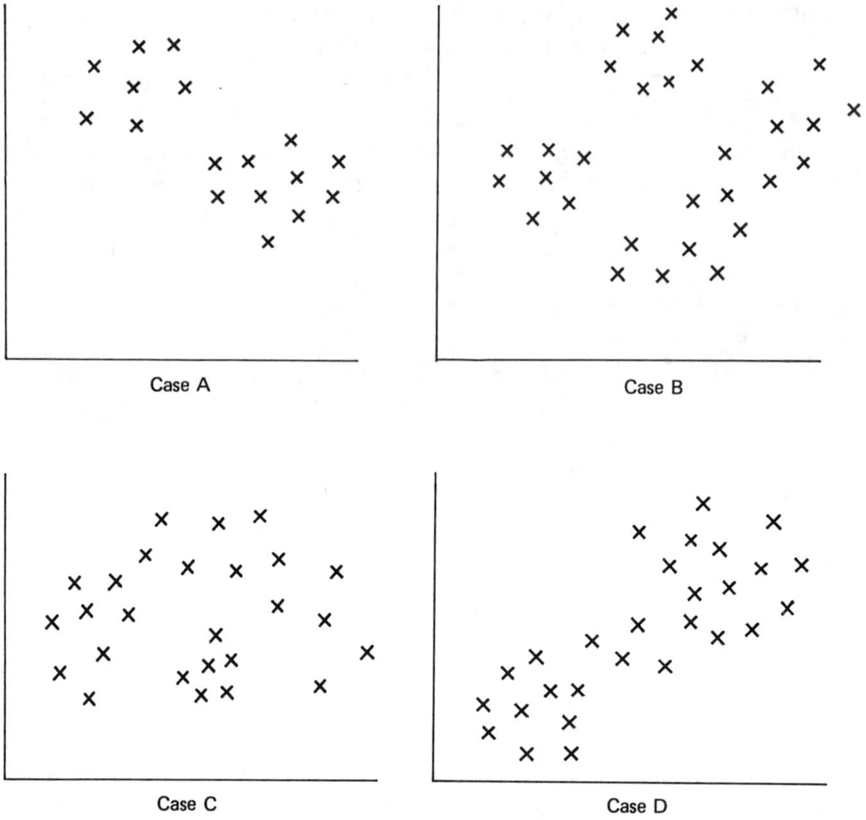


Case A and to a lesser extent case B show two somewhat idealised examples where most non hierarchical analysis procedures would identify the clusters.

Case C is more difficult because one cluster is partially surrounded by the other and the boundary is not linear.

The problem with Case D is one of definition: are there two groups or one?

**Figure 13.5** Some Possible Cluster Configurations

homogeneous groupings, which may not even be there! *If* they exist and if we could find them, appraise their shape and list the individuals whose data values form them, we might learn something new about the process which generated the data or at least generate some questions. Another possibility of course is that clusters may emerge which are well defined statistically, that is they are 'tight', well separated groups, but for which no useful substantive interpretation can be found, or which are simply absurd in the context of the problem. The statistical procedure and their computational implementations will (almost always) work, but they will not guard against such nonsenses.

As we mentioned in Chapter 2, there is always some background, possibly unquantified, information about the individuals themselves. This may serve to tell us at least how many clusters it might be reasonable to search for. It ought also to give some clue as to whether the results of a particular analysis are sensible. Some preliminary data analysis and 'cleaning', to obtain approximate symmetry of the distributions of each of the variables and to remove outliers, may help to improve the compactness of any clusters which do exist. We have already argued for standardisation of the data set, although it is possible that this may blur the information carried by certain variables.

Finally, in interpreting the results of a cluster analysis, a useful ancillary device is that of a plot of the first two sets (or some other pair) of principal component scores, with cluster membership indicated by a suitable choice of plotting symbol. Some programs offer these plots as an option.

There are no easy choices and no quick answers to be had with cluster analysis. Judgment is particularly required.

## 13.8 FACES

The wish to study the nature of multidimensional data has generated a large amount of investigational effort in recent years. One outcome of this effort, which we offer for light relief more than anything else, is the technique of representing each individual by a cartoon face whose features are varied according to the values taken by each of the variables (see Chernoff (1973)). These Chernoff faces are produced by means of a computer controlled digital plotter and can therby achieve a high standard of draughtsmanship.

To illustrate the idea, the faces for the 14 countries of Table 10.1 are reproduced in Figure 13.6. The 7 variables are interpreted in terms of the following features: *POPDEN* upper hair line, *AGEMP* chin shape, *NATINC* lower hair line, *CAPINV* eye size, *INFMOR* mouth size, *ENERGY* space between eyes and *TVSETS* eye slant.

The cheerful looking face of Turkey stands out and is very different from that for example, of Germany; the two countries tend to have variable values at opposite ends of their ranges. Ironically, Turkey's grin is a consequence of its high infant mortality rate which happens to

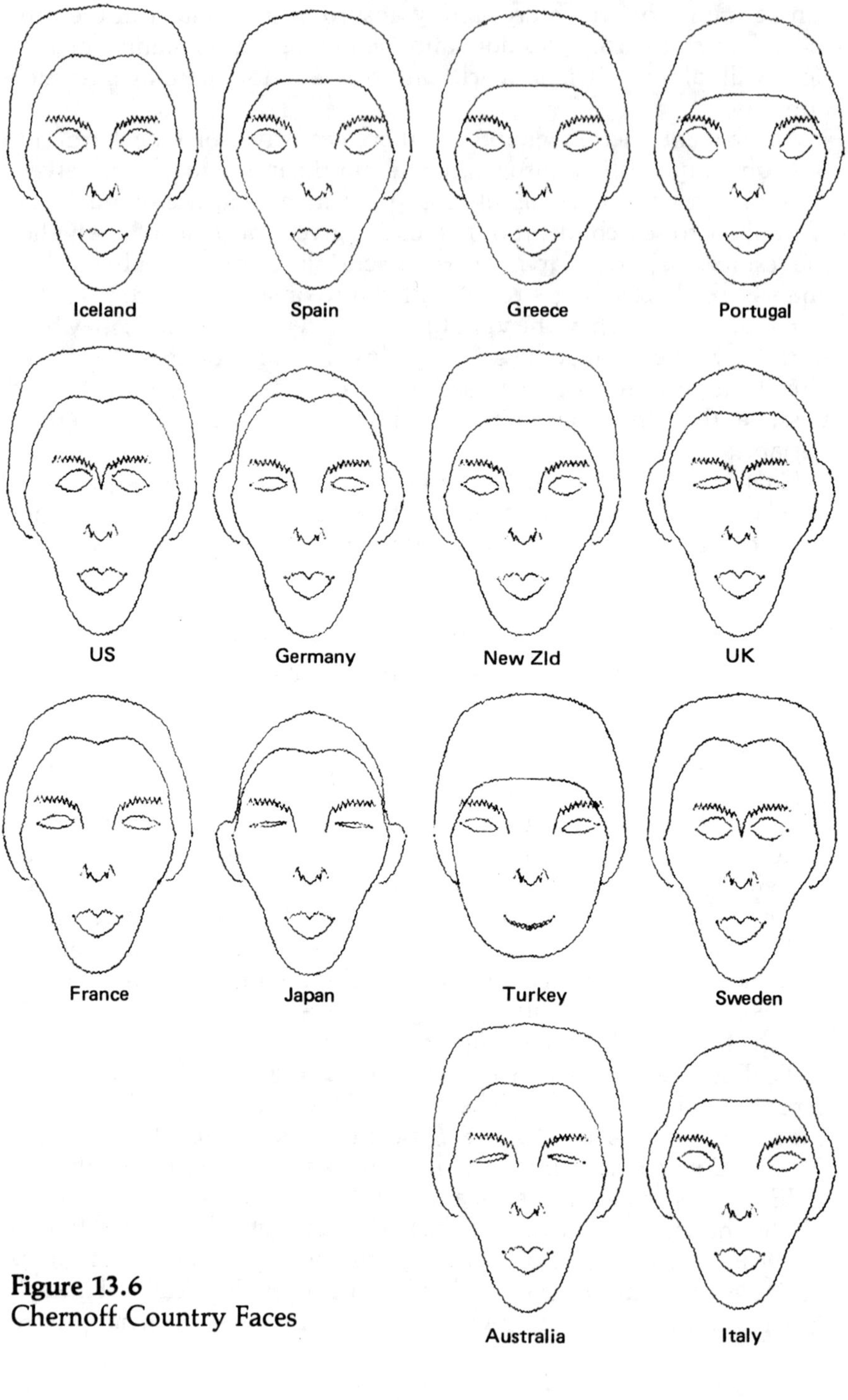


**Figure 13.6**
Chernoff Country Faces

have been visualised in the size of mouth with large values going with a small mouth. Clearly there is plenty of scope for subjective visualisation of the statistics in terms of the emotional impact of the resulting faces.

A purely technical point may be noted. For each feature the FACES program first standardises the corresponding variable by its range. So, for a variable such as *INFMOR* which is highly skew with a large valued outlier (for Turkey), there is little variation in mouth size for most countries: another outlier problem!

The program used to draw these faces was provided by the Biometrics Unit of the Institute of Psychiatry, University of London.

## 13.9 MULTIDIMENSIONAL SCALING

Up to the present point, we have discussed the calculation and interpretation of a proximity matrix from a given data matrix. We have also seen how a proximity matrix may be calculated even if there are categorical variables in the data matrix. Multidimensional scaling (MDS) works in the opposite direction. It is concerned with the problem of finding a data matrix of numerical variables from a given proximity matrix. Thus, if the only information available is a given distance matrix, **D**, MDS would estimate a data matrix which in turn could yield a distance matrix exactly or approximately equal to **D**.

We must first note an inevitable limitation in MDS analysis. It does not and cannot lead to a *unique* data matrix lying behind a proximity matrix, since different versions of transformations of a data matrix can all have the same proximity matrix. Centring a data matrix or transforming to the principal component variables does not change distances between individuals. So a data matrix, its centred version and the matrix of principal component scores will all have the same proximity matrix. We shall in fact see that the MDS solution, the MDS estimated data matrix behind a given proximity matrix, is the principal components version of the underlying data matrix.

There is clearly little point in estimating the numerical variable data matrix which underlies a particular proximity matrix if the lattter has been calculated from an already known numerical variable data matrix. The utility of MDS stems from two other possible situations.

The first is where the *only* data available is a proximity matrix for the $n$ individuals. For example, suppose a social survey has been carried out where respondents have been asked to rate the relative closeness of policies of, say, 10 prominent political figures. This data would yield a $10 \times 10$ proximity matrix for the politicians. Then MDS analysis would estimate a numerical valued data matrix consistent with this proximity matrix.

The second possible situation is where we do start with a data matrix *but some or all of the variables are categorical.* From such a data matrix, a proximity matrix can be calculated, and from this in turn an underlying *numerically valued* data matrix can be estimated by MDS. The possible virtues of this approach, and the MDS procedure itself, will now be illustrated using as an example the UN voting data at Table 13.1.

The starting point for MDS analysis is a proximity matrix. In the case of the UN data it is in fact a 19 × 19 similarity matrix, of which an extract was given in Table 13.3. A detailed description of the MDS solution is beyond the scope of this text but we can offer an outline of what is involved. The key, once more, depends on finding the eigenvalues and the associated eigenvectors of a matrix derived from the given or calculated proximity matrix. This matrix is of order $n$, and will therefore have $n$ eigenvalues and eigenvectors. The MDS underlying data matrix is based upon the eigenvectors (suitably scaled). Schematically, what we have is the following:

| | | | | |
|---|---|---|---|---|
| Eigenvalues (in descending order) | $\lambda_1$ | $\lambda_2$ | .... | $\lambda_{19}$ |
| Eigenvectors (in columns) | $x_{11}$ | $x_{12}$ | .... | $x_{1,19}$ |
| | $x_{21}$ | $x_{22}$ | .... | $x_{2,19}$ |
| | $x_{31}$ | $x_{32}$ | .... | $x_{3,19}$ |
| | .. | .. | .... | .. |
| | $x_{19,1}$ | $x_{19,2}$ | .... | $x_{19,19}$ |

Each column (each eigenvector) represents the values of a variable and each row the variable values for a single individual. The variables are mathematically artefacts and cannot be expected to have any easy interpretative significance; they are known as the principal coordinates, analogous to principal components. The eigenvalues, as usual, stand for the relative effectiveness of the successive variables (eigenvectors) in representing the information in the proximity matrix. In fact, as in PC analysis, each eigenvalue is the variance of the corresponding variable.

For the UN voting data the first two eigenvalues turn out to be 3.3 and 2.2, the total of all eigenvalues is 9.8, and so the first two account for 56% of the whole. A bivariate plot of the two should give a useful impression of the countries' relative positions. This bivariate plot is shown in Figure 13.7 and is based upon the GENSTAT program out-

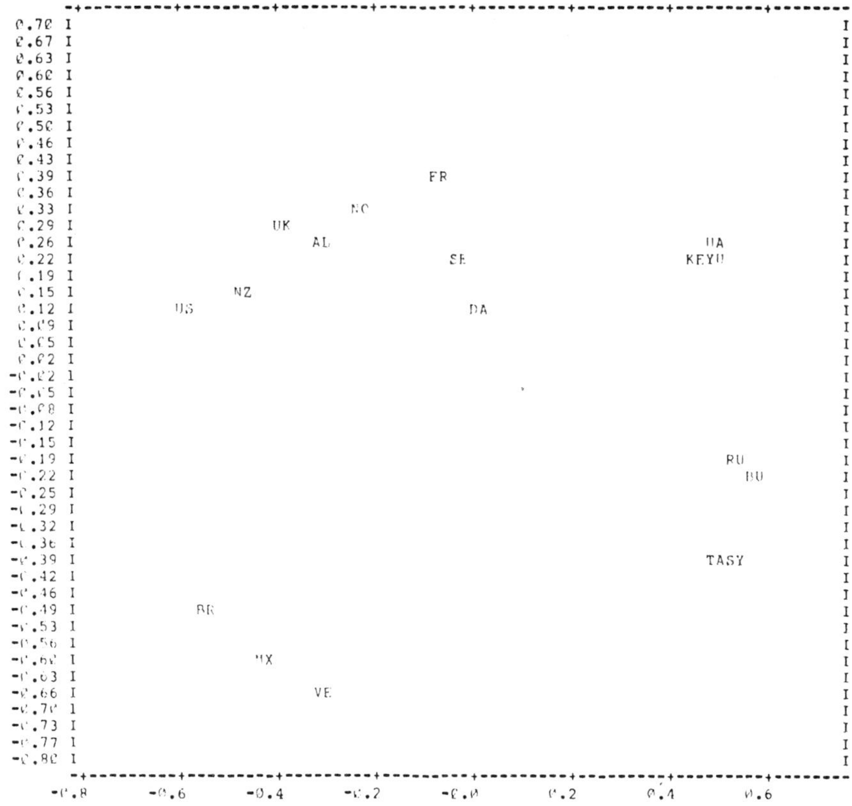


Point coinciding with Point FR: SW

*Source*: GENSTAT output based upon data of Table 13.1

**Figure 13.7** MDS Bivariate Plot of UN Voting Data

put. We leave the reader to assess some fairly obvious features of the plot.

It can sometimes happen that some of the eigenvalues of the MDS solution are negative. Mathematically this has the consequence that the corresponding eigenvectors (variables) cannot be calculated. It implies that the proximity matrix was not truly Euclidean, it being impossible to describe it in terms of an underlying data matrix, whatever the number of artificial variables. Even in this case, a representation in terms of the variables corresponding to the largest (positive) eigenvalues can still be useful.

In this section, we only describe the so-called classical MDS solution. It is a metric method in the sense that it is based upon numerical measures of distances between individuals. There are also non-metric methods of multidimensional scaling where the information about distances between individuals is of a relative nature such as 'Turkey is more different from France than is Spain'. This has its own extensive literature and related computer programs.

## 13.10 CONCLUSION

It will now be clear that most of the techniques discussed in this chapter require access to suitable program packages which may not be widely available. Indeed, without them, there is very little that can be done even at a simple level since the arithmetic quickly becomes very onerous.

## EXAMPLES

**E 13.1** Check selectively a few of the calculations of Gower's similarity coefficient in (i) Table 13.2 and (ii) Table 13.3.

**E 13.2** Figure 13.8 shows the minimum spanning tree derived from a similarity matrix for selected English cathedrals based upon their architectural features. Using similarity levels 85.0, 80.0, 75.0, 70.0 in turn, construct the corresponding dendrogram.

```
CANT . . . 81.2 . . . DURH . . . 69.3 . . . ELY . . . 82.6 . . . PETE
                       (                     (
                       (                     (
                       (                     (  . . . 77.4 . . . RIPON .
                       (                     (
                       (                     (  . . . 81.7 . . . ST. ALB
                       ( . . . 77.9 . . . LINC
```

**Figure 13.8** Minimum Spanning Tree for Example E 13.2

**E 13.3** Compare the MDS bivariate plot at Figure 13.7 with the similarity matrix at Table 13.3 and verify that the greater the dissimilarity between a pair of countries the greater is their distance apart on the plot.

## FURTHER WORK

**F 13.1** Referring to the country data of Table 10.1:

(i) allocate the countries to three or four groups on the basis of your own background knowledge or preconceptions;

(ii) repeat this based upon the purely visual similarities of the faces at Figure 13.6 and;

(iii) compare these allocations with the information in the minimum spanning tree at Figure 13.2.

**F 13.2** If there is access to a suitable cluster analysis program, perform an analysis on the brandy data of Table 12.8 and check the extent to which the cluster analysis separates out the three brandy types.

*This chapter depends upon: Section 0.22, Chapters 1, 2, 4 and 7*

# 14

# Nonparametric Statistics

## 14.1 INTRODUCTION

Nonparametric methods of analysis offer alternative procedures to many of those covered earlier in the text. Nonparametric methods are not descriptive: they are based upon statistical models for the data. Their central distinguishing feature is that they require fewer assumptions to be built into the model, which as we shall see has both advantages and disadvantages. Furthermore, nonparametric methods are not usually restricted in their application to data consisting of values of numerical variables. On the contrary, they are often formulated for ordinal or categorical variables or for numerical variables expressed as rank values.

We can usefully make a fuller assessment of nonparametric methods in the light of two particular examples of their use, which follow in Sections 14.2 and 14.3. A general commentary on aims and methods is in Section 14.4 and later sections discuss further analytical procedures. It should be said that this is a big subject and we can do little more than offer the reader an introduction to its more important aspects. For example, our discussion in the main relates only to two-tailed test procedures though their extension to one-tailed tests is often straightforward. The Bibliography gives references to texts dealing exclusively with nonparametric methods.

## 14.2 COMPARISON OF TWO UNIVARIATE DATA SETS: THE MANN–WHITNEY TEST

In Section 7.10 we considered the problem of testing the significance of the difference between the means of two data sets. The test procedure and the test statistic at (7.10.3) was based upon the assumption that the values in the two data sets were modelled respectively as $y_{1r} = \alpha_1 + u_{1r}$ $(r = 1, \ldots, n_1)$ and $y_{2r} = \alpha_2 + u_{2r}$ $(r = 1, \ldots, n_2)$, where $u_{1r} \sim \mathrm{N}(0, \sigma^2)$ and $u_{2r} \sim \mathrm{N}(0, \sigma^2)$. In other words the two data sets were modelled by normal distributions $\mathrm{N}(\alpha_1, \sigma^2)$ and $\mathrm{N}(\alpha_2, \sigma^2)$ differing, if at all, only as regards their means. The assumption of normality is crucial and, at least for small samples, the test of the null hypothesis that the two distribution means are equal has doubtful validity otherwise.

In this and the following section, we shall look at two non-parametric tests where the normality assumption is no longer necessary. The first is the Mann–Whitney test (alternatively called the Mann–Whitney–Wilcoxon or the Mann–Whitney *U* test). This test assumes that the two data sets are modelled by two distributions whose density functions need not be specified.

The null hypothesis to be tested is that the two distributions are identical. The alternative hypothesis is that the two distributions differ *but only as regards their mean* (or other measure of location). Notationally, the two distributions may be denoted by $f(x, \mu_1)$ and $f(x, \mu_2)$ where the null hypothesis is $\mu_1 = \mu_2$. Figure 14.1 shows the test situation pictorially. Under the null hypothesis, (left-hand diagram) the two distributions are identical. Under the alternative hypothesis, the distribution for one data set is to the left or right of that for the other data set.

The test procedure is as follows. First, rank values are assigned to the values in the two data sets, the ranks being based upon the two data sets *combined*. Table 14.1 shows the resulting rank values for the archaeological jars data which illustrated the test procedure of Section 7.10. A *rank transformation* is said to have been applied to the data as a whole and each value of the numerical variable has been replaced by its rank value. Now if the null hypothesis is false, either the rank values of the first data set will be predominantly small (which would be the case if its data values are modelled by the distribution with the smaller mean), or alternatively they will be predominantly large. Hence the sum of the ranks of the first data set, $S_1$ say, will be unduly small or unduly large if the null hypothesis is false.

The Mann–Whitney test statistic, conventionally denoted by *U*,

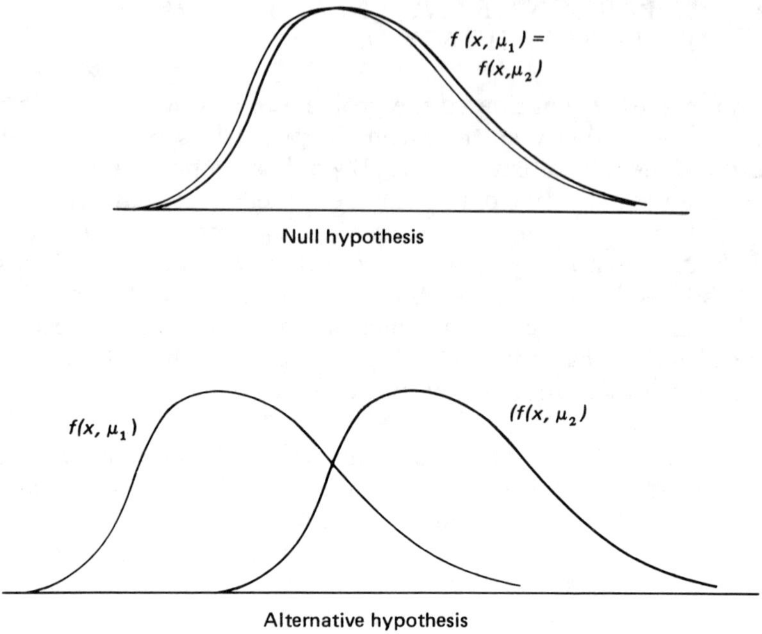


**Figure 14.1** Mann–Whitney Test Distributional Assumptions

depends upon $S_1$ and also upon the number of data values in the first data set $n_1$. Specifically:

$$U = S_1 - \tfrac{1}{2}n_1(n_1 + 1) \qquad \textbf{(14.2.1)}$$

The null hypothesis will be rejected for unduly small values of $U$ (reflecting an alternative hypothesis where the distribution modelling the first data set has the smaller mean) or for unduly large values (where it has the greater mean). The test will therefore be a two-tailed test.

**Table 14.1** Rank Values for Jars Data, Based upon Combined Data Sets

| | | | | | | | | | | |
|---|---|---|---|---|---|---|---|---|---|---|
| Theban | 4 | 8.5 | 10 | 11 | 12 | 15 | 16 | 18.5 | 20 | |
| | 22 | 23.5 | 25 | 26 | 27 | 28 | 29 | 30 | 33 | Total = 378.5 |
| Chania | 1 | 2 | 3 | 5 | 6 | 7 | 8.5 | 13 | | |
| | 14 | 17 | 18.5 | 21 | 23.5 | 31 | 32 | 34 | | Total = 236.5 |

*Source:* Table 4.1

The advantage of the rank transformation is that it can be shown that the distribution of the test statistic $U$ (if the null hypothesis is true) does not depend upon the nature of the common density function. It is in this sense that the test is said to be *distribution-free.* The lower critical value for $U$ is given in published tables of which an extract is in Appendix A. Denoting the lower critical value by $u_{\alpha/2}$ (where $100\alpha\%$ is the level of significance of the test), the upper critical value is calculated by the formula $u_{1-\alpha/2} = n_1 n_2 - u_{\alpha/2}$.

We can now apply the Mann–Whitney test to the archaeological jars data. From Table 14.1, $U = 378.5 - 171 = 207$ (rounded). Also, for a test at the 5% level of significance, $u_{0.025} = 87$ so $u_{0.975} = 288 - 87 = 201$. The observed $U$ therefore lies within the critical region and the null hypothesis is rejected.

## 14.3 COMPARISON OF TWO UNIVARIATE DATA SETS: THE MEDIAN TEST

As before, we start from two probability distribution models, one for each data set. The $t$-test required that the distributions should both be normal; the Mann–Whitney test relaxed this and required that they should be identical though unspecified distributions, at the most differing in their measure of location.

The median test goes further. The distributions can be any distributions at all: one could be, say, normal, and the other lognormal. The null hypothesis which is tested is simply that the median of one equals the median of the other against the alternative hypothesis that the two medians differ. The median is chosen as the measure of location because the test makes use of a property common to any distribution: that if a data value is modelled as a random observation from any distribution with median $M$, say, there is a probability of 0.5 that the value will be less than $M$ (and also a probability of 0.5 that it will be greater than $M$).

To apply the median test, one first calculates the median, $m$ say, of the two data sets combined. Then a table such as Table 14.2 is assembled which shows, for each data set, how many values are below and how many are above $m$. Table 14.2 is easily compiled from the rank transformation data table of Table 14.1 by counting, for each data set, how many rank values are more than 17.5.

If the two data sets are modelled by distributions with the same median, we can look upon the calculated median $m$ as an estimate of it. Furthermore, suppose we calculate for each data set the proportions, $p_1$ and $p_2$ say, of values above the common median. Then if the

**Table 14.2** Jars Data

| | *Theban data* | *Chania data* | *Total* |
|---|---|---|---|
| Above overall median | 11 | 6 | 17 |
| Below overall median | 7 | 10 | 17 |
| Total | 18 | 16 | 34 |

*Source:* Table14.1

null hypothesis is true, one would expect $p_1$ and $p_2$ to be roughly equal to 0.5 and $(p_1 - p_2)$ to be numerically small. If $p_1$ and $p_2$ are very different, this would suggest that the two model distributions have different medians.

The test statistic is

$$T = (p_1 - p_2)/\sqrt{[p(1 - p)(1/n_1 + 1/n_2)]}$$
$$\text{where } p = (n_1 p_1 + n_2 p_2)/(n_1 + n_2) \qquad \textbf{(14.3.1)}$$

Under the null hypothesis, the distribution of $T$ is N(0, 1). (This is an approximate 'large-sample' result but it is satisfactory so long as no entry in Table 4.2 is less than around 5.) For a two-tailed test, the critical values are therefore $\pm$ 1.96. For the archaeological jars data, $p = 0.5$, $(p_1 - p_2) = (11/18) - (6/16) = 0.24$, and $T = 0.24/0.17 = 1.4$. The null hypothesis of a common median is *not* rejected.

## 14.4 THE ROLE OF NONPARAMETRIC STATISTICS IN DATA ANALYSIS

We can now make some general observations in the light of the discussion in the two previous sections.

The first point is that nonparametric analysis has one thing in common with other inferential methods: the analysis depends upon the formulation of a statistical model for the data. The essential difference is that nonparametric statistical methods are less demanding in their model assumptions. For example, in comparing two sets of values of a particular variable, the various tests which have been presented can be placed in sequence as follows:

(i) The *t*-test is the most demanding, being based upon Model B(I) in Chapter 4. This requires that the two data sets are both satisfactorily modelled by normal distributions having identical variances but possibly differing in their means.

(ii) The Mann – Whitney test assumes a model similar to (i) except that the common distribution is not required to be normal.
(iii) The median test is the least demanding in that no distributional assumptions are required.

Given this choice of tests, together with the model specification upon which each is founded, how can we decide which test can be used in practice?

It might be thought that the safest procedure is to use a test which is founded upon the fewest model assumptions. Unfortunately, this is too simple a view. If it is the case that the assumptions implicit in Model B(I) are valid, then conclusions based upon either the *t*-test or the median test will both be valid. However, the latter will be less powerful in the sense that the probability of drawing an incorrect test conclusion is greater. More exactly, in the language of hypothesis testing, for a given level of significance the probability of making a type II error can be expected to be greater for the median test than for the *t*-test. By contrast, if the assumptions required for Model B(I) do not hold, the *t*-test will be less firmly based. In this case, the median test represents a safer choice of test procedure.

For the sake of discussion, we have been contrasting two extremes, the median and the *t*-test. The Mann – Whitney test lies somewhere between the two. On the one hand its model assumptions are less demanding than those of the *t*-test. On the other, it is only a little less powerful than the *t*-test if used in circumstances where the latter is justified. On the whole, unless considerable effort is to be put into the checking of model specification, the Mann – Whitney test represents a useful compromise between underlying model requirements and lack of power should these requirements fail to be met.

Another feature of nonparametric methods is that they often require less 'specific' data. The Mann – Whitney and the median tests are founded upon rank values and, for their application, it would not matter if the numerical values were inaccurately recorded so long as the rank ordering is correct. The sign test to be considered in the next section goes even further and is based solely upon knowledge of the two values of a dichotomous categorical variable. When it is applied to numerical variable data, the numerical values are typically categorised according to whether they fall into one of two categories, such as 'being positive' or 'being negative'.

It is useful therefore to think of a hierarchy of different *levels of measurement*. The highest level is the numerical variable. A lower level of measurement corresponds to the rank transformation of the numerical variable values. The ordinal categorical variable can be said to have the same level of measurement as the rank variable; in both cases the variable values are simply ordered. At the lowest level of measurement is the nominal categorical variable. As we have seen, data measured at one level of measurement can readily be redescribed at a lower level. In nonparametric analysis, this has the implication that a procedure relevant, say, to rank variable data may also be appropriate to numerical variable data. The converse of course is not likely to be true. A procedure which requires ordered data cannot be applied to nominal categorical variable data.

Both of the nonparametric methods discussed so far are distribution-free. It was not necessary to model the data by a stated distribution, such as the normal. This is fairly typical and for this reason *distribution-free statistics* is an alternative term for nonparametric statistics.

A casual glance through any textbook on nonparametric statistics might suggest that nonparametric statistics is solely about hypothesis test procedures. This is not quite true. Some nonparametric methods are concerned with estimation but they are relatively few.

In Sections 14.5 and 14.6, we shall look at the application of some nonparametric methods to univariate data analysis; in Section 14.7 we consider the analysis of bivariate data and particularly where the data is given in the form of rank values. It will be appreciated that, in any application of a nonparametric test, it is important to be clear about:

(i) the level of measurement, bearing in mind that, even if numerical variable data are available, the process by which they were collected may suggest that only a lower level of measurement in terms of ranks can be justified;
(ii) the precision with which the data model is specified, particularly the distributional assumptions; and
(iii) the null and alternative hypotheses and what exactly is implied if the former is rejected.

## 14.5 UNIVARIATE ANALYSIS: THE SIGN TEST

The sign test is a distribution-free test which has a number of variants and we shall look only at its application to univariate data.

In its simplest version, the test requires only data which are values of a dichotomous categorical variable. The two variable values are often indicated by the symbols + and − because, as we shall see subsequently, the test is often based upon the signs of a set of positive or negative numerical variable values. It is a basic model assumption of the test that the data values are mutually independent, that is, that the probability that the value of a particular individual is, say, + does not depend upon the value taken by any other individual. The null hypothesis of the sign test is that the probability of observing + or − equals 0.5.

As an example, consider the rainfall data in Table 14.3. For each month we are in effect told whether the weather was 'wetter than average' (a + sign) or 'drier than average' (a − sign). Taking just the 1980 figures as an example, if that year is typical of past experience, it would be expected that $P(+) = P(-) = 0.5$. However we note that eight of the months were drier than average. It may be asked whether this is evidence of a shift in the weather in that year in the sense that the probability of the monthly rainfall being drier (or wetter) than average was no longer equal to 0.5.

The test procedure is as follows. The null hypothesis is that $P(+) = P(-) = 0.5$ against the alternative hypothesis that $P(+) \neq P(-)$. The test statistic is based upon the proportion of + values, $p$, in the data set. Under the null hypothesis, $p \sim \text{Bi}(n, 0.5)$ (see Section 0.22). For large values of $n$ (say $n > 10$), this may be approximated by the normal distribution $\text{N}(0.5, 0.25/n)$ (see Section 0.22). Hence the test

**Table 14.3** Monthly Rainfall and Sunshine: Differences from Long-Period Averages

| *1979* | | *1980* | |
|---|---|---|---|
| *Rainfall* | *Sunshine* | *Rainfall* | *Sunshine* |
| +12 | + 1 | −31 | +20 |
| +17 | −13 | − 9 | −13 |
| +83 | −25 | +30 | −32 |
| +46 | −37 | −18 | − 1 |
| +57 | + 8 | −34 | +19 |
| −27 | −60 | +14 | −24 |
| −28 | − 9 | +29 | −31 |
| + 5 | −18 | −13 | −26 |
| −43 | +28 | −35 | − 1 |
| +20 | +14 | +37 | +26 |
| −20 | + 1 | −26 | + 6 |
| +63 | + 7 | −27 | +13 |

*Rainfall:* units are mm. per month; the long-run period is 1948–79
*Sunshine:* units are hours per month; the long-run period is 1892–1979
*Source:* Rothamsted weather statistics in Rothamsted Experimental Station *Annual Report*, various issues, Harpenden, Herts.

statistic $z = (p - 0.5)/(0.5/\sqrt{n}) \sim N(0, 1)$. In our example $z = (0.17)/(0.5/\sqrt{12}) = 1.2$. At the 5% level of significance, the null hypothesis is clearly not rejected. In verbal terms, the data give no evidence for any systematic shift in rainfall in 1980.

The sign test may be summed up as follows:

*Data:* Values of a dichotomous categorical variable (indicated by + and −).

*Model assumptions:* The data values are mutually independent each with a constant probability of occurrence.

*Null hypothesis:* $P(+) = 0.5 = P(-)$

*Alternative hypothesis:* $P(+)$ and $P(-)$ are constant but not equal to 0.5.

*Test statistic:* $z = (p - 0.5)/(0.5/\sqrt{n})$ where $p$ = proportion of + values in data set. $z \sim N(0, 1)$ (for large $n$) **(14.5.1)**

The test procedure requires data values at the level of measurement of a dichotomous categorical variable though, as in our example, the categorical variable values may result from numerical variable data. The sign test may also be used to test the null hypothesis that a numerical variable data set is modelled by some distribution having a hypothesised median value, $M_0$, say. The categorical variable values '+' and '−' respectively correspond to data values greater than $M_0$ and data values less than $M_0$.

## 14.6 TIME SERIES AND OTHER ORDERED DATA

The subject matter of this chapter, and indeed of the whole text, is largely concerned with *unordered* data sets. That is to say, the analysis is not dependent upon, nor is there any interest in, the order in which the individual data values are originally presented in the data matrix.

An ordered data set on the other hand is one where not only is the ordering of the values of contextual significance but also there is an analytical interest in how far one value in the ordered sequence is statistically dependent upon other values,—particularly upon neighbouring values. Time series variables offer an important example of ordered data sets; the 'time ordering' of the values of a time series variable (such as those presented in Table 9.1) is a fundamental aspect of the data.

One question may be asked about any set of ordered data: are the values consistent with a data model in which each value is a random observation, from some distribution, which is independent of other observations? That is to say, does the ordering have any statistical significance? Parametric methods of analysis were discussed in Chapter 9; the Durbin—Watson test, for example, is a parametric test of the null hypothesis that the random component values in a regression model are mutually independent. A simple nonparametric test in this area is the *runs test*, (also known as the Wald—Wolfowitz runs test), which we shall now consider.

The rainfall data of Table 14.3 is of course time series data. When the sign test was applied, it was part of the model assumption that the categorical variable values, + and −, were mutually independent and the analysis which followed implicitly ignored the time series ordering. It was based solely upon a total count of the number of occurrences of the value +. This might not have been plausible if there is some persistence in weather conditions which carries over from one month to another: an unduly wet month may tend to be followed by another unduly wet month.

The rationale of the runs test can be understood in terms of the following three idealised sequences of the values + and −. (The reason for the brackets will be explained below.)

(a) (+)(−)(+)(−)(+)(−)(++)(−)(+)
(b) (+)(−−)(+)(−)(++)(−)(++)
(c) (−−−)(++)(−−−)(++)

In case (c), − tends to follow − and + tends to follow +. The sequence of values does not appear to exhibit independence. This is also true of case (a) where, conversely, + tends to follow − and − tends to follow +. Case (b) is an example of a sequence in between these two extremes where an independence model may be plausible. Now these three situations can be distinguished by counting the number of runs. A *run* is defined as a sequence of like values, which is terminated by and/or preceded by unlike values. The runs in the three examples are indicated by the brackets. Thus, case (a) has 9 runs, case (b) has 7 and case (c) has only 4. Lack of independence in the ordered data values is indicated by an unusually small or unusually large number of runs. The number of runs, $r$, is in fact the test statistic of the runs test.

The exact lower and upper critical values of $r$ are tabulated for small values of $n_1$ and $n_2$, the numbers of occurrences of the two categorical values (see references in Appendix A). However, for a two-tailed test at the 5% level of significance, a satisfactory approximation (correct to one) is given by:

$$\text{Lower critical value} = \frac{2n_1 n_2}{(n_1 + n_2)} \left(1 - \frac{2}{\sqrt{(n_1 + n_2)}}\right)$$

$$\text{Upper critical value} = 2 + \frac{2n_1 n_2}{(n_1 + n_2)} \left(1 + \frac{2}{\sqrt{(n_1 + n_2)}}\right) \quad \textbf{(14.6.1)}$$

When $n_1$ and $n_2$ are large, the critical values can be computed from the fact that $r$ has the large sample distribution:

$$N\left\{1 + [2n_1n_2/(n_1 + n_2)], \frac{2n_1n_2(2n_1n_2 - n_1 - n_2)}{(n_1 + n_2)^2\,(n_1 + n_2 - 1)}\right\}$$

The runs test can be used to check the assumption of independence in the time series of rainfall data. Using the data for the period 1979–80, there is the following sequence of signs, with the runs indicated by brackets as before: (+ + + + +) (− −) (+) (−) (+) (−) (+) (− −) (+) (− −) (+ +) (− −) (+) (− −). There are $r = 14$ runs and $n_1 = 12$, $n_2 = 12$. From (14.6.1) the lower critical value = 12(1 − 0.41) = 7 (rounded): the exact value is 7. Similarly, the upper critical value = 2 + 12(1 + 0.41) = 19 (rounded): the exact value is 19. So the null hypothesis is accepted.

## 14.7 BIVARIATE ANALYSIS: RANK CORRELATION

In the present section we shall be particularly concerned with assessing the correlation between two variables where the data values are in the form of rank values. So far as the analytical technique is concerned, it does not matter how the rank values arise. The raw data may originally have been values of two numerical variables from which the rank values have been calculated. Alternatively, the rank values may represent the basic data information in that the individuals have been put in order possibly depending on their values in terms of ordered categorical variables. Table 14.4 falls into the former category; the rank values have in fact been derived from the numerical variable values relating to sensory perceptions which were given in the source cited.

A first step in analysis is to calculate the *rank correlation coefficient* (sometimes known as Spearman's rank correlation coefficient), which is usually denoted by $r_s$. The rank correlation coefficient is nothing more than the ordinary correlation coefficient with the rank values treated as the numerical variable values in the correlation coefficient

**Table 14.4** Sensory Rankings of 12 Varieties of Strawberry and of Jams Made From Them

| | *Strawberries* | | *Jams* | |
|---|---|---|---|---|
| | *Flavour* | *Firmness* | *Flavour* | *Overall impression* |
| Senga Sengana | 10 | 6 | 12 | 12 |
| Bemanil | 5 | 11 | 3 | 7 |
| Jonsok | 6 | 5 | 4 | 4 |
| Tenira | 8 | 8 | 5 | 8 |
| Tamella | 4 | 4 | 7 | 5 |
| Sivetta | 3 | 3 | 1 | 1 |
| Bounty | 10 | 9 | 6 | 6 |
| Tago | 6 | 10 | 8 | 10 |
| Fructana | 8 | 7 | 11 | 11 |
| Totem | 10 | 12 | 10 | 9 |
| Chanil | 2 | 1 | 2 | 2 |
| Fanil | 1 | 1 | 9 | 3 |

*Source:* G. Skrede, 'Strawberry varieties for jam production', *Journal of Science of Food and Agriculture*, 1980, 31. The rank values have been derived from numerical variable evaluations in the reference; high rankings are favourable.

formula (see Section 0.9).[1] So a computer program which computes the correlation coefficient will also compute the rank correlation coefficient. Table 14.5 gives the correlation matrix for the four rank variables of Table 14.4, that is, the rank correlation coefficient for each pair of variables.

A question of primary concern may be whether a value of $r_s$ is 'statistically significant'. In the case of the ordinary correlation coefficient, $r$, we answered this question in Section 7.12 in terms of a regression model for the data which was based upon the normal distribution. For a number of reasons such a data model is entirely inappropriate for ranked data, if only because the rank values are discrete and therefore cannot be satisfactorily modelled by a continous distribution such as the normal. A different approach is therefore needed.

For the purposes of exposition, suppose we have a bivariate data set

1. If there is a substantial number of tied ranks, a correction needs to be made to the standard formula. On this point the reader is referred to specialised texts (see Bibliography).

**Table 14.5** Correlation Matrix for Strawberry and Jam Rank Variables

| | *Strawberries* | | *Jams* | |
|---|---|---|---|---|
| | *Flavour* | *Firmness* | *Flavour* | *Overall impression* |
| *Strawberries* | | | | |
| Flavour | 1.00 | | | |
| Firmness | 0.71 | 1.00 | | |
| *Jams* | | | | |
| Flavour | 0.51 | 0.28 | 1.00 | |
| Overall impression | 0.77 | 0.67 | 0.75 | 1.00 |

*Source:* Table 14.4

in which there are only three individuals. Further suppose that, for the two variables, the individuals' rank values are (1, 1), (2, 2) and (3, 3), that is, there is a perfect correlation between the two sets of ranks with the result that $r_s = 1$. We shall now see how statistical substance can be put to the vague question: is this value of $r_s$ 'statistically significant' or has it come about 'by chance'?

We first examine what other arrangements of rank values could have occurred; we call these the feasible values. For any given rank ordering of the first variable across the three individuals, say ranks 1, 2, 3 respectively, there are six feasible sets of ranks for the second variable: 1, 2, 3; 1, 3, 2; 2, 1, 3; 2, 3, 1; 3, 1, 2; or 3, 2, 1. In other words, the six feasible data matrices (together with the corresponding values of $r_s$) are:

$$\begin{matrix} 1 & 1 \\ 2 & 2 \\ 3 & 3 \end{matrix} \qquad \begin{matrix} 1 & 1 \\ 2 & 3 \\ 3 & 2 \end{matrix} \qquad \begin{matrix} 1 & 2 \\ 2 & 1 \\ 3 & 3 \end{matrix} \qquad \begin{matrix} 1 & 2 \\ 2 & 3 \\ 3 & 1 \end{matrix} \qquad \begin{matrix} 1 & 3 \\ 2 & 1 \\ 3 & 2 \end{matrix} \qquad \begin{matrix} 1 & 3 \\ 2 & 2 \\ 3 & 1 \end{matrix}$$

$$r_s = 1.0 \quad r_s = 0.5 \quad r_s = 0.5 \quad r_s = -0.5 \quad r_s = -0.5 \quad r_s = -1.0$$

(Five further sets could have been written down corresponding to the five other rank orderings for the first variable, but they would repeat the same pattern of $r_s$ values, which is all we shall be concerned with.)

A test procedure can now be formulated. It is based upon the following model assumption: that if the two variables are independent, each of the $6 \times 6 = 36$ feasible data sets has equal probability.

This is plausible enough; under independence, there is no reason to expect any particular rank arrangement to be more common than any other. The test statistic is $r_s$. Its discrete probability distribution (under the null hypothesis of independence) is, taking account of the fact that some values are repeated:

| $r_s$ value | $-1.0$ | $-0.5$ | 0.5 | 1.0 |
|---|---|---|---|---|
| Probability | 1/6 | 1/3 | 1/3 | 1/6 |

The probability of getting a value of $r_s$ as extreme as 1.0 or $-1.0$ is as high as $\frac{1}{3} \simeq 0.33$. A formal two-tailed test of the null hypothesis at the 5% level of significance emphatically does not reject the hypothesis of independence.

The distribution of $r_s$ can be similarly calculated for any value of $n$ and published tables give exact critical values in terms of $n$. When $n$ is large, the distribution of $r_s$ under the null hypothesis of independence is $r_s \sim N(0, 1/n)$. In other words, the distribution of $r_s\sqrt{n}$ is approximately standard normal. (Strictly, $r_s\sqrt{(n-1)}$ is a better approximation, but the arithmetical difference between the two is negligible in practice.) This normal approximation works remarkably well even for small values of $n$ as is clear from the following table which compares the exact upper critical values with the normal approximation critical values (for quartile order 0.975):

| $n$ | 5 | 10 | 15 | 20 | 30 |
|---|---|---|---|---|---|
| Exact | 1.00 | 0.65 | 0.52 | 0.45 | 0.36 |
| Normal | 0.88 | 0.62 | 0.51 | 0.44 | 0.36 |

(The second row of figures is calculated as $1.96/\sqrt{n}$, since $1/\sqrt{n}$ is the approximate standard deviation of $r_s$.) It seems hardly worthwhile bothering with exact tables of the critical values of $r_s$.

The formal test procedure runs as follows, illustrating it in terms of testing the significance of the rank correlation between strawberry flavour and jam flavour of Table 14.5:

*Data:* Values of two rank variables.

*Model assumptions and null hypothesis:* the two variables are independent (in the sense described above) as regards their rank values.

*Alternative hypothesis:* The two variables are positively related. (A negative relation is unreasonable in the context and a one-tailed test procedure is implied.)

*Test statistic:* $r_s\sqrt{n} \sim N(0, 1)$ under the null hypothesis. The critical value of N(0, 1) corresponding to a 5% tail $= 1.64$. This compares to the observed value of $r_s\sqrt{n} = 0.51\sqrt{n} = 1.77$. The null hypothesis is therefore rejected.

The above procedure for testing the significance of $r_s$ is in effect based upon a distribution-free model of the data variables. For numerical variable data, there is therefore a choice between the nonparametric test based upon $r_s$ and the parametric test based upon $r$. The answer to the question 'which is the best?' will throw some light on the advantages and disadvantages of the two approaches.

If the normal bivariate regression model is valid, the parametric test will be more powerful (in the sense explained in Section 14.4) than the nonparametric test. It can however be shown that the difference is not very great, so if there are doubts about the regression model specification, it would be prudent—and nothing very much would be lost—to use the nonparametric test based upon $r_s$.

## 14.8 NONPARAMETRIC METHODS AND BIVARIATE REGRESSION ANALYSIS

In the previous section we have seen that there may be some advantage to be gained in testing for the independence of two numerical variables by applying a rank transformation to the data and using a nonparametric test based upon $r_s$. The advantage stems from the less demanding data model required by the nonparametric procedure. It may be wondered whether this approach may also offer a pay-off in bivariate regression analysis. Regression analysis based upon a normal linear model is particularly sensitive to the presence of outliers and nonparametric procedures typically reduce the impact of outliers either by using a rank transformation of the data or by basing the analysis upon the median rather than the mean.

The first point to be made is that regression analysis based on rank data is not in general a viable procedure. The fact that the dependent variable takes, typically, small integral values makes the assumption of a normal random component implausible.

A different and more useful approach is to require that the data are values of numerical variables, but relax the assumption that the random component values are modelled by the normal distribution. In formal terms, such a model would be specified as:

$$y_r = \alpha + \beta x_r + u_r \; (r = 1, \ldots, n)$$

where the $u_r$'s are independent observations from some fixed (but unspecified) distribution with zero mean.

**(14.8.1)**

Now it is true that the usual least squares procedure can still be used to fit this model and it can be shown that the resulting regression coefficients will be unbiased estimates of the systematic parameters $\alpha$ and $\beta$. However, least squares fitting has the disadvantage that it is unusually sensitive to the presence of outliers. More 'robust' procedures would therefore be desirable.

Theil (1971) describes one such procedure. Theil's model is identical to (14.8.1) except that it is not required that the distribution of the random component should have zero mean. As a consequence the method is concerned only with estimating the slope coefficient $\beta$.

The method depends upon a prior sorting of the data set according to the values of the explanatory variable so that, for example, $x_1$ denotes the lowest X-value and $y_1$ denotes the corresponding value of Y. The data set is then divided into two parts (having deleted the middle observation if $n$ is odd), the first part comprising the observations $(x_1, y_1), (x_2, y_2) \ldots (x_{\frac{1}{2}n}, y_{\frac{1}{2}n})$ and the second part comprising $(x_{\frac{1}{2}n+1}, y_{\frac{1}{2}n+1}), (x_{\frac{1}{2}n+2}, y_{\frac{1}{2}n+2}), \ldots (x_n, y_n)$. The slopes of the lines joining corresponding members of each part are calculated as follows:

$$b_1 = \frac{y_{\frac{1}{2}n+1} - y_1}{x_{\frac{1}{2}n+1} - x_1}, \quad b_2 = \frac{y_{\frac{1}{2}n+2} - y_2}{x_{\frac{1}{2}n+2} - x_2}, \quad \ldots, b_{\frac{1}{2}n} = \frac{y_n - y_{\frac{1}{2}n}}{x_n - x_{\frac{1}{2}n}}$$

Then it can be shown that each $b_r$ has a probability distribution with median $\beta$. This leads to the *median* of the values $b_1, \ldots b_{\frac{1}{2}n}$ as the estimate of $\beta$. Theil also gives a method for calculating what is equivalent to a confidence interval for $\beta$.

Purely descriptive procedures can be looked upon as other instances of distribution-free procedures. For example, Mosteller and Tukey (1977) present a method of fitting a regression relation which avoids any explicit specification for the distribution of the random component and in one sense may therefore be described as distribution-free. Essentially, their method is a weighted least squares procedure (see Section 9.9) in which the smaller weights are assigned to the data points with the larger residuals. The weighting has the effect of making the least squares procedure less affected by outliers. The Mosteller and Tukey approach has the advantage that it is readily applied to multiple regression analysis.

## 14.9 CONCLUSION

Our purpose in this chapter has been to give the reader a flavour of the nonparametric approach to data analysis. The coverage has inevitably been selective and there has been no discussion, for example, of multivariate nonparametric procedures.

As we have seen, the starting point of any nonparametric method is the formulation of a statistical model for the data, but, compared to methods reviewed elsewhere in this text, the nonparametric models are less demanding in their assumptions. As a consequence their conclusions tend to be less informative. However, it may be better to obtain a vague answer which is securely founded than have no answer at all—or have one which is suspect because the data fail to match a more demanding model.

Nonparametric methods are particularly concerned with hypothesis testing and, in consequence, do not usually yield any estimates of the magnitude of 'effects' (differences in medians, for example), if these are found to be statistically significant. In addition it is important to bear in mind the interpretative problems which were reviewed in Section 7.11. In particular, a clear distinction needs to be maintained between statistically significant effects and effects which are significant in the data context.

## EXAMPLES

E 14.1 Referring to Table 12.8, test the significance of the difference in means (for flavour compound C) between German and French grape brandies. Use (i) the $t$-test, (ii) the Mann–Whitney test and (iii) the median test.

E 14.2 Use the sign test to test whether the data in Table 14.3 suggest that the years 1979–80 were 'sunnier than average'.

E 14.3 Use the runs test to determine whether the data on gold prices (Table 9.8) are consistent with a model assuming independence between successive first differences. (Use the data only for 1974–77.)

E 14.4 Calculate the rank correlation coefficient between 'strawberry firmness' and 'jam flavour' (Table 14.4). Test for significance.

*This chapter depends upon: Sections 7.3 and 7.4 and (for Sections 15.6 and 15.7) Chapters 10 and 11*

# 15

# Categorical Variable Analysis

## 15.1 PREVIOUS DISCUSSION

Although the greater part of this text, in common with statistics texts in general, has been concerned with the analysis of numerical variables, a number of aspects of categorical variables has already been covered.

In Chapter 5, we discussed univariate models of frequency data where the data comprised counts of occurrences of a particular value of a categorical variable. Methods of fitting and checking such univariate models were considered in Chapters 6 and 7. In Chapter 9, we saw how categorical variables could be used as explanatory variables in regression analysis by means of the dummy variable technique.

Multivariate analysis (Chapters 10 to 13) was principally concerned with the analysis of data matrices where the information for each of $n$ individuals was in the form of numerical variable values. However, in Chapter 13, which discussed the concept of proximities (that is, distances and similarities) between individuals as determined by their variable values, it was seen how information on categorical variable values could be introduced into proximity calculations. By the method of multidimensional scaling, we saw how a proximity matrix could in turn be represented by a scatterplot of two numerical variables. Discriminant analysis (Chapter 12) also has something to do with categorical variables: although it analyses the overall differences between groups of individuals in terms of values of numerical variables, the group membership is essentially specified in terms of categorical variable values.

Principal components analysis (Chapter 11) is one multivariate technique which does not directly cope with categorical variable data. However, we shall see below that *correspondence analysis*, which is concerned with contingency table data, is based upon a principal components approach.

Finally, in this review of what has gone before, the last chapter showed that a number of nonparametric methods are particularly appropriate for categorical variable data.

The main emphasis of the present chapter will be on the analysis of contingency tables. The earlier sections take a descriptive statistics stance leading in subsequent sections to a consideration of the methods of building and fitting statistical models of contingency table data.

## 15.2 THE ANATOMY OF CONTINGENCY TABLES

The numbers in a contingency table are frequencies based upon counts of numbers of occurrences of categorical variable values. Table 15.1, showing information on road accidents, is typical of the structure of a contingency table. It concerns three categorical variables: (i) the type of road junction (which will be labelled *ROAD*); (ii) the area in which the accident occurred (*AREA*); and (iii) the degree of resulting personal accident severity (*SEVERITY*).

The variables *ROAD*, *AREA*, and *SEVERITY* have respectively 5, 2 and 3 different values, so each individual accident can be described in one of $5 \times 2 \times 3 = 30$ different ways. For example, all the 47 accidents in the top left-hand cell of Table 15.1 had the characteristics—*ROAD*: 'at roundabout', *AREA*: 'urban' and *SEVERITY*: 'fatal'. There are more than 30 cells in Table 15.1 because totals according to the various categorical variable values are also included.

We can also view the accident data in terms of the underlying data matrix which gave rise to this contingency table. In this data matrix, each accident is an individual. The data matrix has three columns, *ROAD*, *AREA* and *SEVERITY* and opposite each individual accident is entered the corresponding variable values. The contingency table is essentially a summary table of the data matrix. It counts up the total number of occurrences of each possible combination of the variable values.

Table 15.1 is the most detailed summary table of the underlying data matrix; indeed, no information is lost as a result of its compilation. At the simplest level of summarisation, we might have looked separately

**Table 15.1** Road Accidents according to (i) Type of Road Junction (*ROAD*), (ii) Area of Occurrence (*AREA*) and (iii) Degree of Severity (*SEVERITY*)

| | *ROAD TYPE* | | | | | |
|---|---|---|---|---|---|---|
| | *Roundabout* | *Cross-roads* | *Other junctions* | *Private drive* | *Non-junction* | *Total* |
| *Urban areas* | | | | | | |
| Fatal | 47 | 340 | 1,263 | 42 | 1,494 | 3,186 |
| Serious | 1,166 | 5,922 | 19,348 | 1,377 | 17,827 | 45,640 |
| Minor | 5,692 | 21,699 | 64,883 | 4,609 | 49,025 | 145,908 |
| Total | 6,905 | 27,961 | 85,494 | 6,028 | 68,346 | 194,734 |
| *Rural areas* | | | | | | |
| Fatal | 22 | 99 | 419 | 76 | 2,019 | 2,635 |
| Serious | 433 | 1,052 | 3,881 | 814 | 15,077 | 21,257 |
| Minor | 1,454 | 1,842 | 7,023 | 1,583 | 24,299 | 36,201 |
| Total | 1,909 | 2,993 | 11,323 | 2,473 | 41,395 | 60,093 |
| *All areas* | | | | | | |
| Fatal | 69 | 439 | 1,682 | 118 | 3,513 | 5,821 |
| Serious | 1,599 | 6,974 | 23,229 | 2,191 | 32,904 | 66,897 |
| Minor | 7,146 | 23,541 | 71,906 | 6,192 | 73,324 | 182,109 |
| Total | 8,814 | 30,954 | 96,817 | 8,501 | 109,741 | 254,827 |

*Source:* Central Statistical Office, *Road Accidents Great Britain 1979*, HMSO, London

at each variable column and counted the numbers of occurrences of the corresponding values. For the *ROAD* variable, this would have given the totals in the bottom row of the table (8,814, 30,954, . . . , 109,741 with total 254,827). The reader will have no difficulty in identifying in the contingency table the corresponding frequencies for *AREA* separately and for *SEVERITY* separately; note that each of these two sets also adds up to the total number of accidents, 254,827.

A summarisation lying in between the more complex Table 15.1 and the summary counts for each variable separately is given by compiling a contingency table based upon just two of the variables, ignoring the information about the third. Such a table is known as a two-way con-

**Table 15.2** Two-Way Contingency Tables

(a) *SEVERITY* × *AREA*

| | *Urban* | *Rural* | *Total* |
|---|---|---|---|
| Fatal | 3,186 | 2,635 | 5,821 |
| Serious | 45,640 | 21,257 | 66,897 |
| Minor | 145,908 | 36,201 | 182,109 |
| Total | 194,734 | 60,093 | 254,827 |

(b) *ROAD* × *SEVERITY*

| | *Fatal* | *Serious* | *Minor* | *Total* |
|---|---|---|---|---|
| R'about | 69 | 1,599 | 7,146 | 8,814 |
| Cross-rds | 439 | 6,974 | 23,541 | 30,954 |
| Other jn | 1,682 | 23,229 | 71,906 | 96,817 |
| Private | 118 | 2,191 | 6,192 | 8,501 |
| Non-jn | 3,513 | 32,904 | 73,324 | 109,741 |
| Total | 5,821 | 66,897 | 182,109 | 254,827 |

*Source:* Table 15.1

tingency table; Table 15.1 is a three-way table. We may, for example, focus attention on only *AREA* (values: Urban, Rural) and *SEVERITY* (values: Fatal, Serious, Minor) and count the number of instances of each possible combination, of which there are 2 × 3 = 6. The resulting contingency table is shown in Table 15.2 (a). Apart from the row and column totals, the table has 6 cells with 3 rows and 2 columns.

There is a simple but important connection between the two-way Table 15.2 (a) and Table 15.1 in that all the entries in Table 15.2 (a) have already been shown in various total rows or columns of Table 15.1. For example, in Table 15.2 (a), the 3,186 fatal accidents in urban areas are also in the total entry of the first row of Table 15.1, where fatal accidents in urban areas are added up for all the road types. Table 15.2 (b), based on the two variables *ROAD* and *SEVERITY* is also a two-way table and is similarly derivable from the three-way Table 15.1.

We have in effect been setting out a hierarchy—a sequence—of contingency tables all based upon the same data matrix. This is set out in Table 15.3. This hierarchy begins with a three-way table and descends, by addition, first to the two-way classifications, then to the one-way classifications and finally to a single total frequency. At the highest level is the three-way Table 15.1, which is said to provide a *cross-classification* of three variables. Symbolically, this table is denoted by *AREA* × *ROAD* × *SEVERITY*, the multiplication signs in this notation reflecting the way the number of cells in the table derive from multiplying 2 × 5 × 3, the respective number of values in the categorical variables. At the next lower level, three two-way tables can be derived, and Table 15.2 shows two of them. These two-way tables are known as the *second order margins* of the contingency table. Finally, at the lowest level, there are the three one-way tables showing, for each

**Table 15.3** Levels of Cross-Classification

| *Level* | *Cross-classification(s)* | *No. of cells* |
|---|---|---|
| 3-way | *AREA* × *ROAD* × *SEVERITY* | 2 × 5 × 3 = 30 |
| 2-way | (i) *SEVERITY* × *AREA* | 3 × 2 = 6 |
| | (ii) *ROAD* × *SEVERITY* | 5 × 3 = 15 |
| | (iii) *ROAD* × *AREA* | 5 × 2 = 10 |
| 1-way | (i) *ROAD* | 5 |
| | (ii) *SEVERITY* | 3 |
| | (iii) *AREA* | 2 |

variable, the frequencies corresponding to the individual values. These are known as the *first order margins*. (The margin of zero order is simply the total frequency.)

It will be clear from the above that although contingency tables conceptually are fairly straightforward, numerically (and indeed algebraically) they can be exceedingly complicated. In the analysis of numerical variables in the multiple regression and multivariate analysis chapters, there were no serious data handling problems due to the multiplicity of variables. However, our contingency table based upon three categorical variables requires as many as 30 frequency numbers and if we had added a fourth, recording say the day of the week of an accident, a table of $7 \times 30 = 210$ frequencies would be needed if a complete cross-classification were to be set out.

In the following Sections 15.3—15.6 we shall concentrate upon analysis of the simpler two-way contingency table and postpone any further consideration of higher order tables.

## 15.3 TWO-WAY CONTINGENCY TABLES: DESCRIPTIVE ANALYSIS

A casual glance at the two-way contingency tables of Table 15.2 gives some information about the overall accident situation. Thus, it can readily be seen from the Total rows or columns that only a small proportion of accidents prove fatal and that relatively few accidents occur at roundabouts. However, the tables as formulated do not tell us, for example, whether an accident is more likely to be fatal if it occurs at, say, a roundabout or a cross-roads, or if it occurs in an urban rather than a rural area. Questions of this kind are about the individual frequencies within the tables and an answer can be approached through the calculation of row and column *proportions*.

Table 15.4 (a) is derived from the contingency table *SEVERITY* × *AREA* at Table 15.2 (a). Operating on each row at a time we calculate proportions using the total frequencies in the last column. Another name for the total frequencies is the *marginal frequencies*. (Thus, 0.55 = 3,186/5,821.) Table 15.4 (b) is similarly constructed from the *columns* of the same contingency table. The latter table tells us for example that urban accidents are on average less serious than rural accidents: 1.6% of the former are fatal compared to 4.4% of the latter.

It will be useful to have some general notation for a contingency table and its row/column proportions. Table 15.5 exhibits the contingency table notation. The two categorical variables are labelled $A$ and $B$ and their values are $A_1, A_2, \ldots, A_r$ and $B_1, B_2, \ldots, B_c$ respectively.

**Table 15.4** Contingency Table *SEVERITY* × *AREA*: Row and Column Proportions

(a) Row Proportions

| | *Urban* | *Rural* | *Total* |
|---|---|---|---|
| Fatal | 0.55 | 0.45 | 1.00 |
| Serious | 0.68 | 0.32 | 1.00 |
| Minor | 0.80 | 0.20 | 1.00 |
| All Severities | 0.76 | 0.24 | 1.00 |

(b) Column Proportions

| | *Urban* | *Rural* | *All areas* |
|---|---|---|---|
| Fatal | 0.016 | 0.044 | 0.023 |
| Serious | 0.23 | 0.35 | 0.26 |
| Minor | 0.75 | 0.60 | 0.71 |
| Total | 1.00 | 1.00 | 1.00 |

*Source:* Table 15.2 (a)

The letters $r$ and $c$ thus denote the number of rows and columns respectively (apart from the row and column of marginal frequencies). The suffix notation for the cell frequencies will be clear; $f_{ij}$ indicates a general frequency for the cell in the $i$th row and $j$th column. This cell corresponds to a conjunction of the two variable values $A_i$ and $B_j$. Finally, we note that, for example, $f_{1.}$ is the total frequency in *row* 1 whereas $f_{.1}$ is the total frequency in *column* 1.

The notation for tables of row proportions or column proportions is set out in Table 15.6. The letter $p$ indicates proportions calculated across rows and $q$ the proportions calculated down columns. By

**Table 15.5** Two-Way Contingency Table Notation

| | | *Values of second categorical variable, B* | | | | | | | *Frequencies T* |
|---|---|---|---|---|---|---|---|---|---|
| | | $B_1$ | $B_2$ | ... | $B_j$ | ... | $B_c$ | | |
| Values of first categorical variable, *A* | $A_1$ | $f_{11}$ | $f_{12}$ | ... | $f_{1j}$ | ... | $f_{1c}$ | | $f_{1.}$ |
| | $A_2$ | $f_{21}$ | $f_{22}$ | ... | $f_{2j}$ | ... | $f_{2c}$ | | $f_{2.}$ |
| | .. | .. | .. | ... | .. | ... | .. | | .. |
| | $A_i$ | $f_{i1}$ | $f_{i2}$ | ... | $f_{ij}$ | ... | $f_{ic}$ | | $f_{i.}$ |
| | .. | .. | .. | ... | .. | ... | .. | | .. |
| | $A_r$ | $f_{r1}$ | $f_{r2}$ | ... | $f_{rj}$ | ... | $f_{rc}$ | | $f_{r.}$ |
| | $T$ | $f_{.1}$ | $f_{.2}$ | ... | $f_{.j}$ | ... | $f_{.c}$ | | $n$ |

**Table 15.6** Row and Column Proportions Notation

(a) Row proportions

| | $B_1$ | $B_2$ | ... | $B_c$ | $T$ |
|---|---|---|---|---|---|
| $A_1$ | $p_{11}$ | $p_{12}$ | ... | $p_{1c}$ | 1 |
| $A_2$ | $p_{21}$ | $p_{22}$ | ... | $p_{2c}$ | 1 |
| .. | .. | .. | ... | .. | .. |
| $A_r$ | $p_{r1}$ | $p_{r2}$ | ... | $p_{rc}$ | 1 |
| All | $p_1$ | $p_2$ | ... | $p_r$ | 1 |

(b) Column proportions

| | $B_1$ | $B_2$ | ... | $B_c$ | *All* |
|---|---|---|---|---|---|
| $A_1$ | $q_{11}$ | $q_{12}$ | ... | $q_{1c}$ | $q_1$ |
| $A_2$ | $q_{21}$ | $q_{22}$ | ... | $q_{2c}$ | $q_2$ |
| .. | .. | .. | ... | .. | .. |
| $A_r$ | $q_{r1}$ | $q_{r2}$ | ... | $q_{rc}$ | $q_r$ |
| $T$ | 1 | 1 | ... | 1 | 1 |

definition, $p_{ij} = f_{ij}/f_{i.}$ and $q_{ij} = f_{ij}/f_{.j}$. Note that, in the last row of Table 15.6 (a), $p_1$ (as an example) is defined as $f_{.1}/n$; it is the proportion of the marginal (or total) frequency of occurrency of the value $B_1$ to the whole total frequency $n$. Similarly $q_1 = f_{1.}/n$.

There is an important and simple pattern of row (or column) proportions which, though rarely met in practice, gives a useful reference point for empirical contingency tables. It is one where the row (or column) proportions in any one row (or column) are the same as the corresponding proportions in any other row (or column). This pattern is exhibited by the left-hand contingency table of Table 15.7. For each row, including the last row of marginal frequencies, the row proportions are the three values $\frac{1}{2}, \frac{1}{6}, \frac{1}{3}$ so if a table such as Table 15.6 (a) were calculated, these figures would be repeated row by row. It is a mathematical consequence that a similar property must also be exhibited by the columns. Specifically:

> If a contingency table shows equality, row by row, of corresponding row proportions, then the corresponding column proportions will also be equal, column by column (and conversely). **(15.3.1)**

The reader can easily confirm that the proportions common to each column of Table 15.7 (a) are $\frac{1}{2}, \frac{1}{4}, \frac{1}{6}, \frac{1}{12}$.

A table of this kind is said to show *no association* between the two variables $A$ and $B$. The proportionate allocation of frequencies to the values $B_1$, $B_2$, and $B_3$ is the same whether we consider individuals with values $A_1$ or $A_2$ or $A_3$ or $A_4$. It is in this sense that the two variables are said to be 'not associated'.

**Table 15.7** Extremes of Association

| | Case $A$ (a) No association | | | | | Case $B$ (b) Strong association | | | |
|---|---|---|---|---|---|---|---|---|---|
| | $B_1$ | $B_2$ | $B_3$ | $T$ | | $B_1$ | $B_2$ | $B_3$ | $T$ |
| $A_1$ | 18 | 6 | 12 | 36 | $A_1$ | 36 | — | — | 36 |
| $A_2$ | 9 | 3 | 6 | 18 | $A_2$ | — | — | 18 | 18 |
| $A_3$ | 6 | 2 | 4 | 12 | $A_3$ | — | 12 | — | 12 |
| $A_4$ | 3 | 1 | 2 | 6 | $A_4$ | — | — | 6 | 6 |
| $T$ | 36 | 12 | 24 | 72 | $T$ | 36 | 12 | 24 | 72 |

*Source:* hypothetical

A 'no association' table such as Table 15.7 (a) has a simple numerical property. Each entry in the table can be calculated by multiplying together its marginal frequencies (those for the row and column in which it lies) and dividing by the total frequency. Thus 18 = (36 × 36)/72 and 9 = (18 × 36)/72. In general, and using the notation of Table 15.5, we have the following result:

For any 'no-association' contingency table

$$f_{ij} = (f_{i.}f_{.j})/n \qquad \textbf{(15.3.2)}$$

By contrast with Table 15.7 (a), Table 15.7 (b) shows a case where there is a very strong association between $A$ and $B$ in the sense that all individuals who have value $A_1$ also have value $B_1$, all who have value $A_2$ have value $B_3$ and so on. The connection is not as straightforward the other way round in that of those with value $B_3$, some have value $A_2$ and some $A_4$. (It is impossible to have a 'perfect' association in this sense if the number of rows is not equal to the number of columns.)

We shall take up the question of measuring the degree of association in Section 15.5.

## 15.4 A PROBABILITY MODEL FOR A TWO-WAY CONTINGENCY TABLE

The model we shall now consider has the following features: it presumes that each value of the variable $A$ has some fixed probability of occurrence, similarly for each value of variable $B$, and, an important assumption, that the occurrences of values of $A$ and $B$ are *independent*. If this model were to be applied to the *SEVERITY* × *AREA* data (Table 15.2 (a)), it would be equivalent to assuming that:

(i) there are fixed probabilities, say, $P$ (Urban) and $P$ (Rural) that an accident occurs in an urban or rural area;
(ii) there are fixed probabilities, say $P$ (Fatal), $P$ (Serious), and $P$ (Minor) that an accident is of one of these three degrees of severity; and
(iii) that the variables *SEVERITY* and *AREA* are independent.

This last assumption implies that, for example, the probability that an accident is fatal is not affected by the kind of area in which the accident occurs.

We now set up the independence model in terms of the general notation of Table 15.5. For the variable $A$, the model associates with the values $A_1, A_2, \ldots, A_r$ the respective probabilities $P(A_1), P(A_2), \ldots, P(A_r)$. Similarly there are the $c$ probabilities $P(B_1), P(B_2), \ldots, P(B_c)$

associated with the $c$ values $B_1, B_2, \ldots, B_c$ of variable $B$. From these two sets of *marginal probabilities*, it is straightforward—using the independence assumption—to calculate the probability that an individual has any particular combination of values of the two variables. For example, the probability of taking the values $A_1$ and $B_1 = P(A_1$ and $B_1) = P(A_1) \times P(B_1)$ because $A_1$ and $B_1$ are independent events (see Section 0.16). In general $P(A_i$ and $B_j) = P(A_i) \times P(B_j)$.

If each of these probabilities (including the marginal probabilities) is multiplied by $n$, the result is a set of *expected frequencies*. The expected frequencies can be looked upon as the long-run average frequencies which would be observed for each cell of the contingency table for a data set with $n$ individuals. Table 15.8 shows the resulting table of expected frequencies for each cell of the contingency table. Note that the rows (and columns) sum across (and down) to the marginal expected frequencies. Thus, for the first row, $nP(A_1)P(B_1) + \ldots + nP(A_1)P(B_c) = nP(A_1)[P(B_1) + \ldots + P(B_c)] = nP(A_1) \times 1 = nP(A_1)$. Also, these expected frequencies exhibit a 'no-association' pattern since $nP(A_i)P(B_j) = [nP(A_i)]\,[nP(A_j)]\,/n$ (see result (15.3.2)).

The table of expected frequencies in Table 15.8 is little more than a restatement of the independence model and its implications. The essential question now is: given an empirical contingency table, how can the model parameters be estimated and the validity of this independence model checked?

The model parameters are defined by the $(r + c)$ marginal expected frequencies, $nP(A_1), \ldots, nP(A_r)$ and $nP(B_1), \ldots, nP(B_c)$. It can be shown that:

> the appropriate estimates of the marginal expected frequencies are the marginal frequencies of the contingency table. **(15.4.1)**

Before we apply this model to a real data set, it will be clearer—and

**Table 15.8** Expected Frequencies for Independence Model

| | $B_1$ | .... | $B_j$ | .... | $B_c$ | *Any* $B_j$ |
|---|---|---|---|---|---|---|
| $A_1$ | $nP(A_1)P(B_1)$ | .. | $nP(A_1)P(B_j)$ | .. | $nP(A_1)P(B_c)$ | $nP(A_1)$ |
| .. | .. | .. | .. | .. | .. | .. |
| $A_i$ | $nP(A_i)P(B_1)$ | .. | $nP(A_i)P(B_j)$ | .. | $nP(A_i)P(B_c)$ | $nP(A_i)$ |
| .. | .. | .. | .. | .. | .. | .. |
| $A_r$ | $nP(A_r)P(B_1)$ | .. | $nP(A_r)P(B_j)$ | .. | $nP(A_r)P(B_c)$ | $nP(A_r)$ |
| Any $A_i$ | $nP(B_1)$ | .. | $nP(B_j)$ | .. | $nP(B_c)$ | $n$ |

**Table 15.9** Data Set Modelled by Independence Probability Model

| (a) Contingency table | | | | | (b) Estimated expected frequencies | | | | | (c) Residuals | | | | |
|---|---|---|---|---|---|---|---|---|---|---|---|---|---|---|
| | $B_1$ | $B_2$ | $B_3$ | $T$ | | $B_1$ | $B_2$ | $B_3$ | $T$ | | $B_1$ | $B_2$ | $B_3$ | $T$ |
| $A_1$ | 22 | 4 | 10 | 36 | $A_1$ | 18 | 6 | 12 | 36 | $A_1$ | 4 | −2 | −2 | 0 |
| $A_2$ | 7 | 4 | 7 | 18 | $A_2$ | 9 | 3 | 6 | 18 | $A_2$ | −2 | 1 | 1 | 0 |
| $A_3$ | 3 | 3 | 6 | 12 | $A_3$ | 6 | 2 | 4 | 12 | $A_3$ | −3 | 1 | 2 | 0 |
| $A_4$ | 4 | 1 | 1 | 6 | $A_4$ | 3 | 1 | 2 | 6 | $A_4$ | 1 | 0 | −1 | 0 |
| $T$ | 36 | 12 | 24 | 72 | $T$ | 36 | 12 | 24 | 72 | **T** | 0 | 0 | 0 | 0 |

*Source:* hypothetical

numerically simpler—to show how the estimation (or fitting) procedure works for a hypothetical data set.

The contingency table is shown in Table 15.9 (a). The column marginal frequencies are 36, 12, 24 and the row marginal frequencies 36, 18, 12, 6. These provide the estimates of the marginal expected frequencies of the independence model, which in turn give the other estimated expected frequencies by applying result (15.3.2). The estimated expected frequencies can alternatively be called the *fitted* frequencies of the independence model (just as in regression analysis, the estimated regression equation gives the fitted dependent variable values). Table 15.9 (c) shows the residuals, defined—as usual—as the difference between the data value and the fitted value.

This fitting procedure can be described in terms of general notation. We start from the contingency table with $f_{ij}$ representing the general frequency, $f_{i.}$ the general row marginal frequency and $f_{.j}$ the general column marginal frequency. We denote the corresponding entries in the table of fitted values (that is, the estimated expected frequencies) by capital $F$'s. So $F_{i.} = f_{i.}$ (for $i = 1, \ldots, r$) and $F_{.j} = f_{.j}$ (for $j = 1, \ldots, c$). Also $F_{ij}$, the fitted frequency for the $i$th value of $A$ and the $j$th value of $B$, equals $(F_{i.}F_{.j})/n = (f_{i.}f_{.j})/n$. Denoting the residuals by the label $e$, we have $e_{ij} = f_{ij} - F_{ij} = (f_{ij} - (f_{i.}f_{.j})/n)$.

The validity of the independence model is easily checked. The test statistic, known as the $\chi^2$-statistic, and often simply labelled $\chi^2$, is $\Sigma(f_{ij} - F_{ij})^2/F_{ij}$, where the summation is taken over all $rc$ cells in the contingency table. (This $\chi^2$-statistic is a two-way table extension of the $\chi^2$ goodness of fit statistic of Section 7.4.) It can be shown that, if the independence model is valid, the test statistic $\chi^2$ has the distribution $\chi^2(\nu)$ where the degrees of freedom $\nu = (r-1)(c-1)$. The quantity $(r-1)(c-1)$ arises because of the condition that the marginal expected frequencies should be equal to those given by the data. As a result, given any $(r-1)$ cell frequencies in any row, the remaining cell frequency can be calculated by subtraction. A similar argument holds for the columns.

The test is a one-tailed test since only large values of the test statistic favour the alternative hypothesis of association. Applying this test to the hypothetical data of Table 15.8, $\chi^2 = 4^2/18 + \ldots + (-1)^2/2 = 6.7$. Since $\chi^2_{0.95}(6) = 12.6$, the independence model is not rejected. (The warning in Section 7.4 about analysis with cell frequencies less than 3 has been ignored for simplicity of exposition.)

Textbook examples typically apply this $\chi^2$-test to contingency table data, real or hypothetical, with relatively small frequencies—as we have just done. However, an important model-building aspect is brought out if we see how it can work out for the larger frequencies such as in the accident data. We take, as an example, the contingency table for the cross-classification *ROAD* × *SEVERITY* (Table 15.2 (b)). For this table, $F_{11} = (8,814 \times 5,821)/254,827 = 201.3$ so the residual $e_{11} = 69 - 201.3 = -132.3$ and $(f_{11} - F_{11})^2/F_{11} = 87.0$. A similar computation for all the other cells gives a $\chi^2$-value which is very large indeed: 2,550.

This is another example of a feature of hypothesis testing to which we have referred on a number of occasions. Given a sufficiently large value of $n$, any null hypothesis will almost certainly be rejected. Whether such a conclusion has any contextual significance depends upon the objectives of the analysis.

## 15.5 MEASURES OF ASSOCIATION

The ordinary correlation coefficient, $r$, can be looked upon as a measure of association between two numerical variables. In the present section, we shall be concerned with extending this idea to the case of two categorical variables in terms of the frequencies in the contingency table.

It is tempting to think that the $\chi^2$-statistic might be a suitable measure of association. It has some desirable properties. It takes the value 0 if the contingency table shows no association and, for a given total frequency, takes higher values as the frequencies approach the maximum degree of association. The problem is that the $\chi^2$-statistic is affected by the value of $n$, the total frequency. We saw at the end of the previous section how a data set with numerically large frequencies can have exceedingly large $\chi^2$-statistic values. The point can be put slightly differently by observing that if the frequencies in the contingency table of Table 15.9 (a) were each increased by a factor of 10, the $\chi^2$-statistic would increase from 6.7 to 67, with no change in the row or column proportions.

There is no shortage of measures of association. The SPSS program (as part of the subprogram CROSSTABS) lists some 15 of them (to 5 decimal places)! The problem is that they each serve their own particular purpose depending upon the orientation of the analysis in hand. We shall discuss only a few of them.

One of the simplest of the association measures is the *Pearson coefficient of contingency* defined as:

$$P = \frac{\chi^2/n}{\sqrt{(1 + \chi^2/n)}} \qquad (15.5.1)$$

This coefficient has the property that $P = 0$ for a no-association contingency table but it is inevitably less than its maximum possible value of unity however strong the association. Cramer's $V$ statistic, defined as

$$V = \frac{\chi^2/n}{\sqrt{[\min(r-1,\ c-1)]}}, \tag{15.5.2}$$

overcomes the latter difficulty; it equals unity in the case of complete association.

These two measures are purely descriptive and a particular numerical value of either has significance only in comparative terms. A more purposeful approach is offered by measures which take an asymmetric view of the table and base their measure of association on an answer to the question: how far does knowledge of one variable allow a better prediction of the value of the other variable? The *uncertainty coefficient* takes this approach. It is based upon the concept of the *entropy*.

Suppose first we have a discrete probability distribution for the $r$ values $A_1, A_2, \ldots, A_r$ of a categorical variable $A$. The respective probabilities are, say, $\pi_1, \pi_2, \ldots, \pi_r$. The entropy of this distribution is defined as:

$$H = \sum_{i=1}^{r} \pi_i \log(1/\pi_i) \tag{15.5.3}$$

where the logarithms are to the base $e$.[1] $H$ can be shown to have a maximum value when all the $\pi_i$'s are the same and therefore equal $1/r$: this maximum value is $r(1/r)\log(r) = \log(r)$. For any other probability distribution $H$ is smaller. For example in the extreme, when $\pi_1 = 1$ and $\pi_2 = \ldots = \pi_r = 0$, $H$ equals its minimum value zero. (When $\pi = 0$ $\pi \log(1/\pi)$ is defined as equal to zero.) The entropy can be looked upon as measuring the degree of uncertainty of some future observed value of $A$. If the probabilities are equal, $H$ is at its maximum and uncertainty about the value is at its greatest. On the other hand if the probabilities are more heavily weighted on some values of $A$, these are the more likely, uncertainty is less and entropy is lower.

Empirically, entropies are estimated from observed relative frequencies. For example, in the first row of each table at Table 15.7, we have two separate frequency distributions. For table (a) first row, the relative frequencies are $\frac{1}{2}, \frac{1}{6}, \frac{1}{3}$ and $H = \frac{1}{2}\log 2 + \frac{1}{6}\log 6 + \frac{1}{3}\log 3 = 1.01$. For the other table (b) first row, $H = 1 \log 1 = 0$. If the cell entries had each been equal to 12, the entropy would have taken the maximum value of $3(\frac{1}{3}\log 3) = 1.10$.

1. We use the letter $H$ to denote entropy in conformity with Theil's (1971) usage.

The uncertainty coefficient $U$ depends upon comparing two entropy calculations based upon proportions in the contingency table as estimates of the underlying probabilities. The first, which we label $H(A)$, is the entropy associated with the marginal proportions of $A$. In terms of our general notation, the marginal proportions associated with the values $A_1, \ldots, A_r$ of $A$ are respectively $f_{1.}/n, f_{2.}/n, \ldots, f_{r.}/n$. The entropy $H(A) = \Sigma(f_{i.}/n) \log (n/f_{i.})$. For the contingency table at Table 15.9, $H(A) = \frac{1}{2}\log 2 + \frac{1}{4}\log 4 + \frac{1}{6}\log 6 + \frac{1}{12}\log 12 = 1.20$. The second entropy calculation is based upon a weighted average of the entropies for each *column* of the contingency table, that is, the entropy of $A$ given that one of the values of $B$ has recurred. The weights are the marginal proportions of the variable $B$. The formula for this is then:

$$H(A \mid B) = \sum_{j=1}^{c} (f_{.j}/n) \left[\sum_{i=1}^{r} (f_{ij}/f_{.j}) \log(1/(f_{ij}/f_{.j}))\right] \qquad \textbf{(15.5.4)}$$

Applying (15.5.4) to Table 15.9, the three column entropies are 1.07, 1.29 and 1.20. The weighted average $H(A \mid B) = 1.15$.

Thus $H(A \mid B)$ (= 1.15) is less than $H(A)$ (= 1.20) and it can be shown that this is always so unless the contingency table exhibits no association, when $H(A \mid B) = H(A)$. The uncertainty coefficient is defined as:

$$U(A \mid B) = [H(A) - H(A \mid B)]/H(A) \qquad \textbf{(15.5.5)}$$

It lies between 0 and 1 and equals 0 if there is no association. For our example $U(A \mid B) = 0.04$. This is small and it will be recollected from Section 15.4 that an independence model for this hypothetical data was not rejected.

The uncertainty coefficient is asymmetric in that it focuses attention on one variable. Similarly $U(B \mid A)$ can be written down by interchanging $A$ and $B$ in (15.5.5).

The above is necessarily a brief account of some measures of association and the ideas behind them. As with the correlation coefficient, we think that patterns of association are often best explored in the context of model building, a topic which will be taken up in Section 15.8.

A fuller account of measures of association, including measures for the case where the contingency table is based upon ordinal categorical variables, is given in Everitt (1977).

## 15.6 TWO-WAY CONTINGENCY TABLE PLOTS

In this section, we consider methods of plotting two-way contingency tables having no more than three columns (or no more than three rows). This is rather restrictive, but in Section 15.7 it will be seen how plotting can be extended to larger tables.

A two-way contingency table is usually symmetrical in the sense that it is a matter of arbitrary choice which of the two categorical variables is ascribed to the rows and which to the columns of the table. Analysis of the table is typically symmetrical as well: the fitting of the independence model in Section 15.4 treated one variable exactly the same as the other. Sometimes attention may be focused upon one categorical variable as when we looked at tables of row proportions in Section 15.3 (for example, Table 15.4 (a)). But symmetry is restored in the sense that the same methodology can be applied to column proportions. In the discussion below we shall in fact be describing the plotting and interpretation of row proportions, but it should be borne in mind that the methods can be applied, with appropriate changes, to columns.

Table 15.10 shows the contingency table for the cross-classification *ROAD* × *AREA* and we suppose that the focus of interest is on how the various kinds of road hazards differ in their accident experience as regards whether they are in urban or rural areas. The relevant row proportions are shown in brackets. We note that cross-roads and non-junction accidents are at the two extremes: the former have the highest proportion of accidents in urban areas, the latter the lowest.

**Table 15.10** Contingency Table *ROAD* × *AREA* with Row Proportions

| | *Urban* ($p_1$) | *Rural* ($p_2$) | *Total* ($p_1 + p_2$) |
|---|---|---|---|
| Roundabout | 6,905<br>(0.78) | 1,909<br>(0.22) | 8,814<br>(1.0) |
| Cross-roads | 27,961<br>(0.90) | 2,993<br>(0.10) | 30,954<br>(1.0) |
| Other junctions | 85,494<br>(0.88) | 11,323<br>(0.12) | 96,817<br>(1.0) |
| Private drive | 6,028<br>(0.71) | 2,473<br>(0.29) | 8,501<br>(1.0) |
| Non-junction | 68,346<br>(0.62) | 41,395<br>(0.38) | 109,741<br>(1.0) |
| All types | 194,734 | 60,093 | 254,827 |

*Source:* Table 15.1

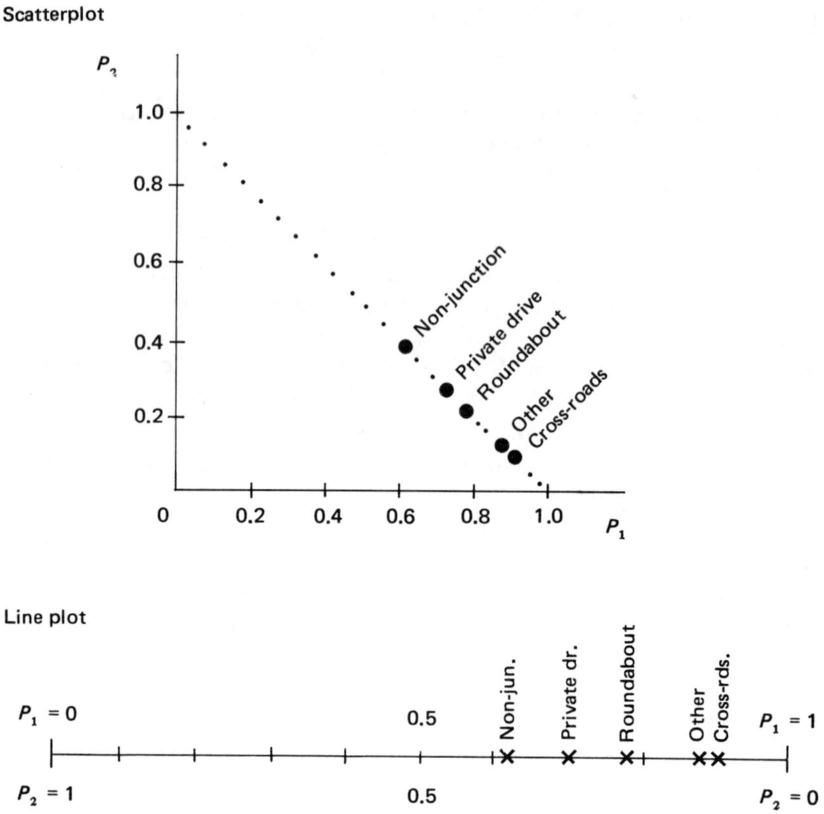


*Source*: Table 15.10

**Figure 15.1** Plots of Row Proportions for a Two-Column Contingency Table

The information may be plotted in a conventional scatterplot in which the individual data points correspond to the road features and the two 'variables' are the proportions $p_1$ and $p_2$. Figure 15.1 shows the scatterplot and it will be observed that each point lies on a straight line. The reason of course is that, for each road hazard, $p_1 + p_2 = 1$ and each data point lies on the straight line with this equation. As a consequence, to plot this data, all we need is the line plot shown beneath the scatterplot. The degree of scatter of the points on the line plot reflects the degree of association between the two variables. All the points would be coincident if the row proportions were all equal.

**Table 15.11** Contingency Table and Row Proportions Table: Three-Column Case

(a) Contingency table

| | *a* | *b* | *c* | *T* |
|---|---|---|---|---|
| *A* | 2 | 3 | 5 | 10 |
| *B* | 8 | 10 | 22 | 40 |
| *C* | 4 | 3 | 13 | 20 |
| *D* | 21 | 21 | 18 | 60 |
| *E* | 1 | 18 | 1 | 20 |
| *T* | 36 | 55 | 59 | 150 |

(b) Row proportions table

| | $p_1$ | $p_2$ | $p_3$ | *T* |
|---|---|---|---|---|
| *A* | 0.2 | 0.3 | 0.5 | 1.0 |
| *B* | 0.2 | 0.25 | 0.55 | 1.0 |
| *C* | 0.2 | 0.15 | 0.65 | 1.0 |
| *D* | 0.35 | 0.35 | 0.30 | 1.0 |
| *E* | 0.05 | 0.90 | 0.05 | 1.0 |

*Source:* hypothetical

A similar effect is shown if there are three columns in the contingency table rather than two. Again, each row concerns the values of one of the categorical variables and these are thought of as the 'individuals' of the data matrix and the three columns $p_1$, $p_2$, and $p_3$ are treated as if they were the sets of variable values. A hypothetical contingency table of this kind is shown in Table 15.11 (hypothetical, in order to bring out better some methodological points). To help focus ideas, one might think of the table as being derived from a population census. The frequencies are in units of thousand people, the row categories *A*, *B*, *C*, *D*, *E*, represent local areas in a city and the column categories numbers of people in three broad socio-economic groups labelled *a*, *b*, *c*. If we are interested in comparing these local areas the frequencies are not very informative—they merely reflect the varying sizes of the area. The row proportions are the more relevant data. They show, for example, that areas *A* and *B* are very similar and that areas *C* and *E* are very different.

These similarities and dissimilarities can be plotted, at least notionally, in a three dimensional scatterplot (see Section 10.7 and Figure 10.3). A drawing of such a scatterplot is at Figure 15.2. However, the points of the scatterplot are not distributed in three dimensional space but in two dimensions. They all lie in a plane and indeed within the triangle outlined on the diagram. The reason is that, for each data point, $p_1 + p_2 + p_3 = 1$. In three dimensional coordinate geometry, this connection between the three variables determines a plane.

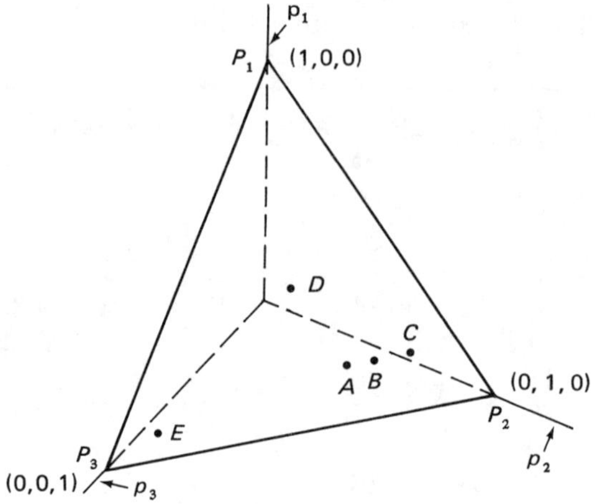


**Figure 15.2** Plot (Three Dimensional) of Row Proportions

The triangle and its plotted points are shown as a two dimensional plot in Figure 15.3. The triangle is equilateral. The vertices $P_1$, $P_2$, $P_3$ correspond to any rows with respectively the extreme proportions (1, 0, 0) (0, 1, 0) and (0, 0, 1). The scale of the diagram is such that any one vertex is at unit distance from its opposite side. A particular row such as $D$ (with $p_1 = 0.35$, $p_2 = 0.35$, $p_3 = 0.30$) is plotted by the

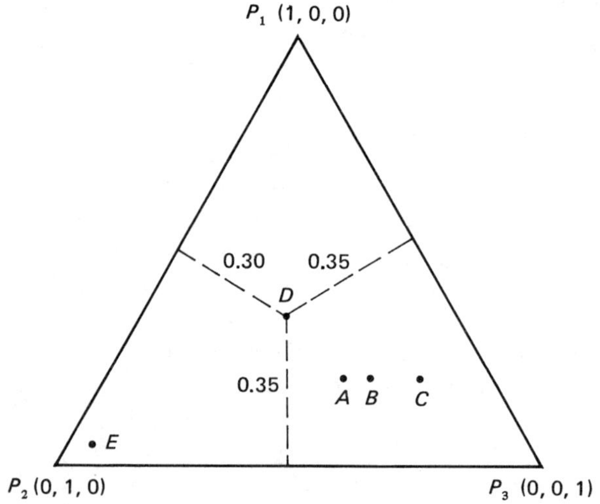


*Source*: Table 15.11(b)

**Figure 15.3** Triangular Plot of Row Proportions

point with vertical distances 0.35 from $P_2P_3$, 0.35 from $P_3P_1$ and 0.30 from $P_1P_2$.

The positions of the points give a ready visual indication of which local areas are similar to each other as regards their socio-economic composition. Areas *A, B, C* are very similar and *E* and *A* are very different.

## 15.7 PRINCIPAL COMPONENTS AND CORRESPONDENCE ANALYSIS

We now turn to a consideration of how the rows of a contingency table may be described and compared in the presence of any number of columns. As before, it will be presumed that the focus of interest is the row proportions rather than the row frequencies. Also as before, we note that the various procedures can be applied to columns as well as to rows.

Correspondence analysis is closely related to principal components analysis (which was considered in Chapter 11). So it will be helpful to look first at the implications of applying PC analysis to a table of row proportions, regarding it as a data matrix. We start with the simple case where, as in the previous section, there are just three columns (variables) in the row proportions data matrix. Table 15.11 (b) will again be used as an example, together with its scatterplot at Figure 15.2.

PC analysis is carried out on the *centred* data matrix. Geometrically this amounts to shifting the three axes $p_1$, $p_2$, $p_3$ parallel to themselves so that the origin lies at the centroid of the data. (This is an extension of the two variable case described in Section 11.2.) This new origin lies in the plane of the triangle containing the data points and the axes themselves now stick out of the plane. PC analysis derives a new set of axes which we call $Z_1$, $Z_2$ and $Z_3$ (consistent with the notation in Chapter 11). The direction of $Z_1$ is along the greatest elongation of the scatter of the five points *A, B, C, D,* and *E* and $Z_1$, not surprisingly, lies within the triangular plane, and so does $Z_2$. As a consequence, $Z_3$—which is at right angles to both $Z_1$ and $Z_2$—is perpendicular to the triangular plane. These geometrical features are depicted in the two diagrams of Figure 15.4.

These geometrically expressed ideas have important implications. First, because $Z_3$ is at right angles to the triangular plane, the coordinates of the data points in its direction are all zero. All the third principal component scores are therefore identical and this principal component is redundant. Secondly, the first two principal component

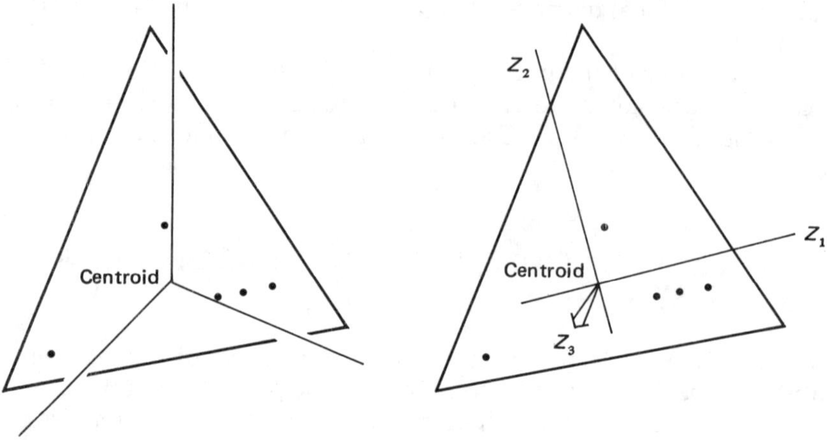


**Figure 15.4** PC Axes for Row Proportions

axes $Z_1$ and $Z_2$ are in exactly the same directions as if we had simply applied a two variable PC analysis to the two-dimensional scatterplot of Figure 15.3. So the PC scores will be also identical in the two cases.

To sum up then, the application of PC analysis to the row proportions data matrix confirms what we saw in the last section: only two variables are required to describe the similarities and dissimilarities of the 'individuals' (the row proportions) in a contingency table with three categories. The PC analysis, as it were, automatically converts the three dimensional scatterplot of Figure 15.2 into the two-dimensional version of Figure 15.3.

Applying PC analysis to the row proportions data matrix of Table 15.11 (b) gives the following results. The eigenvalues are $\lambda_1 = 2.34$, $\lambda_2 = 0.66$ and $\lambda_3 = 0$. We recall that the eigenvalues are the variances of the PC component scores (Section 11.6). The third eigenvalue being zero simply reflects the fact that the scores for the third principal component are identical—it is another way of saying that the third PC is redundant. The two sets of PC scores are given in Table 15.12. The reader may care to plot these. It will be found that the pattern of the points is the same as that of Figure 15.3.

As we saw in Chapter 11, when PC analysis is applied to a multivariate data matrix it may be possible to describe the individuals in terms of a small number of PC variables. The same will be true if the data matrix is a row proportions table with many columns and the smallest eigenvalue will again be zero. A bivariate plot of the first two PC scores may give an informative picture of the similarities and differences between the categories defining the rows, if the first two eigenvalues are a large proportion of total variance.

**Table 15.12** PC Scores for Row Proportions of Table 15.11 (b)

| Row name | *A* | *B* | *C* | *D* | *E* |
|---|---|---|---|---|---|
| PC 1 score | 0.42 | 0.65 | 1.12 | 0.50 | −2.69 |
| PC 2 score | −0.24 | −0.38 | −0.65 | 1.41 | −0.14 |

*Source:* GENSTAT output based upon correlation matrix

*Correspondence analysis,* to which we now turn, also carries out a PC analysis but on a transformed version of the row proportions table. So it is a little more complicated. We illustrate it in terms of Table 15.13: a contingency table of road casualties according to hours of the day (the rows) and days of the week (the columns).

The first stage is to calculate the row proportions as before. Two new features now arise. First, the values in each *column* of the row proportions data matrix are divided by the square root of the marginal relative frequency (the marginal frequency divided by the total frequency) of the same column of the contingency table. For example, for the casualty data, the first column total for Monday in the contingency table is 43,792, the overall total frequency is 334,513 and so the divisor is $\sqrt{0.131} = 0.362$. A PC analysis is then carried out, but here the second new feature arises. Each individual row (each hour of the day) is given a weighting in the PC analysis equal to the total row relative frequency in the contingency table. Thus, the weight for the 16.00 hours row is the row total, which is 28,368, divided by 334,513 and so the weight equals 0.085. This weighting procedure works in the following fashion. Suppose that one row turns out to have twice the weight of another row. Then the weighted PC analysis has the effect of including two (identical) individuals rather than one in computing the principal components.

The reasons for these special features in correspondence analysis, the column transformation and the row weighting, will be picked up below. Meanwhile we pass on to the correspondence analysis results as applied to the casualty data. The output has the same structure as PC output. With $p$ variables (that is $p$ columns to the contingency table) there are $p$ eigenvalues each with associated 'PC type' coefficients which in turn allow computation of the scores for each individual.

The correspondence analysis eigenvalues are 0.068, 0.006, 0.0022, 0.0003, 0.0002, 0.0001. These $(p - 1) = 6$ eigenvalues exclude the eigenvalue corresponding to the redundant component—which occurs in the correspondence analysis solution just as in the pure PC

**Table 15.13** Road Casualties by Hour and Day of Week

*Number*

| *Hour from* | *Mon* | *Tues* | *Wed* | *Thurs* | *Fri* | *Sat* | *Sun* |
|---|---|---|---|---|---|---|---|
| Midnight | 731 | 637 | 601 | 661 | 992 | 2,433 | 2,731 |
| 1.00 | 489 | 390 | 420 | 447 | 781 | 1,876 | 1,940 |
| 2.00 | 224 | 287 | 247 | 329 | 613 | 1,478 | 1,663 |
| 3.00 | 162 | 145 | 140 | 148 | 231 | 516 | 756 |
| 4.00 | 109 | 135 | 86 | 107 | 139 | 335 | 411 |
| 5.00 | 247 | 220 | 213 | 172 | 218 | 291 | 339 |
| 6.00 | 516 | 490 | 566 | 436 | 579 | 396 | 304 |
| 7.00 | 1,875 | 2,037 | 2,096 | 2,136 | 2,249 | 721 | 398 |
| 8.00 | 3,285 | 3,837 | 3,563 | 3,683 | 3,649 | 1,127 | 498 |
| 9.00 | 1,897 | 2,092 | 1,904 | 1,983 | 2,130 | 1,501 | 800 |
| 10.00 | 1,755 | 1,748 | 1,696 | 1,711 | 2,036 | 2,327 | 1,463 |
| 11.00 | 2,043 | 1,909 | 1,850 | 2,055 | 2,307 | 3,110 | 1,998 |
| 12.00 | 2,424 | 2,259 | 2,230 | 2,428 | 2,798 | 3,406 | 2,298 |
| 13.00 | 2,320 | 2,304 | 2,370 | 2,411 | 2,702 | 3,228 | 2,201 |
| 14.00 | 2,242 | 2,105 | 2,152 | 2,350 | 2,741 | 3,611 | 3,628 |
| 15.00 | 2,978 | 2,777 | 2,726 | 3,119 | 3,987 | 3,879 | 3,143 |
| 16.00 | 4,057 | 3,870 | 4,029 | 4,437 | 5,600 | 3,444 | 2,931 |
| 17.00 | 4,274 | 3,993 | 4,176 | 4,415 | 4,783 | 3,297 | 2,589 |
| 18.00 | 2,621 | 2,559 | 2,650 | 2,882 | 3,243 | 2,428 | 2,213 |
| 19.00 | 2,139 | 2,238 | 2,347 | 2,483 | 2,982 | 2,264 | 2,097 |
| 20.00 | 1,805 | 1,703 | 1,705 | 2,022 | 2,573 | 2,123 | 1,770 |
| 21.00 | 1,725 | 1,646 | 1,735 | 1,912 | 2,431 | 2,121 | 1,648 |
| 22.00 | 1,773 | 1,952 | 2,049 | 2,491 | 2,968 | 2,388 | 2,232 |
| 23.00 | 2,100 | 2,112 | 2,354 | 3,076 | 5,784 | 5,374 | 3,194 |

*Source:* Central Statistical Office, *Road Accidents Great Britain 1979* HMSO London

solution. For a technical reason which is unimportant in practical terms, the redundant correspondence analysis eigenvalue has the value 1.0 and not zero as for the PC solution.[2] The first two non-redundant eigenvalues account for as much as 96% of the total, so a bivariate plot of the individual correspondence analysis scores will

2. It is a consequence of the correspondence analysis 'PC solution' being based on the *non-centred* data matrix. This complication does not affect the numerical output of the analysis but it simplifies the theory.

well represent the similarities and dissimilarities of the rows in the contingency table as regards the proportions.

It is a virtue of the correspondence analysis approach that the theoretical set-up allows the analysis of *column* proportions to be carried out simultaneously with the row proportions. (This is a consequence of the transformation and weighting procedures referred to above.) There are in fact straightforward mathematical relations between the row scores and the column scores. For this reason, it is usual in correspondence analysis to plot both the row scores and the column scores on the same scatterplot. This is shown in Figure 15.5. (The symbol : indicates a coincidence of the points for the hours 20, 21, and 22.) The pattern looks plausible in that hours which could be expected to have similar experience over the week are grouped

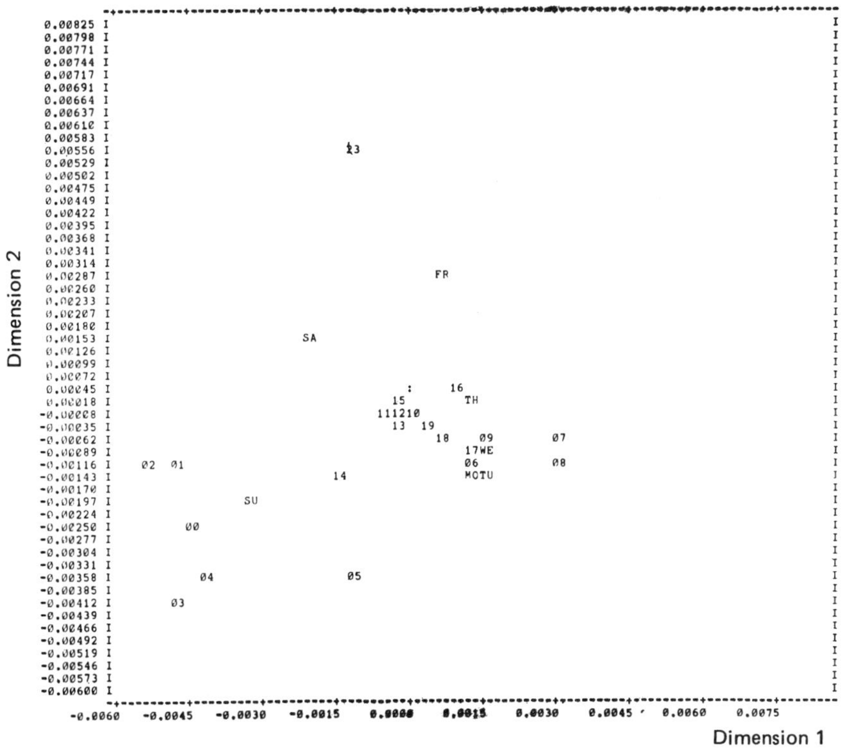


*Source*: GENSTAT analysis of Table 15.13 data

**Figure 15.5** Correspondence Analysis Plot of Contingency Table Data together. The same goes for the days in that the weekday points MO, TU, WE, TH, are fairly close to each other; SA and SU are separate and so indeed is FR presumably reflecting the start of week-end traffic (and casualty) patterns on Friday evenings.

It is *sometimes* possible to identify particular features of the contingency table from a simultaneous consideration of the correspondence analysis row and column plots. For example, the row data point for 23.00 (23) is well separated from the rest as are the column data points for Friday (FR) and Saturday (SA). Examination of the contingency table shows that one particular feature is that Friday and Saturday show relatively large frequencies for the hour 23.00. Similarly, Sunday shows relatively large frequencies in the early hours of the day and the column point SU is close to the row points 00, 01, 02. Near coincidences of row and column points do not necessarily carry significance, but at the very least they direct attention to particular parts of a possibly large contingency table for further study.

Finally we note (without attempting an explanation) that the transformation and weighting procedures of correspondence analysis have the useful consequence that if the frequencies of some rows (or columns) are aggregated, this does not affect the analytical results for the remaining rows (or columns). Suppose, for example, that the contingency table were a population census cross-classification of local areas' × occupations; the scores for larger areas would not be affected by grouping together the frequencies for the smaller areas.

## 15.8 LOG-LINEAR MODELS FOR TWO-WAY CONTINGENCY TABLES

In Section 15.4 we showed how a model for a two-way contingency table could be derived by assuming independence between the two sets of categorical variable values. In particular, we described how the fitted values could be derived from the marginal frequencies using result (15.3.2). Specifically:

the fitted frequencies for the independence model are,

$$F_{ij} = (f_{i\cdot} f_{\cdot j})/n \qquad (15.8.1)$$

Let us now approach this independence model from a different stance, by setting up a model in which a dependent variable is expressed as the sum of a systematic component and a random component. A simple assumption which will yield a formulation for the systematic component is provided by the assumption of independence between the row and column variables. We start by recalling from

Table 15.8 that the expected frequency in row $i$ and column $j$ is given by $nP(A_i)\ P(B_j)$. Therefore the logarithm of the expected frequency is $\log(n) + \log P(A_i) + \log P(B_j)$ where it is convenient to take logarithms to the base $e$. In practice, even if the independence model is valid, the data frequencies will differ from the expected frequencies and this difference may be represented by a random component. Thus, we may formally write the independence model as

$$\log(f_{ij}) = \log(n) + \log P(A_i) + \log P(B_j) + \text{random component}$$
$$\text{where } i = 1, \ldots, r \text{ and } j = 1, \ldots, c \qquad \textbf{(15.8.2)}$$

The model consists of a set of equations, one for each cell, and the dependent variable is the logarithm of the cell frequency.

The probabilities $P(A_i)$ and $P(B_j)$ can, as we saw in Section 15.4, be estimated from the first order margins: the appropriate row and column marginal frequencies. The estimated systematic component and the consequential fitted frequency are therefore:

$$\log(F_{ij}) = \log(n) + \log(f_{i\cdot}/n) + \log(f_{\cdot j}/n)$$
$$i = 1, \ldots, r \qquad j = 1, \ldots, c$$
$$\text{where } F_{ij} \text{ is the fitted value, under the assumed model.} \qquad \textbf{(15.8.3)}$$

This formulation (15.8.3) expresses the logarithm of each fitted frequency, $F_{ij}$, as the sum of:

(i) a constant term, $\log(n)$;
(ii) a row effect, $\log(f_{i\cdot}/n)$, a term which depends solely upon the row in which the fitted value lies; and
(iii) a column effect, $\log(f_{\cdot j}/n)$.

Simple algebra shows that the fitted values themselves may alternatively be expressed in a multiplicative fashion:

$$F_{ij} = n(f_{i\cdot}/n)(f_{\cdot j}/n) \qquad \textbf{(15.8.4)}$$

which is of course identical to (15.8.1). In practice, only $(r-1)(c-1)$ fitted values need to be calculated, since the fitted values sum to the given marginal frequencies in the contingency table. Furthermore, for the purpose of calculating the fitted values, it does not matter whether the multiplicative form (15.8.4) is used or whether the values for $\log F_{ij}$ are first obtained from (15.8.3) and then transformed by the exponential function.

It will help understand the more complicated set-up for three-way contingency tables if we set out the general procedure whereby the fitted values of the independence model are obtained, starting with the $rc$ frequencies of the contingency table. The procedure operates in two quite separate stages. At the 'pre-fitting' stage the row first order margins (the marginal row frequencies $f_{1\cdot}, f_{2\cdot}, \ldots, f_{r\cdot}$ and the marginal column frequencies $f_{\cdot 1}, f_{\cdot 2}, \ldots, f_{\cdot c}$) are calculated. Summing either of these gives in turn the total frequency, $n$, the zero order margin.

The fitting stage works the opposite way round, as it were. Here, we start with the given first order and zero order margins and by applying (15.8.4) arrive at the fitted values.

Having arrived at the fitted values the next step is to assess the goodness of fit by comparing them with the data frequencies, as was done in Section 15.4 using the $\chi^2$-statistic. If the independence model is rejected, the implication is that there is some association between the variables $A$ and $B$. In other words, the cell frequencies cannot be adequately represented simply by knowledge of the marginal frequencies. When this is the case there is said to be *interaction* between the two categorical variables $A$ and $B$. If interaction is present then it is possible in principle to include in the model extra parameters known as interaction terms (denoted by $u_{ij}$). With such an addition, the log-linear independence model (15.8.3) takes the form:

$$\log(F_{ij}) = \log(n) + \log(f_{i\cdot}/n) + \log(f_{\cdot j}/n) + u_{ij}$$
$$i = 1, \ldots, r \qquad j = 1, \ldots, c \qquad (15.8.5)$$

The term $u_{ij}$ is specific to each cell and, given these $u_{ij}$ values, the fitted frequencies inevitably exactly equal the data frequencies. For a two-way table, it is the case that there is no half-way house between the simple independence model and a model which fits the data in every respect.

## 15.9 THREE-WAY CONTINGENCY TABLE NOTATION AND STRUCTURE

In Section 15.2, we saw that successive totalling procedures lead to what are called the margins of a contingency table. Any three-way table has three second-order margins and three first order margins (see Table 15.3). If we add up any of the first order margins we get the total frequency which is also called the zero order margin The three-way contingency table itself (excluding any totals) can be called the third order margin. (This latter term may seem superfluous at this stage, but it illuminates later work by allowing a consistent terminology based upon the term 'margin', for all levels of classification.)

We now turn to the notation for a three-way table and its margins. The three-way contingency table of Table 15.1 will continue to be used for illustrative purposes. However, for purposes of clarity, the number of *ROAD* categories will be reduced from five to three (cross-roads, non-junction and other) and frequencies will be expressed in units of a thousand.

In a two-way contingency table, the *rows* set out frequencies of the $r$ values of the first 'row variable' and the *columns* frequencies of the $c$ values of the second 'column variable'. In a three-way table, the frequencies relating to the third variable are named the *layer* frequencies for a reason which will emerge shortly; its number of values is denoted by $l$. Altogether there are $rcl$ cells in a three-way contingency table.

Two of the second order margins—*SEVERITY* × *AREA* and *ROAD* × *SEVERITY*—have already been displayed in Table 15.2. The third, *ROAD* × *AREA*, is shown in Table 15.14. The variable SEVERITY will be regarded as the 'row variable'; it has three values or *levels* as they are more usually called: fatal, serious, minor. So the number of rows, $r$, equals 3. *ROAD* is the column variable, with $c = 3$ and the levels cross-roads, non-junction, other. *AREA* is the layer variable. The symbol $l$ denotes the number of levels of the layer variable: here $l = 2$ and the levels are urban, rural.

Altogether there are $rcl = 3 \times 3 \times 2 = 18$ cell frequencies in our three-way contingency table. These frequencies together with various totals (which correspond to the various margins) are contained in Table 15.1.

To describe a particular cell in a three-way table, we now need three subscripts, say, $i$, $j$, $k$ where $i$ refers to the row variable *SEVERITY*, $j$ to the column variable *ROAD*, and $k$ to the layer variable *AREA*. Thus $f_{132}$ indicates the frequency in the cell corresponding to the first level of *SEVERITY* (fatal), the third level of *ROAD* (other) and the second level of *AREA* (rural). Table 15.1 shows that $f_{132}$ equals (in

**Table 15.14** *ROAD* × *AREA* *thousands*

| | *Cross-roads* | *Non-junction* | *Other* | *Total* |
|---|---|---|---|---|
| Urban | 28.0 ($f_{.11}$) | 68.3 ($f_{.21}$) | 98.4 ($f_{.31}$) | 194.7 ($f_{..1}$) |
| Rural | 3.0 ($f_{.12}$) | 41.4 ($f_{.22}$) | 15.7 ($f_{.32}$) | 60.1 ($f_{..2}$) |
| Total | 31.0 ($f_{.1.}$) | 109.7 ($f_{.2.}$) | 114.1 ($f_{.3.}$) | 254.8 ($f_{...} = n$) |

*Source:* Table 15.1

thousands) 0.022 + 0.419 + 0.076 = 0.52 (rounded). Similarly $f_{131}$ (fatal, other, urban) = 1.35.

The second-order margins such as *ROAD* × *AREA* (Table 15.14) are derived by adding together appropriate frequencies in the three-way contingency table. Thus, in Table 15.14 the figure of 68.3 is the result of adding the frequencies $f_{121}$ (fatal, non-junction, urban), $f_{221}$ (serious, non-junction, urban) and $f_{321}$ (minor, non-junction, urban). The notation $f_{\cdot 21}$ for this cell indicates that it aggregates frequencies for all levels of the row variable *SEVERITY*. Careful inspection of Table 15.14 will make clear the notation for all other entries in this second order margin table. Entries in the second order margin *AREA* × *SEVERITY* will likewise have notation of the form $f_{i \cdot k}$ and in *ROAD* × *SEVERITY* the form $f_{ij \cdot}$.

A schematic three-dimensional diagram of the additive procedure leading from the contingency table through the second order and first order margins to the total frequency is shown in Figure 15.6. We call this the cube diagram. The three axes correspond to the three variables and their values or levels are indicated on the axes. The first order margins (the marginal frequencies for each level of each variable) are denoted by a notation which is an extension to three variables of that for two variables. Thus $f_{\cdot\cdot 2}$, the total rural frequency, equals 60.1, $f_{\cdot\cdot 1}$, total urban, equals 194.7, and $f_{\cdot\cdot 2} + f_{\cdot\cdot 1} = 254.8 = n$ (sometimes denoted by $f_{\cdot\cdot\cdot}$).

The second order margins, abbreviated to $R \times A$, $A \times S$ and $R \times S$, are shown on the faces of the axis system. Thus, the symbols for the $R \times A$ cells correspond to those in Table 15.14. Furthermore, the diagram represents the additive connections between the second order and first order margins. Addition along any line of a second order margin gives the first order margin on the axis (for example $f_{\cdot 32} + f_{\cdot 22} + f_{\cdot 12} = f_{\cdot\cdot 2}$).

The 18 cells of the three-way contingency table are shown as a cube in space in relation to the axes. It can be looked upon as two layers of nine points, the top layer corresponding to cells having the category 'rural' and the bottom layer to the category 'urban'. Alternatively, it can be thought of either as three slices each of six points defined in terms of the levels of *SEVERITY*, (in one direction) or as three slices for the levels of *ROAD* (in the other direction).

If we take any line of points in the cube (such as $f_{131}$, $f_{231}$, $f_{331}$) the addition of their frequencies gives one cell entry in one of the second order margins (in this case $f_{\cdot 31}$ of $R \times A$). Following the direction of sets of points in the same line of the cube leads to the corresponding second order (and lower order) margins.

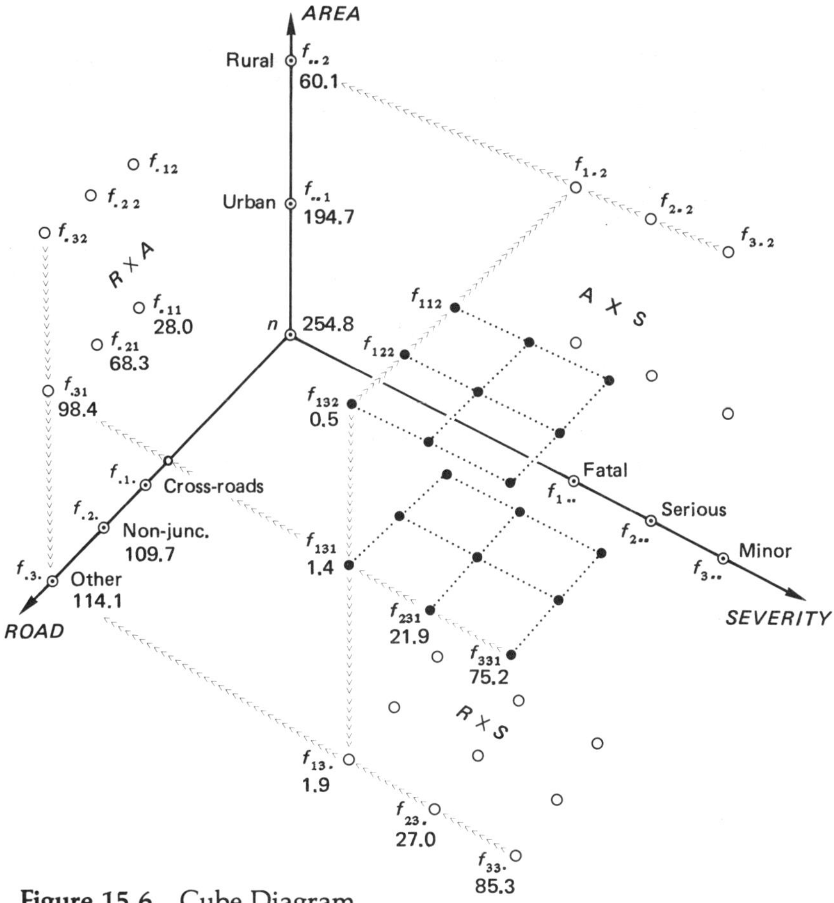


**Figure 15.6** Cube Diagram

## 15.10 THE FITTED VALUES IN A THREE—WAY TABLE FOR VARIOUS MODELS

The simplest model for a three-way contingency table is to assume that the three variables are mutually independent. This parallels the independence model for a two-way table (Section 15.8). The fitted value $F_{ijk}$, may be derived by a simple extension of the two-way formulation at (15.8.4), namely:

$$F^{I}_{ijk} = n(f_{i..}/n)(f_{.j.}/n)(f_{..k}/n) \quad \textbf{(15.10.1)}$$

The $I$ indicates a fitted value for the independence model. This may be justified by appeal to the rule for the joint probability of three independent events, say, $A_i$, $B_j$, $C_k$ where the probability of all three = $P(A_i$ and $B_j$ and $C_k) = P(A_i)P(B_j)P(C_k)$. The terms in brackets in (15.10.1) may be regarded as estimates of these probabilities $P(A_i)$, $P(B_j)$, $P(C_k)$. The factor $n$, on the right-hand side, results in 'estimated' expected frequencies on the assumption of independence.

Expression (15.10.1) yields the fitted values in terms of the first order margins. In order to develop fitting procedures for more complicated models, we first note that the fitted values for this independence model could alternatively be obtained by a two-stage procedure. The first is to determine the fitted values for each of the two-way tables forming the second order margins. Consider, for example, *ROAD* × *AREA*. Since mutual independence is assumed for all three variables, then *ROAD* and *AREA* are independent. The fitted values of *ROAD* × *AREA* are therefore given by applying (15.8.4) with appropriate notation. Thus the fitted value of the cell whose observed frequency is $f_{.jk}$ is:

$$F^I_{.jk} = n(f_{.j.}/n)\,(f_{..k}/n) \qquad \textbf{(15.10.2)}$$

(The $I$ again indicates independence.)

Referring to the cube diagram at Figure 15.6, repeated application of (15.10.2) allows *fitted* frequencies to be calculated for all the second order margins, (which will in future be denoted by the brief notation $R \times A$, $A \times S$, $R \times S$). Furthermore, each of these tables of fitted frequencies will constitute what we have called no-association tables (Section 15.4).

Expression (15.10.1) can now be rewritten in the following form:

$$F^I_{ijk} = n(F^I_{ij.}\, F^I_{.jk}\, F^I_{i.k}) \,/\, (f_{i..} f_{.j.} f_{..k}) \qquad \textbf{(15.10.3)}$$

This may be easily verified by substituting for $F^I_{ij.}$, $F^I_{.jk}$, $F^I_{i.k}$ their values in terms of the expression (15.10.2) and cancelling.

As an example, we calculate the fitted value $F^I_{132}$(for the cell fatal, other, rural). The data first order margins are $f_{1..} = 5.82$, $f_{.3.} = 114.1$ and $f_{..2} = 60.1$. Hence $F^I_{13.} = (5.82)\,(114.1)/254.8 = 2.61$. Similarly, $F^I_{1.2} = 1.37$ and $F^I_{.32} = 26.9$. Applying (15.10.3):

$$F^I_{132} = 254.8\ [(2.61)(1.37)(26.9)]\ /\ [(5.82)(114.1)(60.1)] = 0.62$$

The reader can easily verify that the formula at (15.10.1) gives the same answer. Having calculated the fitted values for all cells, the goodness of fit can be tested but we reserve a discussion of this until later.

The independence model may prove inadequate in failing to take account of association between pairs of variables. The simplest model assumption, and one we shall examine first, is that there is association (interaction) between only one pair of variables. We choose *ROAD* and *AREA* as an example, but the methods of fitting will equally apply to either of the other two pairs.

The basic formula (15.10.3) will provide a starting point for calculating the set of fitted frequencies which will be indicated by $F^{RA}_{ijk}$. The formula calls for, among others, the fitted frequencies for the second order margin $R \times A$. Under the assumption of association between two variables, however, the fitted frequencies must now be set equal to the actual frequencies of $R \times A$, and not those calculated under the assumption of independence. Notationally, if we denote the fitted frequencies as $F^{RA}_{.jk}$, then $F^{RA}_{.jk} = f_{.jk}$, the actual frequencies.

The other terms in (15.10.3) remain unchanged as we are still assuming independence between *ROAD* and *SEVERITY* and between *SEVERITY* and *AREA*. Hence:

$$\begin{aligned} F^{RA}_{ijk} &= n\,(F^{I}_{ij.}\,F^{RA}_{.jk}\,F^{I}_{i.k})/(f_{i..}\,f_{.j.}\,f_{..k}) \\ &= n\,(F^{I}_{ij.}\,f_{.jk}\,F^{I}_{i.k})/(f_{i..}\,f_{.j.}\,f_{..k}) \end{aligned} \qquad \textbf{(15.10.4)}$$

Expression (15.10.4) can be simplified, using the relationships for $F^{I}_{ij.}$ and $F^{I}_{i.k}$ of the form shown at (15.10.2). After substitution and cancelling, we have:

$$F^{RA}_{ijk} = (1/n)\,f_{i..}\,f_{.jk} \qquad \textbf{(15.10.5)}$$

To illustrate, we calculate $F^{RA}_{132}$, the fitted value for the cell (fatal, other, rural). It is simplest to use (15.10.5) and we have:

$F^{RA}_{132} = (1/n)\,f_{1..}\,f_{.32} = (1/254.8)(5.8)(15.7) = 0.36$ (thousands), where $f_{1..} = 5.8$ from Table 15.2 and $f_{.32} = 15.7$ from Table 15.14.

We now turn to the calculation of fitted frequencies where the model assumption is that there is interaction between two pairs of variables, such as $R \times A$ and $A \times S$. This may readily be done by adapting (15.10.4). The term $F^{I}_{i.k}$ is based upon the assumption of independence between *AREA* and *SEVERITY*. This assumption is now dropped and so $F^{I}_{i.k}$ is replaced by $f_{i.k}$. Specifically, the formula for the fitted values, now denoted by $F^{RA+AS}_{ijk}$, can be written

$$F^{RA+AS}_{ijk} = n(F^{I}_{ij.}f_{.jk}f_{i.k})/(f_{i..}f_{.j.}f_{..k}) \qquad \textbf{(15.10.6)}$$

The simplified version (corresponding to (15.10.5) for $F^{RA}_{ijk}$) is:

$$(f_{.jk}f_{i.k})/f_{..k} \qquad \textbf{(15.10.7)}$$

Finally, we may fit a model which assumes association between all three pairs of variables, $R \times A$, $A \times S$ and $S \times R$. Perhaps surprisingly, there is no algebraic formula for the fitted frequencies in this case, but they can be produced by methods which require the use of a computer program.

The various models we have discussed can be summed up in terms of the equality between certain data margins and the corresponding margins of the table of fitted frequencies. Working back from the last model considered, all three second order margins, ($R \times A$, $A \times S$ and $S \times R$) are identical for the fitted frequencies and the data frequencies. The reason is that fitting procedure takes the relevant second order margins in the data and determines the fitted frequencies consistently with them. If a model is fitted which assumes fewer interactions then a similar constraint applies, but only to those pairs of variables which specify the interaction term or terms.

Finally, for the independence model, only the first order margins are preserved in the fitted frequencies; by contrast, the second order margins of the contingency table and the fitted frequencies differ, since the latter are now expected values under the assumption of independence.

## 15.11 DEVIANCE AND MODEL CHOICE

It will now be seen that the modelling of three-way contingency tables yields a set of possible models which may be arranged in order of increasing complexity. At the lowest level is the independence model. Next, in order of complexity, come three models each of which models the interaction between one pair of variables: (i) $A \times R$ (ii) $A \times S$ and (iii) $R \times S$. Proceeding further, we have models assuming two interactions: (i) $A \times R$ and $A \times S$ (ii) $A \times S$ and $R \times S$ and (iii) $R \times S$ and $A \times R$. Finally, there is the single model which contains all three interactions: $A \times R$ and $A \times S$ and $R \times S$. With seven possible models there is clearly a problem of choice and the main purpose of this section is to discuss how such choice should be made.

We may begin by recalling the way in which a similar problem is resolved in regression analysis as regards the choice and number of explanatory variables. The starting point there is that the total sum of squares (TSS) equals the explained sum of squares (ESS) plus the residual sum of squares (RSS). The obvious effect of adding explanatory variables is to increase ESS and reduce RSS and the question of choice may be resolved by considering whether any reduction in RSS is statistically significant. The appropriate ANOVA test procedure is presented in Section 8.6.

As an aid to settling the choice of contingency table model, the *deviance* plays a similar role to the various sums of squares in regression analysis. The *total deviance* is analogous to the total sum of squares in that it is a measure of the variation in the cell frequencies of the contingency table. Having fitted a particular model, we may calculate the *residual* (or *unexplained*) *deviance* which is based upon a comparison between fitted and actual frequencies. The *explained deviance* is simply the difference between the total and the residual deviance. These values are easily computed from the sets of actual and fitted frequencies and in any case, will often be part of a standard computer output. Their formulae are:

$$\begin{aligned} \text{Total deviance} &= 2\Sigma f \log(f/\bar{f}) \\ \text{Residual deviance} &= 2\Sigma f \log(f/F) \\ \text{Explained deviance} &= 2\Sigma f \log(F/\bar{f}) \end{aligned} \qquad \textbf{(15.11.1)}$$

where, to simplify the notation, $f$ stands for $f_{ijk}$, $F$ for $F_{ijk}$, $\bar{f}$ is the mean of all cell frequencies and the summation is taken over all values of $i$, $j$ and $k$ (that is, over all cells).

We begin by noting that two quantities may be associated with any model, namely, the residual deviance for that model and its *degrees of freedom*, the latter depending upon the 'complexity' of the model as we will explain below. A first step towards model selection, therefore, is based upon identifying a hierarchy of models with successively smaller residual deviances. Starting from the independence model, we then search for (i) the best model with just one interaction, (ii) the best with two, leading to, (iii) the only one with three interactions. It is easily established that if a particular interaction enters at stage (i) it will also enter at stage (ii). In fact, as we shall now see, the optimal sequence for our data is: (i) $R \times A$ (ii) $R \times A + A \times S$ and (iii) all three interactions.

The information in Table 15.15 shows how this optimal sequence of models is arrived at. The table is arranged according to the number of interactions included in each model (from zero to three). Of these models with one first order interaction, $R \times A$ is the best in the sense that it has the smallest residual deviance. At the next level, the model $R \times A + A \times S$ is optimal with a deviance of only 840. It is obvious from the column of residual deviances that there is no better set of models.

Table 15.15 also shows the degrees of freedom appropriate to each model. The degrees of freedom are typically set out in computer programs for this type of analysis but they may be readily calculated in terms of the change in DF from one model in the hierarchy to the next. For example, the degrees of freedom for the independence model (12) equals $(rcl - 1 - (r - 1) - (c - 1) - (l - 1))$. The reduction of 2 in

**Table 15.15** Residual Deviances for Various Models

| *Model* | *Residual deviance* | *Change in deviance** | *DF* | *Change in DF** |
|---|---|---|---|---|
| I | 2,740 | . . | 12 | |
| One first order interaction | | | | |
| $R \times A$ | 5,920 | 21,820 | 10 | 2 |
| $A \times S$ | 22,660 | | 10 | |
| $S \times R$ | 25,410 | | 8 | |
| Two first order interactions | | | | |
| $R \times A + A \times S$ | 840 | 5,080 | 8 | 2 |
| $A \times S + S \times R$ | 20,330 | | 6 | |
| $S \times R + R \times A$ | 3,590 | | 6 | |
| $R \times A + A \times S + S \times R$ | 24 | 816 | 4 | 4 |

*Change from previous value of chosen hierarchy
*Source:* Table 15.1 and GENSTAT output

passing to $R \times A$ equals $(c - 1)(l - 1)$ and the further reduction of 2 as a result of adding $A \times S$ equals $(l - 1)(r - 1)$.

Up to this point we have simply set out a hierarchy of possible models of increasing complexity and with better fit. However, as the model specification becomes increasingly complex, so does interpretation. A first general principle may therefore be invoked, namely that of *parsimony.* This may be translated into practice by proposing that the move through the sequence of complexity—from independence to the inclusion of three interaction effects—should stop when the subsequent *reduction* in the residual deviance is 'relatively small'. Formally, 'relatively small' may be taken to mean 'statistically not significant'. It can be shown that, under the appropriate null hypothesis, the reduction in residual deviance from one model to the next has approximately a $\chi^2$-distribution with degrees of freedom equal to the *change* in DF between the two models. In the case of a contingency table with a very large total frequency, as in our example, all reductions are likely to be significant, implying that all models are, strictly speaking, statistically inadequate. In such circumstances, less formal criteria must be used such as the choice of a model whose residual deviance is some small proportion of the total, or a model which is capable of interpretation in its context.

## 15.12 LOG-LINEAR MODEL INTERPRETATION

In the discussion so far, it has been possible to rely almost exclusively upon the multiplicative formulation rather than the log-linear version introduced (for a two-way table) at (15.8.5). However, the latter is particularly informative when one comes to assess the magnitudes of various interactions in a model. We demonstrate this for a model with just one first order interaction, choosing, by way of illustration, the model with the single interaction $R \times A$ which, as was seen in the last section, was the preferred model of this type.

The fitted values are given by (15.10.5) and taking logarithms gives

$$\log F_{ijk} = -\log n + \log f_{i..} + \log f_{.jk} \qquad \textbf{(15.12.1)}$$

The second term on the right-hand side $f_{i..}$ reflects the proportionate effect on the fitted values of the various marginal frequencies of *SEVERITY*. It is analysis of the final term, $\log f_{.jk}$, which leads to a measure of the $R \times A$ interaction effect.

Table 15.16 shows: (i) the log-frequencies for the *ROAD* × *AREA* margin and (ii) in brackets, the interaction terms whose calculation will now be explained. Since the *ROAD* variable and *AREA* variable

**Table 15.16** *ROAD* × *AREA*: Log-Frequencies and Interaction Effects

| | *Cross-roads* | *Non-junction* | *Other* | *Layer Mean* | *Deviation from m* |
|---|---|---|---|---|---|
| Urban | 10.24 | 11.13 | 11.50 | 10.96 | 0.76($l_1$) |
| | (0.36) | (−0.51) | (0.16) | | |
| Rural | 8.00 | 10.63 | 9.66 | 9.43 | −0.76($l_2$) |
| | (−0.36) | (0.51) | (−0.16) | | |
| Column mean | 9.12 | 10.88 | 10.58 | Overall Mean $m = 10.19$ | |
| Deviation from *m* | −1.07 | 0.69 | 0.39 | | |

*Source:* Table 15.14

are respectively the column and layer variables, we shall describe Table 15.16 in terms of 'columns' and 'layers' (rather than columns and rows). The first stage is to calculate the layer means (shown in the penultimate column), the column means (shown in the penultimate row), and the mean of the log-frequencies for all cells together (the overall mean $m$). The *layer effects* are now calculated as the difference between each layer mean and the overall mean and, correspondingly, for the *column effects*. Each log frequency ($\log f_{.jk}$) can now be written as the sum of the overall mean, a layer effect ($l_k$), a column effect ($d_j$) and an interaction $u_{.jk}$, that is:

$$\log (f_{.jk}) = m + c_j + l_k + u_{.jk} \qquad \textbf{(15.12.2)}$$

For example, $\log (f_{.11}) = 10.24 = 10.19 - 1.07 + 0.76 + u_{.11}$, so $u_{.11} = 0.36$, (calculated by differencing).

For any no association table, the interactions are all zero, as may easily be verified. So the size of the calculated interactions identifies a conjunction of variable levels (in the context of the model) which are the sources of association. For example, the interaction terms for (urban, other) and (rural, other) (equal to $\pm$ 0.16) are relatively small which implies that the proportion of accidents at 'other junctions' is roughly the same in urban and rural areas. At the other extreme, the interactions for (urban, non-junction) and (rural, non-junction) are numerically large. In particular, the *negative* interaction ($-0.51$) for (urban, non-junction) tells us that there are *fewer* accidents in this category than would be expected if the variables *ROAD* and *AREA* were not associated.

If we use expression (15.12.2) to replace $\log f_{.jk}$ in (15.12.1), then, after a little algebra, it can be shown that $\log F_{ijk}$ can be finally expressed as the sum of an overall mean, the row, column and layer effects, and the column $\times$ layer interaction effect:

$$\log F_{ijk} = m + r_i + c_j + l_k + u_{jk} \qquad \textbf{(15.12.3)}$$

The overall mean ($m$) is the average over all cells of the log-frequencies. The terms $r_i$, $c_j$ and $l_k$ are known as the main effects.

The formulation at (15.12.3) can be extended to a model with two two-variable interactions such as $R \times A$ and $A \times S$. We do not go into detail, but such a model may be written:

$$\log F_{ijk} = m + r_i + c_j + \mathrm{l}_k + u_{ik} + u_{jk} \qquad \textbf{(15.12.4)}$$

A further obvious extension for a model with all three two variable interactions is to add to (15.12.4) the term $u_{ij}$ reflecting the measure of the interaction between the row variable (*SEVERITY*) and the column variable (*ROAD*).

The terms on the right-hand side of expressions such as (15.12.3) and (15.12.4) may be regarded as estimates of the systematic parameters of some prespecified model. If assumptions are made about the distribution of the error term, then standard errors of these parameter estimates may also be calculated and are shown in computer program output, but a discussion of this area takes us beyond the scope of this text.

## 15.13 COMPUTATION AND THE LOG-LINEAR MODEL

As we have shown, the calculation of fitted frequencies and the deviances are straightforward (except for the case where all three interactions are included in the model). They are, however, arithmetically extremely burdensome since a separate calculation is necessary for each cell and, furthermore, one needs to examine several possible models. A suitable computer program is therefore almost essential. There are a number available; for our own numerical work we have used GENSTAT (see Appendix B).

In interpreting computer output, a complication arises from there being interrelationships between the estimated parameters: for example, it will be seen from Table 15.16 that $l_1 + l_2 = 0.76 + (-0.76) = 0$. These interrelationships can be handled in different ways: '$l_1$' could be given the value zero and '$l_2$' the value 1.52. Either approach is valid since we are simply comparing the effect of 'urban' with that of 'rural'. Different computer programs handle these relationships in different ways and an understanding of the particular procedure which a program has used is an essential prerequisite for interpretation of the estimated parameters. However, the fitted frequencies (and therefore the deviances) are not affected by the treatment of the parameter interrelationships.

## 15.14 CATEGORICAL VARIABLES IN REGRESSION ANALYSIS

In Section 9.5 we saw how categorical variables may be used as explanatory variables in regression analysis, by means of the dummy variable technique. It will be recalled that, where, in particular, the categorical variable is dichotomous (having just two values), the corresponding dummy variable takes the numerical values 0 or 1. Now we turn to the case where a categorical variable is the *dependent* variable and, to facilitate the discussion, we suppose that the categorical variable is dichotomous. Unfortunately, the standard

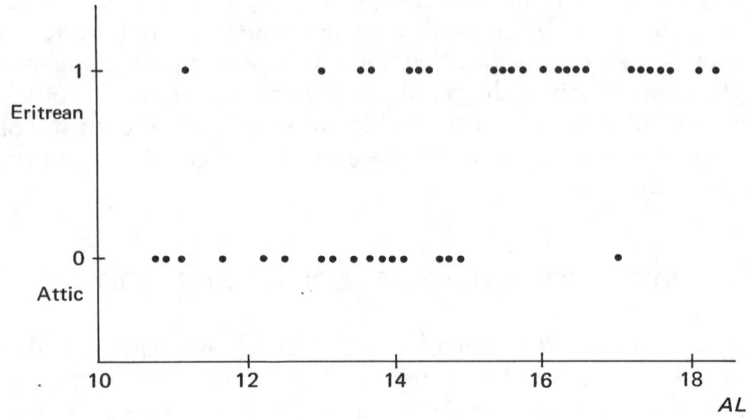


**Figure 15.7** Scatterplot of $AL$ according to Origin of Pottery Fragment

regression analysis is no longer a viable or indeed a sensible procedure and the reasons will be clear from a simple bivariate example.

For illustration, we draw upon the pottery fragment data of Table 12.1. That table shows the values of a number of numerical variables for two groups of individuals: one in the category 'Attic' and the other in the category 'Eritrean'. (For simplicity, the doubtful groups are included, for this exposition, in the corresponding main groups.) Now one might try to set up a bivariate regression framework if one asks: how far does the value of $AL$, say, 'determine' the category of a pottery fragment, Attic or Eritrean? Following from the dummy variable technique it might seem sensible to have a dependent variable with value 1 (Attic) or 0 (Eritrean) and the explanatory variable is the numerical variable $AL$.

The resulting scatterplot is shown in Figure 15.7 and it will be readily appreciated why fitting a conventional regression line to these data points would not be appropriate: (i) the residuals could not be assumed to be approximately normally distributed; and (ii) the fitted values of the dependent variable would have no meaning since they would have uninterpretable decimal values.

Clearly a different approach is needed and although a full exposition would take us beyond the level of this text, some intuitive idea of a viable procedure can be given. We note from Figure 15.7 that there is a relation in the sense that as the value of the explanatory variable increases, the *proportion* of individuals in the Eritrean group also tends to increase. Thus, if we group the data in terms of class intervals for $AL$, we have:

| $AL$ interval | 10.5−12.4 | 12.5−14.4 | 14.5−16.4 | 16.5−18.4 |
|---|---|---|---|---|
| Proportion of Eritrean to total* | 0.2 | 0.4 | 0.7 | 0.9 |

*Within each class interval
*Source:* Table 12.1

This suggests that we might set up a statistical model in which the *probability* of being Eritrean (rather than Attic) depends upon the value of the explanatory variable $AL$. Such a model is called the logistic model. It relates the probability $\pi$, that an individual is in one of two categories, to the value of an explanatory variable, $x$, by means of a relationship of the form:

$$\log[\pi/(1-\pi)] = \alpha + \beta x + \text{random component} \quad \mathbf{(15.14.1)}$$

The logit transformation is used on the left-hand side for the reason given in Section 8.11. The procedure for fitting this model and estimating $\alpha$ and $\beta$ (and their standard errors) uses the information for each data point, and arbitrary grouping, such as is done above, is not needed. This is not a standard regression set-up if only because the left-hand side of (15.14.1) is not observable and a specific program is required.

The logistic model sketched above is readily extendable to several explanatory variables and, as in ordinary regression analysis, some of these may be categorical.

## EXAMPLES (all based upon Table 15.17)

**E 15.1** Table 15.17 is a three-way contingency table. Identify and derive the three second order margins and the three first order margins. What is the zero order margin?

**E 15.2** For the second order margin *Voting intention* × *House ownership* ($V \times H$): (i) derive the corresponding expected frequencies under the assumption of no association; (ii) compute the $\chi^2$-statistic; and (iii) test the hypothesis of no association.

**E 15.3** Calculate, for the second order margin ($V \times H$): (i) the entropy of the total frequency allocation to the three levels of 'voting intentions'; (ii) the corresponding entropies given 'house-owner' and 'other'; and (iii) the uncertainty coefficient.

**E 15.4** Use the formula at (15.10.1) to calculate the fitted values for the independence model. Hence calculate the residual deviance (see (15.11.1)).

**E 15.5** Under the assumption that the independent model is valid, the residual deviance has a $\chi^2$-distribution with degrees of freedom $(rcl - r - c - 1 + 2)$. Check the validity of this model.

**Table 15.17** Voting Intentions according to Earnings and House-Ownership

*Number*

| | *Labour* | *Conservative* | *Other* |
|---|---|---|---|
| *High earnings* | | | |
| House-owner | 54 | 10 | 10 |
| Other | 27 | 5 | 4 |
| *Low earnings* | | | |
| House-owner | 34 | 6 | 5 |
| Other | 42 | 6 | 3 |

*Source:* adapted from sample survey data at Table 20, in Goldthorpe, J. H. *et al.*, *The Affluent Worker: Political Attitudes and Behaviour*, Cambridge University Press (1968)

# *Appendix A*

# Statistical Tables

A.1 Standard Normal Distribution
A.2 $t$-Distribution
A.3 $\chi^2$-Distribution
A.4 $F$-Distribution
A.5 Geary Test
A.6 Kolmogorov – Smirnov Test
A.7 Durbin – Watson Statistic
A.8 Mann – Whitney Test

The following pages present statistical tables referred to in the text. Their primary purpose is expositional. They have been kept as simple as possible, though for many purposes they may be sufficient and accurate enough.

There are many compilations of statistical tables. One which is particularly appropriate to the level and coverage of this text is: H. R. Neave, *Elementary Statistics Tables,* Allen and Unwin, 1981. The same author's *Statistics Tables,* Allen and Unwin, 1978 are more comprehensive. Textbooks on nonparametric statistics often include a full range of tables; see, for example, Daniel (1978).

**Table A.1** Standard Normal Distribution

| *z value* | *Cumulative area* | *Central area* | *Two-tailed area* | *One-tailed area* |
|---|---|---|---|---|
| 0 | 0.500 | 0.000 | 1.000 | 0.500 |
| 0.2 | 0.579 | 0.159 | 0.841 | 0.421 |
| 0.4 | 0.655 | 0.311 | 0.689 | 0.344 |
| 0.6 | 0.726 | 0.451 | 0.549 | 0.274 |
| 0.8 | 0.788 | 0.576 | 0.424 | 0.212 |
| 1.0 | 0.841 | 0.683 | 0.317 | 0.159 |
| 1.2 | 0.885 | 0.770 | 0.230 | 0.115 |
| *1.28* | *0.900* | *0.800* | *0.200* | *0.100* |
| 1.4 | 0.919 | 0.838 | 0.162 | 0.081 |
| 1.6 | 0.945 | 0.890 | 0.110 | 0.055 |
| *1.64* | *0.950* | *0.900* | *0.100* | *0.050* |
| 1.8 | 0.964 | 0.928 | 0.072 | 0.036 |
| *1.96* | *0.975* | *0.950* | *0.050* | *0.025* |
| 2.0 | 0.977 | 0.954 | 0.046 | 0.023 |
| *2.33* | *0.990* | *0.980* | *0.020* | *0.010* |
| *2.58* | *0.995* | *0.990* | *0.010* | *0.005* |
| 3.0 | 0.999 | 0.997 | 0.003 | 0.001 |

*Reference:* Section 3.8

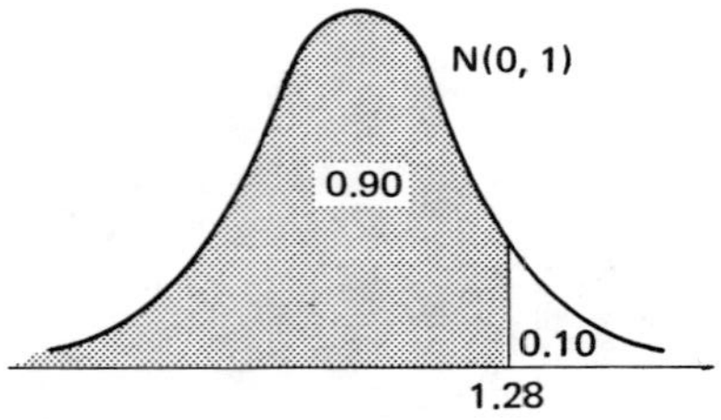

Cumulative and one-tailed areas

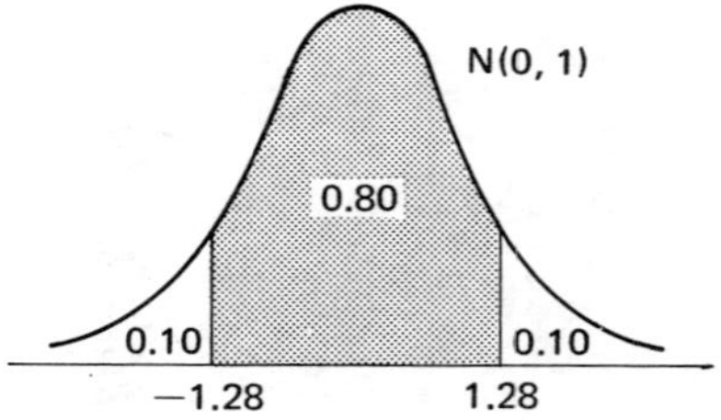

Central and two-tailed areas

**Table A.2** *t*-Distribution

| | | | | | |
|---|---|---|---|---|---|
| Cumulative area | 0.90 | 0.95 | 0.975 | 0.99 | 0.995 |
| Central area | 0.80 | 0.90 | 0.95 | 0.98 | 0.99 |
| Two-tailed area | 0.20 | 0.10 | 0.05 | 0.02 | 0.01 |
| One-tailed area | 0.10 | 0.05 | 0.025 | 0.01 | 0.005 |
| *ν values* | | | | | |
| 4 | 1.53 | 2.13 | 2.78 | 3.75 | 4.60 |
| 6 | 1.44 | 1.94 | 2.45 | 3.14 | 3.71 |
| 8 | 1.40 | 1.86 | 2.31 | 2.90 | 3.36 |
| 10 | 1.37 | 1.81 | 2.23 | 2.76 | 3.17 |
| 12 | 1.36 | 1.78 | 2.18 | 2.68 | 3.05 |
| 14 | 1.35 | 1.76 | 2.15 | 2.62 | 2.98 |
| 16 | 1.34 | 1.75 | 2.12 | 2.58 | 2.92 |
| 18 | 1.33 | 1.73 | 2.10 | 2.55 | 2.88 |
| 20 | 1.33 | 1.72 | 2.09 | 2.53 | 2.85 |
| 30 | 1.31 | 1.70 | 2.04 | 2.46 | 2.75 |
| ∞ | 1.28 | 1.64 | 1.96 | 2.33 | 2.58 |

*Reference:* Section 4.4

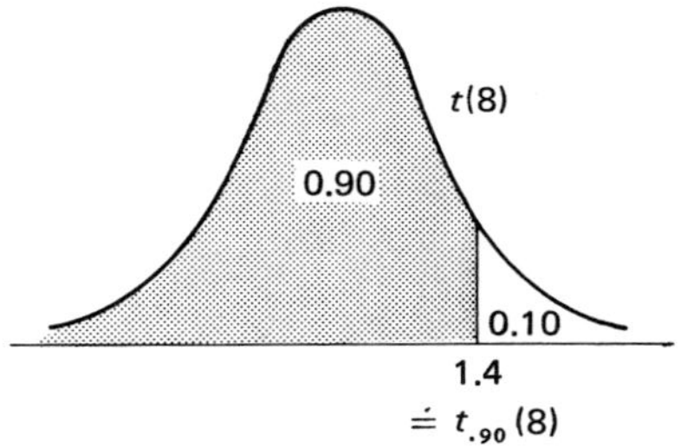


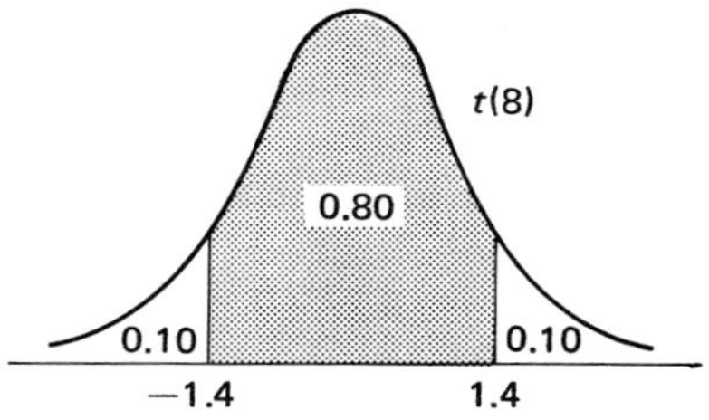

**Table A.3** $\chi^2$-Distribution

| | RH tail area | | | |
|---|---|---|---|---|
| $\nu$ | *0.10* | *0.05* | *0.025* | *0.01* |
| 1 | 2.71 | 3.84 | 5.02 | 6.64 |
| 2 | 4.61 | 5.99 | 7.38 | 9.21 |
| 3 | 6.25 | 7.82 | 9.35 | 11.35 |
| 4 | 7.78 | 9.49 | 11.14 | 13.28 |
| 5 | 9.24 | 11.07 | 12.83 | 15.09 |
| 6 | 10.65 | 12.59 | 14.45 | 16.81 |
| 7 | 12.02 | 14.07 | 16.01 | 18.48 |
| 8 | 13.36 | 15.51 | 17.54 | 20.09 |
| 9 | 14.68 | 16.92 | 19.02 | 21.67 |
| 10 | 15.99 | 18.31 | 20.48 | 23.21 |
| 12 | 18.55 | 21.03 | 23.34 | 26.22 |
| 14 | 21.06 | 23.68 | 26.12 | 29.14 |
| 16 | 23.54 | 26.30 | 28.85 | 32.00 |
| 18 | 25.99 | 28.87 | 31.53 | 34.81 |
| 20 | 28.41 | 31.41 | 34.17 | 37.57 |
| 25 | 34.38 | 37.65 | 40.65 | 44.31 |
| 30 | 40.26 | 43.77 | 46.98 | 50.89 |
| 40 | 51.80 | 55.76 | 59.34 | 63.69 |
| 50 | 63.17 | 67.51 | 71.42 | 76.15 |

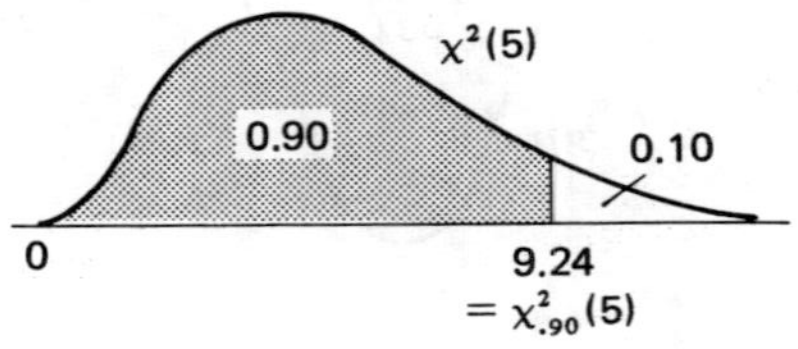


*Reference:* Section 7.3

**Table A.4** *F*-Distribution

$F_{.95}$ values (for right-hand tail area = 0.05)
$\nu_1$ = degrees of freedom of numerator
$\nu_2$ = degrees of freedom of denominator

| $\nu_2$ | $\nu_1$: 1 | 2 | 3 | 4 | 6 | 8 | 10 | 20 | ∞ |
|---|---|---|---|---|---|---|---|---|---|
| 5 | 6.61 | 5.79 | 5.41 | 5.19 | 4.95 | 4.82 | 4.74 | 4.56 | 4.36 |
| 6 | 5.99 | 5.14 | 4.76 | 4.53 | 4.28 | 4.15 | 4.06 | 3.87 | 3.67 |
| 8 | 5.32 | 4.46 | 4.07 | 3.84 | 3.58 | 3.44 | 3.35 | 3.15 | 2.93 |
| 10 | 4.96 | 4.10 | 3.71 | 3.48 | 3.22 | 3.07 | 2.98 | 2.77 | 2.54 |
| 15 | 4.54 | 3.68 | 3.29 | 3.06 | 2.79 | 2.64 | 2.54 | 2.33 | 2.07 |
| 20 | 4.35 | 3.49 | 3.10 | 2.87 | 2.60 | 2.45 | 2.35 | 2.12 | 1.84 |
| ∞ | 3.84 | 3.00 | 2.60 | 2.37 | 2.10 | 1.94 | 1.83 | 1.57 | 1.00 |

*Reference:* Section 8.5

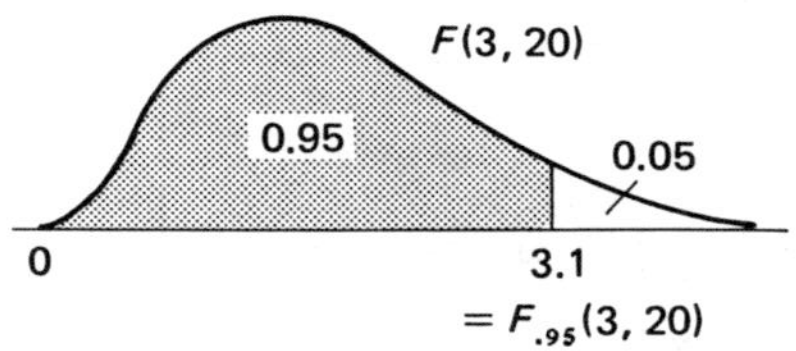

**Table A.5** Geary Test, 5% Critical Values

| *n* | *Lower critical value* | *Upper critical value* |
|---|---|---|
| 11 | 0.68 | 0.92 |
| 16 | 0.70 | 0.90 |
| 21 | 0.71 | 0.89 |
| 31 | 0.72 | 0.88 |
| 41 | 0.73 | 0.87 |
| 51 | 0.74 | 0.86 |
| 101 | 0.75 | 0.84 |

*Reference:* Section 7.7

**Table A.6** Kolmogorov–Smirnov Test, 5% Critical Values

| *n* | *Critical value* | *n* | *Critical value* |
|---|---|---|---|
| 5 | 0.56 | 16 | 0.33 |
| 6 | 0.52 | 18 | 0.31 |
| 7 | 0.48 | 20 | 0.29 |
| 8 | 0.45 | 25 | 0.26 |
| 9 | 0.43 | 30 | 0.24 |
| 10 | 0.41 | 40 | 0.21 |
| 12 | 0.38 | 50 | 0.19 |
| 14 | 0.35 | 100 | 0.13 |

*Reference:* Section 7.8

**Table A.7** Durbin–Watson Statistic
Upper ($d_U$) and lower ($d_L$) critical values for a test at the 5% level of significance

| | *Number of explanatory variables* | | | | | | | | | |
|---|---|---|---|---|---|---|---|---|---|---|
| | *1* | | *2* | | *3* | | *4* | | *5* | |
| *T* | $d_L$ | $d_U$ | $d_L$ | $d_U$ | $d_L$ | $d_U$ | $d_L$ | $d_U$ | $d_L$ | $d_U$ |
| 15 | 1.08 | 1.36 | 0.95 | 1.54 | 0.82 | 1.75 | 0.69 | 1.97 | 0.56 | 2.21 |
| 16 | 1.10 | 1.37 | 0.98 | 1.54 | 0.86 | 1.73 | 0.74 | 1.93 | 0.62 | 2.15 |
| 17 | 1.13 | 1.38 | 1.02 | 1.54 | 0.90 | 1.71 | 0.78 | 1.90 | 0.67 | 2.10 |
| 18 | 1.16 | 1.39 | 1.05 | 1.53 | 0.93 | 1.69 | 0.82 | 1.87 | 0.71 | 2.06 |
| 19 | 1.18 | 1.40 | 1.08 | 1.53 | 0.97 | 1.68 | 0.86 | 1.85 | 0.75 | 2.02 |
| 20 | 1.20 | 1.41 | 1.10 | 1.54 | 1.00 | 1.68 | 0.90 | 1.83 | 0.79 | 1.99 |
| 25 | 1.29 | 1.45 | 1.21 | 1.55 | 1.12 | 1.66 | 1.04 | 1.77 | 0.95 | 1.89 |
| 30 | 1.35 | 1.49 | 1.28 | 1.57 | 1.21 | 1.65 | 1.14 | 1.74 | 1.07 | 1.83 |
| 40 | 1.44 | 1.54 | 1.39 | 1.60 | 1.34 | 1.66 | 1.29 | 1.72 | 1.23 | 1.79 |
| 50 | 1.50 | 1.59 | 1.46 | 1.63 | 1.42 | 1.67 | 1.38 | 1.72 | 1.34 | 1.77 |
| 75 | 1.60 | 1.65 | 1.57 | 1.68 | 1.54 | 1.71 | 1.51 | 1.74 | 1.49 | 1.77 |
| 100 | 1.65 | 1.69 | 1.63 | 1.72 | 1.61 | 1.74 | 1.59 | 1.76 | 1.57 | 1.78 |

*Reference:* Section 9.3

**Table A.8** Mann–Whitney Test, 5% Lower Critical Values

| | $n_2$ | | | | | | | |
|---|---|---|---|---|---|---|---|---|
| $n_1$ | *6* | *8* | *10* | *12* | *14* | *16* | *18* | *20* |
| 6 | 6 | 9 | 12 | 15 | 18 | 22 | 25 | 28 |
| 8 | 9 | 14 | 18 | 23 | 27 | 32 | 37 | 42 |
| 10 | 12 | 18 | 24 | 30 | 37 | 43 | 49 | 56 |
| 12 | 15 | 23 | 30 | 38 | 46 | 54 | 62 | 70 |
| 14 | 18 | 27 | 37 | 46 | 56 | 65 | 75 | 84 |
| 16 | 22 | 32 | 43 | 54 | 65 | 76 | 87 | 99 |
| 18 | 25 | 37 | 49 | 62 | 75 | 87 | 100 | 113 |
| 20 | 28 | 42 | 56 | 70 | 84 | 99 | 113 | 128 |

*Reference:* Section 14.2

# *Appendix B*

# Statistical Computer Programs

The idea that statistical analysis requires a flexibility of approach and a need to experiment—to try out models of more or less complexity and to obtain necessary diagnostic information, for example—is one which appears in most of the chapters of this book. Some of the analytical steps will be matters of straightforward, if sometimes tedious, arithmetic; others require calculations which go well beyond what can be achieved with pencil, paper and a pocket (or desk top) calculator. The availability of appropriate computer programs which will cope with both the straightforward and the complicated steps of an analysis is essential.

We cannot hope to provide a consumer's guide to statistical computer programs in the space of a brief Appendix. What can be done is to refer to those which we have used and mention in passing a few others which offer similar facilities.

Most of our routine work in Chapters 2, 4, 8 and 9 was done using IDA. IDA is a conversational style program. The user is prompted, through the terminal, for the information which the program requires and the results are similarly returned to the terminal. IDA enables most of the standard summary statistics to be obtained and has good facilities for regression and time series analysis. Diagnostic information of the kind discussed in Chapters 8 and 9 and graphical displays are easily obtained.

There are a number of other conversational style programs available. One example is Minitab. This offers most of the conventional analyses and also some of the more novel displays, such as stem and leaf plots, and computes the 'influence statistics' described in Chapter 8. IDA does not have these latter capabilities.

Most conversational style programs stop short of offering anything for performing multivariate analyses or for fitting models to contingency tables. To perform the calculations for Chapters 10, 11 and 12, we made use of SPSS and GENSTAT. SPSS has good facilities for handling large data files but, in our view, it is somewhat 'rigid' in the way if offers options to the user and some of the more recent developments in exploratory statistical techniques cannot easily be done; one cannot for example, directly make plots of individuals according to their principal component scores.

GENSTAT, by contrast, is a very impressive program. Because of its great flexibility and coverage, practically everything discussed in this book can be done with it. Its drawback is that it requires an investment of time to become familiar with its procedures. Furthermore, not all computer installations which possess GENSTAT allow it to run 'on-line', that is, the user cannot initiate a job from a terminal and expect to have the results sent back to the terminal while the user waits. We were fortunate in that the Computing Service of the University of York allows GENSTAT to be used in this way. Without GENSTAT, we would have found it difficult to complete the calculations and graphical work required by Chapters 13 and 15.

We should also refer to three other programs. BMDP offers a very wide range of statistical procedures and the reader is referred to Frane, J. W., The 1981 version of BMDP, *The American Statistician*, November 1981 for a summary of its capabilities. CLUSTAN is primarily directed towards cluster analysis and some of the analyses in Chapter 13 could have been carried out using this program. TSP (Time Series Processor) is a program for the econometric analysis of time series data and it particularly concentrates upon the subject matter of Chapter 9. It does not work in a conversational manner.

Finally, GLIM (Generalised Linear Interactive Modelling) offers an integrated system for fitting linear and log-linear models such as those discussed in Chapters 8, 9 and 15. GLIM is designed to be run on-line but is not conversational as are, for example, IDA and Minitab. It requires a higher level of theoretical knowledge of the user than this book presumes.

The following lists the publishers of the reference manuals for the programs mentioned above:

BMDP Reference Manual: University of California Press.
CLUSTAN User Manual: Program Library Unit, University of Edinburgh.
GENSTAT Manual: Rothamsted Experimental Station, Harpenden, Hertfordshire.
GLIM Manual: Numerical Algorithms Group, 7 Banbury Road, Oxford.
IDA User's Manual: Scientific Press, Palo Alto, California.
Minitab Manual: Pennsylvania State University.
SPSS Manual: Mc-Graw Hill Book Company.
TSP Manual: TSP International, Junipero Serra Blvd, Stanford, California, CA 94305.

# Answers to Selected Examples

## CHAPTER 1

E 1.1 (i) Numerical (discrete) though practically continuous;
(ii) (iii) and (iv) categorical (nominal);
(v) numerical (continuous).

## CHAPTER 3

E 3.1 1.96, 2.33, −1.64, 0.95, 0.02, 0.995.
E 3.2 9.88, 10.99, −0.92, 0.975, 0.22, 0.16, 0.256.
E 3.3 (−0.92, 8.92), (−1.88, 9.88), (−3.74, 11.74)

## CHAPTER 4

E 4.1 (a) (i) 2.45, 2.23, 2.09, 1.96; (ii) 3.14, 2.76, 2.53, 2.33;
(b) 0.95, 0.995, 0.95, 0.95.
E 4.2 2.18, 2.12, 2.09.
E 4.3 Set 1: $3.0 \pm 2.02$, $3.0 \pm 2.87$; Set 2: $2.5 \pm 2.02$, $2.5 \pm 2.87$; Set 3: $2.5 \pm 1.66$, $2.5 \pm 2.30$; Set 4: $2.5 \pm 1.10$, $2.5 \pm 1.48$.
E 4.4 $0.5 \pm 2.2$.
E 4.5 $0.8 \pm 0.84$.
E 4.6 83, 68, 0.82, 0.91 (N.B. TSS $= 10 \times 8.3$).

## CHAPTER 5

E 5.1 (iv) 0.024, 1.009, LN(0.024, 1.009).
E 5.3 Po(2): 0.135, 0.271, 0.271, 0.181, 0.090, 0.036, 0.012;
Po(3.5): 0.030, 0.106, 0.185, 0.216, 0.189, 0.132, 0.077.
E 5.4 $k = 4$; 0.197, 0.165, 0.110, 0.064, 0.034.

## CHAPTER 6

E 6.2 (i) LN(0.02, 1.01); (ii) LN(0.10, 0.84).

## CHAPTER 7

E 7.1 $A : G = 0.71$, $B : G = 0.75$. Both consistent with normality at 5%.
E 7.3 2.5 ± (2.23)(1.0); 0.8 ± (2.23)(0.3).

## CHAPTER 8

E 8.1 0.18 ± 0.26; 1.41 ± 2.06; −2.12 ± 4.52.
E 8.3 3,480; 3,407; 3,409; 3,596.
E 8.4 (i) (2, 18), (6, 14); (ii) 3.55, 2.85; (iii) both significant.
E 8.5 $R_0^2 = 0.661$, $R_1^2 = 0.680$, $k_1 = 3$, $k_0 = 2$, $F$-ratio = 0.987.

## CHAPTER 9

E 9.1 (i) inconclusive; (ii) not significant; (iii) significant.
E 9.2 (i) inconclusive; (ii) not significant; (iii) significant.
E 9.3 2.7; $F_{.95}(20, 20) = 2.12$; significant.
E 9.5 $T = 24$, $k = 3$, $\nu = 20$; (i) 8,620 ± (2.09)(292); (ii) 8,620 ± (2.09)(178).

## CHAPTER 10

E10.1 6.6, 13.0, 4.8, 6.6.
E10.2 See Table 11.1 for $X_3$ and $X_6$. For each column, centred values equal SD times standardised values.
E10.3 See columns $Z_1$ and $Z_2$ of Table 11.1.
E10.4 For Portugal—Turkey: (a) 51, (b) 0.55.

## CHAPTER 11

E11.3 Variances are: (i) 1.83, 0.17; (ii) 1.01, 0.99.
E11.4 The correlation is greater for $X_3$ and $X_6$ than for $X_4$ and $X_6$.
E11.5 Changing the directions of the axes does not alter relative positions and, therefore, distances between data points are unchanged.
E11.6 The distances between two data points can be calculated by formula (10.8.4) using either the original data values for all variables or the PC scores for all PC's. In the latter case, the contribution to the calculation from the scores on the first two PC's will often account for a large proportion of the answer.

## CHAPTER 13

E13.2 At level 80 or more, link (ELY, PETE, ST.ALB); Link (CANT, DURH). At level 75 or more link LINC to (CANT, DURH); link RIPON to (ELY, PETE). At level 69.3 link (CANT, DURH, LINC) to (RIPON, ELY, PETE, ST.ALB).

## CHAPTER 14

**E14.1** (i) $t = 0.87$ (not sig), $U = 41$ (not sig), $T = 1.0$ (not sig).
**E14.2** $P(+) = 11/24$, $z = 0.4$ (not sig).
**E14.3** No. of runs = 25; $n_1 = 25$, $n_2 = 22$; (not sig).
**E14.4** $r_s = 0.28$, (not sig).

## CHAPTER 15

**E15.1** Using obvious notation, the second order margins are $V \times H$, $H \times E$, $E \times V$. The first order margins are $V$ (157, 27, 22), $H$ (119, 87), $E$ (110, 96). Zero order margin = 206.
**E15.2** The $V \times H$ expected frequencies are: First row 90.7, 15.6, 12.7, second row 66.3, 11.4, 9.3. $\chi^2 = 1.2$ ($\nu = 2$). Not significant.
**E15.3** $H(V) = 0.712$; $H(V \mid H) = 0.709$; $U(V \mid H) = 0.004$. Virtually no association.
**E15.4** Fitted values are, row 1: 48.4, 8.3, 6.8; row 2: 35.4, 6.1, 5.0; row 3: 42.3, 7.3, 6.0; row 4: 30.9, 5.3, 4.3. Residual deviance = 11.1.
**E15.5** DF = 7. Critical value = 14.1. Independence accepted.

# References and Bibliography


## CHAPTERS 0–7 GENERALLY

There are many texts which cover the introductory material of Chapter 0. One example is:

Fuller, M.F. and Lury, D.A., *Statistics Workbook for Social Science Students*, Philip Allan, 1977.

Examples of texts covering similar ground to Chapters 0 to 7, though laying more emphasis on theory (and less on application) are:

Dixon, W.J. and Massey, F.J., *Introduction to Statistical Analysis*, McGraw-Hill, 3rd edn, 1969.

Sanders, D.H., Murph, A.F. and Eng, R.J., *Statistics: A Fresh Approach*, McGraw-Hill, 1980.

Wonnacott, T.H. and R.J., *Introductory Statistics for Business and Economics*, Wiley, 2nd edn, 1977.

## CHAPTER 2 UNIVARIATE DESCRIPTIVE ANALYSIS

There are a number of texts which provide a guide to methods of univariate descriptive analysis. Two books which discuss stem and leaf plots and some other, more recent developments of this type are:

Erickson, B.H. and Nosanchuk, T.A., *Understanding Data*, Open University Press, 1979.

Velleman, P.F. and Hoaglin, D.C., *Applications, Basics, and Computing of Exploratory Data Analysis*, Duxbury Press, Boston, 1981.

These texts are inspired by the methods of John Tukey, for which see:

Tukey, J.W., *Exploratory Data Analysis*, Addison-Wesley, 1979.

All of these cover a wider field than univariate analysis.

The references cited in this chapter are:

Dixon, W.J. and Massey, F.J., *Introduction to Statistical Analysis*, McGraw-Hill, 1969.

Doane, D.P., Aesthetic frequency classifications, *The American Statistician*, Vol. 30, No. 4, November 1976.

Smith, G.L., Michie, C. and Pope, J.A., Robust estimators of location, *Bulletin in Applied Statistics*, Vol. 5, No. 1, 1978.

## CHAPTER 4 LINEAR MODELS

The reference cited is:

Hoel, P.G., *Introduction to Mathematical Statistics*, Wiley, 3rd edn, 1962.

## CHAPTER 5 THEORETICAL FREQUENCY AND PROBABILITY DISTRIBUTIONS

## CHAPTER 6 FITTING DISTRIBUTIONS TO DATA

A convenient source reference on the properties of theoretical distributions is:

Hastings, N.A.J. and Peacock, J.B., *Statistical Distributions*, Wiley, 1974.

Two works giving examples of the use of the negative binomial and the lognormal distributions are, respectively:

Mosteller, F. and Wallace, D.L., *Inference and Disputed Authorship: The Federalist*, Addison-Wesley, 1964.

Williams, C.B., *Style and Vocabulary*, Griffin, 1970.

## CHAPTER 7 TESTING THE VALIDITY OF MODELS

As regards problems in the use and interpretation of test procedures, see:

Henkel, R.E. and Morrison, D.E. (eds), *The Significance Test Controversy*, Butterworths, 1970.

The reference cited in this chapter is:

Pearson, E.S. and Hartley, H.O., *Biometrika Tables for Statisticians Vol. I*, Cambridge University Press, 1970.

## CHAPTERS 8 AND 9 MULTIPLE REGRESSION ANALYSIS AND TIME SERIES ANALYSIS

There are a number of books on regression analysis at about the level of this text. One with a strongly practical emphasis is:

Chatterjee, S. and Price, B., *Regression Analysis by Example*, Wiley, 1978.

A more theoretical treatment may be found in the book by Wonnacott and Wonnacott (see above) which also includes material on the analysis of time series data. A more advanced text, which may be useful on a selective basis is:

Draper, N.R. and Smith, H., *Applied Regression Analysis*, Wiley, 2nd edn, 1981.

The references cited in these chapters are:

*Chapter 8*

Hoaglin, D.C. and Welsch, R.E., The hat matrix in regression and ANOVA, *The American Statistician*, 32, 1978.

Knox, E.G., Foods and diseases, *British Journal of Preventive and Social Medicine*, 31, 1977.
Segall, J.J., Is milk a coronary health hazard?, *British Journal of Preventive and Social Medicine*, 31, 1977.

*Chapter 9*
Brown, P. and Payne, C., Election night forecasting (with discussion), *Journal of the Royal Statistical Society, Series A*, Part 4, 1975.
Johnston, J., *Econometric Methods*, McGraw-Hill, 1972.
Mosteller, F. and Tukey, J.W., *Data Analysis and Regression*, Addison-Wesley, 1977.

## CHAPTERS 10–13 MULTIVARIATE STATISTICAL METHODS

There are very few texts which deal with multivariate analysis at the level which we have set. One which comes close, although concentrating upon descriptive and graphical methods, is:

Everitt, B., *Graphical Techniques for Multivariate Data*, Heinemann, 1978.

Two texts which are at a slightly higher technical level are:

Hope, K., *Methods of Multivariate Analysis*, University of London Press, 1968.
Bennett, S. and Bowers, D., *An Introduction to Multivariate Techniques for Social and Behavioural Sciences*, Macmillan, 1976.

Case studies of the application of multivariate methods include:

Davenport, M. and Studdert-Kennedy, G., The statistical analysis of aesthetic judgment: an exploration, *Applied Statistics*, 21, 1972.
Jeffers, J.N.R., Two case studies in the application of principal components analysis, *Applied Statistics*, 16, 1967.
Willmott, A.J. and Grimshaw, P.N., Cluster analysis in social geography, in Cole, A.J. (ed.), *Numerical Taxonomy*, Academic Press, 1969.

Some further discussion of minimum spanning trees and their use in multivariate analysis (with a brief reference to a practical situation) may be found in:

Gower, J.C. and Ross, G.J.S., Minimum spanning trees and single linkage cluster analysis, *Applied Statistics*, 18, 1969.

A very useful guide to the ideas of cluster analysis and its uses and limitations is:

Everitt, B., *Cluster Analysis*, Heinemann, 2nd edn, 1980.

The other references cited in these chapters are:

Chatfield, C. and Collins, A.J., *Introduction to Multivariate Analysis*, Chapman and Hall, 1980.
Chernoff, H., Using faces to represent points in *k*-dimensional space graphically, *Journal of the American Statistical Association*, 68, 1973.
Gower, J.C., A general coefficient of similarity and some of its properties, *Biometrics*, 27, 1971.
Hartigan, J.A., *Clustering Algorithms*, Wiley, 1975.

## CHAPTER 14 NONPARAMETRIC STATISTICS

There are several introductory texts on nonparametric statistical methods. One with a practical orientation is:

Daniel, W.W., *Applied Nonparametric Statistics*, Houghton Mifflin, 1978.

The references cited are:

Mosteller, F. and Tukey, J.W., *Data Analysis and Regression*, Addison-Wesley, 1977.

Theil, H., *Principles of Econometrics*, North-Holland, 1971.

## CHAPTER 15 CATEGORICAL VARIABLE ANALYSIS

The analysis of contingency tables is a subject which has attracted much research in recent years and a number of textbooks, of varying levels of complexity, have appeared. A useful start is:

Everitt, B., *The Analysis of Contingency Tables*, Chapman and Hall, 1977.

The subject of correspondence analysis is one which has been developed in recent years particularly by French statisticians. One of the few practical introductions to correspondence analysis written in English is:

Greenacre, M.J., Practical correspondence analysis, in Barnett, V. (ed.), *Interpreting Multivariate Data*, Wiley, 1981.

## FURTHER REFERENCES

Three useful sources of empirical and expository papers, at about the level of this text (or somewhat higher), are:

*The American Statistician*, published by the American Statistical Association.

*Bulletin in Applied Statistics* (BIAS), published by the Department of Mathematics and Statistics, Sheffield Polytechnic.

*The Statistician*, published by the Institute of Statisticians.

There is, of course, a vast number of published sources of data of the kind which has been considered in this text. The most generally available sources for further practical study are the official statistical publications of the British and United States governments and the United Nations Statistical Office.

# Index

*The main references are to sections, e.g. 8.5; minor or passing references are to pages, e.g. 167. Other references are to tables (e.g. T5.1), figures (e.g. F9.5), or chapters (e.g. Ch 15).*